Biodegradable Polymers, Blends and Biocomposites

Biobased biodegradable polymers are emerging as an alternative to fossil fuel-based plastics. *Biodegradable Polymers, Blends and Biocomposites: Trends and Applications* discusses trends in the development of microbial/other renewable source-based bioplastic products, their blends and biocomposites applications in various industrial fields. It covers biodegradable polymeric materials preparation, extraction, formulation, modification of properties, product development and applications and end-of-life options. Furthermore, the book discusses topics like bioplastic resources, isolation procedures, utilization at commercial level and markets and economy.

Features:

1. Explains emerging application possibilities of biobased biodegradable polymers.
2. Provides detailed application notes on agricultural waste-based bioplastics.
3. Covers microbial and agro-based biocomposites and their applications.
4. Summarizes bioplastic degradation and blending research.
5. Discusses application possibilities of biobased biodegradable polymers.

The book is aimed at researchers and graduate students in polymers and composites.

Emerging Materials and Technologies

Series Editor:
Boris I. Kharissov

The *Emerging Materials and Technologies* series is devoted to highlighting publications centered on emerging advanced materials and novel technologies. Attention is paid to those newly discovered or applied materials with potential to solve pressing societal problems and improve quality of life, corresponding to environmental protection, medicine, communications, energy, transportation, advanced manufacturing, and related areas.

The series takes into account that, under present strong demands for energy, material, and cost savings, as well as heavy contamination problems and worldwide pandemic conditions, the area of emerging materials and related scalable technologies is a highly interdisciplinary field, with the need for researchers, professionals, and academics across the spectrum of engineering and technological disciplines. The main objective of this book series is to attract more attention to these materials and technologies and invite conversation among the international R&D community.

Smart Micro- and Nanomaterials for Drug Delivery
Edited by Ajit Behera, Arpan Kumar Nayak, Ranjan K. Mohapatra, and Ali Ahmed Rabaan

Smart Micro- and Nanomaterials for Pharmaceutical Applications
Edited by Ajit Behera, Arpan Kumar Nayak, Ranjan K. Mohapatra, and Ali Ahmed Rabaan

Friction Stir-Spot Welding
Metallurgical, Mechanical and Tribological Properties
Edited by Jeyaprakash Natarajan and K. Anton Savio Lewise

Phase Change Materials for Thermal Energy Management and Storage
Fundamentals and Applications
Edited by Hafiz Muhammad Ali

Nanofluids
Fundamentals, Applications, and Challenges
Shriram S. Sonawane and Parag P. Thakur

MXenes
From Research to Emerging Applications
Edited by Subhendu Chakroborty

Biodegradable Polymers, Blends and Biocomposites
Trends and Applications
Edited by A. Arun, Kunyu Zhang, Sudhakar Muniyasamy and Rathinam Raja

For more information about this series, please visit www.routledge.com/Emerging-Materials-and-Technologies/book-series/CRCEMT

Biodegradable Polymers, Blends and Biocomposites

Trends and Applications

Edited by
A. Arun, Kunyu Zhang, Sudhakar Muniyasamy
and Rathinam Raja

First edition published 2025
by CRC Press
2385 NW Executive Center Drive, Suite 320, Boca Raton FL 33431

and by CRC Press
4 Park Square, Milton Park, Abingdon, Oxon, OX14 4RN

CRC Press is an imprint of Taylor & Francis Group, LLC

Funded by the Scheme for Promotion of Academic and Research Collaboration (SPARC) (Project No. SPARC/2018-2019/P485/SL; Dated: 15 March 2019), Ministry of Education, Government of India.

ISBN: 9781032302492 (hbk)
ISBN: 9781032302508 (pbk)
ISBN: 9781003304142 (ebk)

DOI: 10.1201/9781003304142

Typeset in Times
by codeMantra

Dedication

The editors would like to dedicate this book to Padmabushan Dr. R.M. Alagappa Chettiar and Professor Emo Chiellini.

Padmabushan Dr. R.M. Alagappa Chettiar was a great philanthropist, eminent educationist, and a multi-faceted versatile personality, who harbored a firm conviction that education is an absolute must for a human being to become productive, wholesome and humane. In 1947, at Dr. Annie Besant's centenary celebrations, the Vice-Chancellor of Madras University called industrialists to start colleges to educate India. Spontaneously, Dr. Chettiar answered the call in the same function, and within three days, Alagappa Arts College started functioning at Gandhi Maleghai in Karaikudi.

His generous donations led to the establishment of a string of educational institutions, which formed the basis for the establishment of Alagappa University in 1985 by the Government of Tamil Nadu. Alagappa University is located in Karaikudi, Tamil Nadu, India. Alagappa University is accredited by the National Assessment and Accreditation Council with an A+ Grade (CGPA: 3.64), Graded as Category-1 & Granted Autonomy; BRICS India Rank: 20, BRICS Rank: 104 and Asia Rank: 216, is located at Karaikudi in Tamil Nadu, India.

To bring prominence to the educational institutions, he had the vision to convince Prime Minister Nehru to house one of the Government's National Research Institutes in the heart of the Alagappa campus.

Dr. Radhakrishnan, the Vice President of India, expressed his opinion during the opening of the Central Electrochemical Research Institute (CECRI)on 14 January 1953: "The magnificent gift of 300acres of land, 1.5 million Rupees by Dr. Alagappa Chettiar helped the Government of India to select Karaikudi as the seat of the Electrochemical Research Institute. Being a businessman himself, Dr. Alagappa Chettiar is aware of the industrial possibilities of our country and the need for scientific, technical, and technological education; in his lifetime, he has built a monument for himself, you have only to look around".

As editors, we would like to dedicate this book to the deceased Professor Emo Chiellini (died on 21 August 2020), the research supervisor for Dr. Sudhakar Muniyasamy who is one of the editors of this book. Professor Chiellini was Emeritus Professor of Chemistry at Pisa University in Italy and developed plastic with his colleagues in the post-war period. They realized that the material durability they had designed into plastic would cause an environmental problem if discarded into the open environment. They, therefore, found a way to cause plastic to degrade and then biodegrade entirely in the natural environment and called it "Environmental Degradable/Compatible

Polymeric Materials." They have contributed considerably to protecting our environment, which will be long remembered.

Prof. Chiellini received the prestigious Giulio Natta Award in 2018, named after Professor Giulio Natta, who discovered polypropylene. Each year, this award is given to one senior researcher in the chemical industry whose work has reached outstanding achievements, contributed to advancing the quality of life and has left a significant mark on the community. Prof. Emo Chiellini was in charge of research contracts financed by industries and the European Community. He has been the Founder and President of the Doctorate Course in Biomaterials of the Doctorate School in Biological and Molecular Sciences, "now, it is called BioLab," at the University of Pisa, Italy. He has been the Founder and Director of the Interdisciplinary Laboratory of Bioactive Polymeric Materials for Biomedical and Environmental Applications, where researchers had competencies in Materials Science and Technology, Organic Chemistry, Chemical Engineering, Materials Science, Pharmaceutical Technologies, Microbiology and Environmental Chemistry. He has been among the promoters of the foundation of the "Inter-university National Consortium of Materials Science & Technology (INSTM)" (1992), of which he was honored to serve as a Member of the Scientific Council. He is widely published in the areas of polymer science and technology and packaging materials.

Contents

Preface

Biobased biodegradable polymers are emerging as an essential component for global sustainability as alternatives to fossil fuel–based plastics. This book aims to discuss the recent trends in the development of microbial/other renewable source–based bioplastic products, their blends, and biocomposite applications in various industrial fields.

Interestingly, biopolymers and biodegradable plastics are a hot topic in the plastics industry. They are used in many sectors that rely on plastics, such as packaging, medical devices, construction, life cycle assessment, and economical mode. Moreover, biopolymers have been given much consideration because they play a significant role in habitual life due to their unique tunable characteristics and applications in all fields. Recent research aims to find solutions to reduce plastic usage and produce economically friendly plastic products to reach society.

This book is arranged into various chapters to provide an in-depth overview of bioplastics and the economics of the current situation, which plays a predominant part and gives the reader a logical and expressive portrayal. Chapter 1 comprises biopolymers' and biodegradable polymers' past, present, and future availability. Classification, extraction, biobased monomers, biopolymers from bacteria, and an overview of biodegradable products are explored in Chapter 2. Chapter 3 analyzes the global plastic waste generation, indoor and outdoor wastes, dietary exposure, and health impacts of microplastics and nanoplastics. Chapter 4 compiles bioplastic's role, sustainability, principles, and applications. Chapters 5–7 are on biopolymer production from microbes, agricultural wastes, and biobased monomers. Chapter 8 contains fascinating content comprising the blending techniques to produce cost-effective biopolymers. Chapters 9–13 clarified the end-of-life options, economy, circular economy, and life cycle assessment study of bioplastics and bio-nanocomposites. Finally, Chapter 14 finalizes the impact of microplastics and their toxicological effect on aquatic environments.

Overall, this book is proposed to have the best knowledge about biopolymers' economic value and situation. The editors and authors are well-known persons who have worked in this field for many years. On behalf of the Scheme for Promotion of Academic and Research Collaboration (SPARC), we thank all the authors for contributing to this book.

Acknowledgments

We would like to acknowledge the Scheme for Promotion of Academic and Research Collaboration (SPARC) (No. SPARC/2018-2019/P485/SL; Dated: 15 March 2019) and the authorities of the Alagappa University for their dedicated support in the compilation of this handbook.

About the Editors

A. Arun is a Professor and the Head of the Department of Microbiology, Alagappa University, Karaikudi, Tamilnadu, India. His research team is working on bioplastic, bioenergy, microbial fuel cells, bioremediation, biomass and bioactive compounds. He has vast experience in the field, having published several research articles and book chapters in peer-reviewed journals. In addition, he received many significant research projects and two consultancy projects, including one Indo-China project funded under SPARC by MHRD, India, for around five crores of Indian rupees. Moreover, he has a proven track record of technical achievements with over 100 journal publications with a cumulative impact factor of 415, 1959 citations, a 26 h-index, a 50 i10 index, and 1 filed patent application.

Kunyu Zhang is a Senior Scientist at Petrochemical Research Institute, Petro China Company Limited, China. Before this position, he was an Associate Professor at the Tianjin University, China. He has 20 years of experience in polymer science: chemistry/process/application and has contributed to research in the following topics: sustainable polymer, biocomposite, high-performance polymer synthesis and processing. He has a proven track record of technical achievements with 38 journal publications, receiving over 1400 citations and ten filed patent applications.

Sudhakar Muniyasamy is a Principal Scientist currently leading research and development on sustainable biobased biodegradable plastics at the CSIR-Chemicals, South Africa. He holds a Ph.D. in Chemical Science from the University of Pisa, Italy. He is a research associate in the Department of Chemistry at Nelson Mandela University, South Africa. Before this position, he was a post-doctoral researcher at the University of Guelph, Canada.

His current research activity mainly focuses on developing sustainable, biodegradable plastics from renewable resources and studying end-of-life options such as recyclability, biodegradation and composting and verifying the environmental claims for biobased products. He has authored and co-authored numerous peer-reviewed papers in the fields of biobased polymers and biodegradation.

Rathinam Raja is working as a Professor and Director since June 2024 at Centre for Integrated Medical Research, Sree Balaji Medical College and Hospital (SBMCH), BIHER, Chromepet, Chennai, India. He pursued his Ph.D. (microalgae) at the University of Madras in 2004. After that, he completed his post-doctoral research at IIT Madras for two years. He also worked for a few years as a Lead Scientist at Aquatic Energy LLC, Louisiana. Previously, he worked in Taiwan, South Korea and Portugal for over a decade. As a result, Dr. Raja has published 38 international articles and 14 book chapters and edited four books.

Contributors

Selvaraj Anitha
Toxicogenomics and Systems Toxicology Laboratory, Department of Animal Health and Management
Alagappa University
Karaikudi, Tamil Nadu, India

A. Arun
Department of Microbiology
Alagappa University
Karaikudi, Tamil Nadu, India

Muniyandi Biruntha
Vermitechnology Laboratory, Department of Animal Health and Management
Alagappa University
Karaikudi, Tamil Nadu, India

Abhispa Bora
Bioenergy and Bioremediation Laboratory, Department of Microbiology
Alagappa University
Karaikudi, Tamil Nadu, India

Orebotse Joseph Botlhoko
CSIR Chemical Cluster
Nanostructured and Advanced Materials,
Pretoria, South Africa

A. Gada
Department of Chemistry
Nelson Mandela University
Gqebera, South Africa

C. Hoyo
Department of Biological Sciences
North Carolina States University
Raleigh, North Carolina

Muthusamy Govarthanan
Department of Environmental Engineering
Kyungpook National University
Deagu, Republic of Korea

Sivaprakash Gurusamy
Bioenergy and Bioremediation Laboratory, Department of Microbiology
Alagappa University
Karaikudi, Tamil Nadu, India

S.P. Hlangothi
Department of Chemistry
Nelson Mandela University
Gqebera, South Africa

Letwaba John
CSIR Chemical Cluster
Nanostructured and Advanced Materials
Pretoria, South Africa
and
Department of Chemical
Metallurgical and Materials Engineering, Tshwane University of Technology
Pretoria, South Africa

Nachimuthu Karmegam
Department of Botany
Government Arts College (Autonomous)
Salem, Tamil Nadu, India

Mandla V. Khumalo
Department of Chemistry
Nelson Mandela University
Gqebera, South Africa

Mohanrasu Kulanthaisamy
Bioenergy and Bioremediation Laboratory, Department of Microbiology
Alagappa University
Karaikudi, Tamil Nadu, India

Ponnuchamy Kumar
Toxicogenomics and Systems Toxicology Lab, Department of Animal Health and Management
Alagappa University
Karaikudi, Tamil Nadu, India

Kulanthaivel Langeswaran
Department of Biomedical Science
Alagappa University
Karaikudi, Tamil Nadu, India

Rakgoshi Lekalakala
CSIR Chemical Cluster
Nanostructured and Advanced Materials
Pretoria, South Africa

P. Melariri
Faculty of Health Sciences
Nelson Mandela University
Gqebera, South Africa

Raphaahle Caroline Mphahlele
Centre for Nanostructures and Advanced Materials
Council for Scientific and Industrial Research (CSIR)
Pretoria, South Africa

Sudhakar Muniyasamy
CSIR Chemical Cluster
Nanostructured and Advanced Materials
Pretoria, South Africa
and
Department of Chemistry, Faculty of Science
Nelson Mandela Metropolitan University
Port Elizabeth, South Africa

Kathiresan Nachammai
Department of Biotechnology
Alagappa University
Karaikudi, Tamil Nadu, India

Nomadolo Nomvuyo
CSIR Chemical Cluster
Nanostructured and Advanced Materials
Pretoria, South Africa
and
Department of Chemistry
Nelson Mandela University
Gqebera, South Africa

Dada Omotola
Department of Biological Sciences, Faculty of Basic and Applied Sciences
Elizade University
Ondo State, Nigeria

Thavamani Palanisami
School of Engineering
The University of New Castle
New South Wales, Australia

Logeshwaran Panneerselvan
School of Engineering
The University of New Castle
New South Wales, Australia

Mothoa Pertunia
Department of Chemical, Metallurgical and Materials Engineering
Tshwane University of Technology
Pretoria, South Africa

Dong PoSong
School of Materials Science and Engineering
Tianjin University
Tianjin, China

R. Guru Raj Rao
Department of Bioinformatics,
Alagappa University
Karaikudi, Tamil Nadu, India

Nithya Rathinavel
Bioenergy and Bioremediation Laboratory, Department of Microbiology
Alagappa University
Karaikudi, Tamil Nadu, India

Durgadevi Sabapathi
Toxicogenomics and Systems Toxicology Lab, Department of Animal Health and Management
Alagappa University
Karaikudi, Tamil Nadu, India

Pandi Sangavi
Department of Bioinformatics
Alagappa University
Karaikudi, Tamil Nadu, India

Gnanasekaran Sivakumar
College of Education
Alagappa University
Karaikudi, Tamil Nadu, India

S. Vishnu Suba
Department of Economics
Mannar Thirumalai Naicker College
Madurai, Tamil Nadu, India

Andri Swanepoel
CSIR Chemical Cluster
Nanostructured and Advanced Materials
Pretoria, South Africa

T. Angelin Swetha
Bioenergy and Bioremediation Laboratory, Department of Microbiology
Alagappa University
Karaikudi, Tamil Nadu, India

Anathanarayanan Yuvaraj
Department of Zoology
Periyar University
Salem, Tamil Nadu, India

Kunyu Zhang
Advanced Materials Research Center, Petrochemical Research Institute
PetroChina Company Limited
Beijing, China
and
School of Materials Science and Engineering
Tianjin University
Tianjin, China

1 Biopolymers and Biodegradable Polymers

Past, Present and Future

Andri Swanepoel, Sudhakar Muniyasamy and Omotola Dada

1.1 INTRODUCTION

Polymers or plastics derived from fossil fuels such as poly(ethylene) (PE), poly(propylene) (PP) and poly(ethylene terephthalate) (PET) are increasingly replacing glass and metal in various sectors such as the packaging and automotive industries due to the flexibility and low cost of these materials. The ever-increasing demand for plastic products and the high rates of plastic production required to meet this demand results in vast accumulation of disposable plastic waste in the environment (Pan et al., 2020). According to the United Nations Environment Programme report, only 9% of all plastic is recycled and 12% is incinerated. The remaining plastic pollutes the environment or ends up in landfills due to high costs of collection and a lack of adequate infrastructure required for alternative solutions. Fundamentally, plastic pollution has a direct and indirect economic impact on industries such as fishing, shipping and tourism. Moreover, unmanaged, non-biodegradable conventional plastic waste has the potential to generate greenhouse gases when incinerated and exposed to solar radiation that contributes to the growing threat of global warming (Luckachan & Pillai, 2011).

Increased awareness worldwide of the negative environmental impact of overconsumption of plastics has resulted in several countries opting to prohibit the sale of plastic bags and to impose fees or levies on customers towards minimizing the quantity of waste generated (Convery et al., 2007). Increasing concern regarding the environmental impact of plastic waste and plastic-related emissions is motivating greater efforts towards implementing a circular plastic economy. In a circular plastic economy, the primary focus is on minimization of the usage of non-renewable resources and plastic waste production via the reuse and recycling of materials (Rosenboom et al., 2022). This approach is complicated due to difficulties in collecting and handling of materials for recycling, as well as the technical challenges related to recycling of material composites, including the need for sophisticated technology (Allred & Busselle, 2000). Biodegradable plastics that degrade naturally over time have emerged as a viable option to mitigate the environmental impact of conventional plastics (Zhu & Wang, 2020). The unique characteristics of biodegradable plastics such as their excellent mechanical, physicochemical and

DOI: 10.1201/9781003304142-1

degradation properties make for an attractive alternative to conventional plastics in many industrial applications (Moshood et al., 2022).

Several biodegradable polymers such as poly(lactic acid) (PLA) and poly(hydroxy alkanoates) (PHAs), including poly(hydroxybutyrate) (PHB), poly(hydroxy valerate) (PHV) and poly(hydroxy butyrate-co-valerate) (PHBV), are commercially available (Zhu & Wang, 2020). These biodegradable polymers are utilised in various applications but are three to five times more expensive than conventional plastics. Such high costs prevent wider uptake by customers (Yaguchi et al., 2020). This lack of uptake as well as a lack of infrastructure for handling and composting of bioplastic waste remains a global issue (Mehta et al., 2021). There thus exists a need to balance the cost and desired properties of biodegradable plastics for use in specific applications. From an environmental standpoint, the adoption of biodegradable plastics as (partial) replacements for fossil fuel–based plastics is extremely beneficial. Biodegradable plastics are chemically converted into natural components upon disposal, reducing stress on already saturated landfill sites and limiting long-term plastic pollution (Moshood et al., 2022). The usage of biodegradable plastics in various applications can also decrease greenhouse gas emissions (Widiastuti, 2014). To maximise benefit to the environment, however, biodegradable polymers must be disposed of correctly.

1.2 ENVIRONMENTAL IMPACT OF PLASTICS

Plastic pollution affects the natural environment on both macro- and micro-scales. The impact of larger pieces of litter on marine life, as evidenced by images of birds and other larger animals entangled in plastic waste, as well as the well-documented floating trash heaps in the oceans, are well known. Larger animals ingest plastic or become entangled in discarded fishing equipment (Gregory, 2009). The ubiquitous presence and the effects of microplastics have become an important area of research in the last decade or so. Microplastics are plastic fragments or fibres, generally smaller than 5 mm, that are generated via the degradation of larger pieces of plastic pollution, from shedding from clothing manufactured from synthetic fibres, or from products containing microplastics such as cosmetics (Solomon Ogunola et al., 2018). It is estimated that by 2100, there will be between 2.5×10^7 and 1.3×10^8 tonnes of floating microparticles in the Earth's marine environment (Everaert et al., 2018).

While not as visible as macro plastic pollution, the accumulation of microplastics in both the terrestrial (Ng et al., 2018, de Souza Machado et al., 2018) as well as the marine environments (Everaert et al., 2018, Bohdan, 2022) have been documented. The possibility of inhalation of airborne microplastics has also been raised (Gasperi et al., 2018).

Research has been conducted on the effects of microplastic ingestion on different marine species (Browne et al., 2008, Wright et al., 2013, Fossi et al., 2014), as well as on the distribution of microplastics in the different stratas of the world's oceans (Spindola Vilela et al., 2022). A major concern is the introduction of persistent organic pollutants originating from plastic additives into marine food systems via microplastics (Ivar Do Sul & Costa, 2014). Researchers are still working towards a thorough understanding of the impact of the accumulation of microplastics in the biosphere on human health as well as the broader ecosystem, with definitive conclusions not yet available (Kontrick, 2018, Everaert et al., 2018, Huang et al., 2022).

Meanwhile, several approaches are proposed to reduce plastic pollution. Where conventional plastics are concerned, the implementation of a circular economy is proposed, whereby the recycling and reuse of plastics are promoted to approach a closed-loop system (United Nations Environment Programme, 2024). The replacement of conventional plastics with alternative products is also proposed. Simulations have shown that reduced demand as well as better management of plastic waste is required to enable significant changes in the current projected outlooks for plastic pollution (Lau et al., 2020). The replacement of conventional plastics with biodegradable polymers is also proposed as a solution. The use of biodegradable polymers from sustainable sources (other than petroleum-based) can potentially address the issues of pollution as well as sustainable consumption concurrently.

Currently, commercially biodegradable polymers make up a fraction of the total plastics production (around 1% of the total global production) (Plastics Europe, 2022). Implementation of biodegradable plastics is hampered by the lack of mechanical strength and ductility of many biodegradable polymers, as well as the high costs and complicated synthesis processes required to produce many biodegradable polymers.

1.3 BIOPOLYMER SYNTHESIS AND APPLICATIONS

Biodegradable polymers from renewable sources are used in various fields, including biomedical applications, food packaging and agricultural films (Song et al., 2018, Auras et al., 2004). Since the 1990s, more emphasis has been placed on developing biodegradable polymers for all purposes and engineering applications, rather than simply focusing on biodegradability (Nakajima et al., 2017).

The application of biodegradable polymers is often hampered by restrictive processing conditions, and once processed, the materials often exhibit poor mechanical as well as thermal and electrical properties (La Mantia & Morreale, 2011). The high cost of biodegradable polymer production is of course a very important factor and one of the key reasons why conventional plastics are still preferred.

Composites must be used where characteristics of individual biodegradable polymers alone simply cannot meet the demands required of the final product. Fillers can also be incorporated to alter characteristics of biodegradable polymers depending on the final application without impacting on the biodegradability of the final product (Sun et al., 2018). As this chapter deals with biopolymers from renewable sources, the composites and fillers utilised along with biodegradable polymers mentioned here are also from renewable sources.

The characteristics, synthesis and applications of PLA, thermoplastic starch (TPS), PHAs and their copolymers of PHB, PHV and PHBV, cellulose and other natural materials and the composites of these materials will be discussed in the next few sections.

1.3.1 Polylactic Acid (PLA) and Its Composites

Amongst biodegradable polymers, PLA is arguably the most attractive alternative to conventional petroleum-based materials. PLA has comparable mechanical and physical properties to those of thermoplastics, particularly PP, high-density polyethylene (HDPE), PET and polystyrene (PS) (Almeida, 2011, Auras et al., 2005), as well as being biocompatible and biodegradable (Ljungberg & Wesslen, 2003). This appeal is evident in the large amount of commercially available PLA products. One of the most well-known commercial PLA suppliers is Nature Works, supplying PLA under the trademark Ingeo (Nakajima et al., 2017). Other commercial suppliers include Unitika (Terramac®), Corbion/Total Energies (Luminy®), Zhejiang Hisun Biomaterials (Revode) and Futerro (Renew™).

PLA can be used in different applications, such as the automotive and packaging industries, and in agricultural and biomedical applications, amongst others (Auras et al., 2004, Alexandre & Dubois, 2000, Ahmed & Varshney, 2011, Baiardo et al., 2003, Arakawa et al., 2006, Castro-Aguirre et al., 2016). PLA is widely used in biomedical applications due to its biocompatibility and the ability to be reabsorbed safely by the body. Due to the versatility of PLA and other biodegradable polymer compounds, they are rapidly replacing the use of other materials such as ceramics, metals and alloys in this field (Song et al., 2018). The use of PLA in various biomedical applications has been studied, including for use in drug delivery (Baimark & Srisa-Ard, 2012, Maji et al., 2015) and implants and scaffolds (Amnael Orozco-Díaz et al., 2020, Chen et al., 2006, Leiggener et al., 2006, Llorens et al., 2015, Nazhat et al., 2001). Commercially available products based on biodegradable polymer technology are mainly made from PLA produced by companies such as Conmed, DePuy, Arthrex, Linvatec and Innovasive Devices (Song et al., 2018), with product offerings largely confined to soft tissue engineering (resorbable sutures) and bioresorbable implants. Several reviews on the use of PLA in biomedical applications have been published to provide further insight into the topic (Tyler et al., 2016, DeStefano et al., 2020).

PLA is produced via the ring-opening polymerization of lactides or via polycondensation of lactic acid (Martinez Villadiego et al., 2022, Jacobsen et al., 2000). Lactic acid in turn is produced via fermentation of carbohydrates by lactobacillus bacteria. When the carbohydrates required for bacterial fermentation are obtained from agricultural waste, the PLA can be produced in a fully renewable cycle.

PLA can be processed via conventional processing technologies (Castro-Aguirre et al., 2016), as well as other technologies such as melt spinning (Panichsombat et al., 2019), electrospinning (Llorens et al., 2015), 3D printing (Amnael Orozco-Díaz et al., 2020) and foaming (Mihai et al., 2010).

As mentioned, PLA exhibits good mechanical properties, depending on the degree of crystallinity of the polymer, the molecular weight and the stereochemical configuration of the backbone of PLA (Perego et al., 1996). PLA is, however, not highly compostable and degradation residues affect the quality of soil by increasing the acidity of the soil (Barnes et al., 2009). Its brittleness and low toughness limit its application in areas where flexible load-bearing is required (Hamad et al., 2018). PLA also has greater permeability as compared to other commercial polymers such as PP and PE, PET and PS (Petersen et al., 2001, Auras et al., 2003). Such poor barrier properties may hinder the adoption of PLA in the packaging industry, especially in applications where the sensitivity of the contents to moisture or oxygen is crucial. Its thermal stability is relatively low, which makes PLA unsuitable for use in high-temperature applications (Auras et al., 2003). PLA is also hydrophobic and lacks reactive functional groups. Modification of PLA is thus required to tailor it to applications where greater toughness and hydrophilicity are required (Farah et al., 2016).

PLA is more expensive than petrochemical plastics (Auras et al., 2011). An effective approach towards reducing the costs of PLA products is to reinforce PLA with abundant and inexpensive renewable precursors such as sawdust, calcium carbonate ($CaCO_3$), chiton and algae, all lightweight, high-strength and biodegradable compounds (Dimonie & Rapa, 2010). However, there are limitations to this approach, including reduced mechanical properties of the reinforced PLA due to incompatibility between the precursor and the PLA matrix. Improvements in the compatibility between PLA and fillers can be made by introducing maleic anhydride poly(lactic acid) (PLA-g-MA) to the formulation. Chemical treatment is also an effective method to improve interfacial adhesion between PLA and precursors.

Substantial investigations have been conducted on the reinforcement of PLA with sawdust to reduce costs and improve properties such as mechanical and thermal properties while maintaining biodegradability (Kuciel et al., 2020, Mazur et al., 2022, Qiang et al., 2012). Sawdust is an inexpensive renewable material with relatively low density, high tensile strength and minimal abrasiveness for processing equipment. PLA-based sawdust is earning a growing share of the market due to its renewable and biodegradable nature (Qiang et al., 2012). Mazur and co-workers have examined the mechanical and biodegradable properties of biocomposites of PLA with wood fibres and basalt fibres (Mazur et al., 2022). Their results showed that the rate of biodegradation can be controlled, and the tensile properties of the composites can be tailored by varying the type and amount of fibres used.

Often, wood flour (from sawdust) must be chemically treated as weak interfacial adhesion exists between the hydrophilic wood flour and PLA. The resulting poor dispersion of the filler throughout the PLA matrix negatively affects the mechanical properties of the composite, especially the impact strength (Okubo et al., 2009, Molnar et al., 2009, Lee & Kim, 2009). Wood flour chemically treated with silane was used to improve the interfacial adhesion between PLA and recycled wood flour (RWF) (Yu et al., 2022). It was found that the tensile strength of the composites was independent of the RWF, whereas the toughness and elongation at break decreased with increased RWF content. It is therefore imperative to examine the effect of the particle size of sawdust in combination with chemical treatment via grafted maleic anhydride on the compatibility between PLA and sawdust.

Among other fillers, an alga is also a promising candidate for production of renewable polymer composite matrices due to its low cost and its abundance in nature. Algae biomass is a source of hydrosoluble polysaccharides (agarose, carrageenan and alginate) and is mainly used in sectors such as agriculture, food and cosmetics industries (Bulota & Budtova, 2015). Incorporation of algae in PLA composites allows for lowering of the costs and carbon emissions resulting from synthesizing the polymer composites while also acting as reinforcing filler in the biocomposites (Huang et al., 2020, Barghini et al., 2010, Lannace et al., 1999). An added environmental benefit of utilizing algae is the possible mitigation of the negative impact of algae overgrowth on the ecosystem, further motivating researchers to invest in utilising algae in the field of biocomposites (Huang et al., 2020).

The dynamic mechanical and thermal properties of biocomposites reinforced with bleached red algae fibre (BRAF) were measured (Sim et al., 2009). It was found that the dynamic mechanical and thermomechanical properties of the biocomposites were improved with increasing BRAF loading. In another study, the mechanical properties of PLA-algae as a function of algae type (treated with salt), concentration and particle size were investigated (Bulota & Budtova, 2015). Young's modulus of the composite was optimised at an algae loading of 40%, although the strain at break and tensile strength decreased. Their results also indicated that the algae type used had little to no effect on the mechanical properties of the composite.

Currently, a limited number of studies exist regarding the mechanical and thermal properties of PLA-algae composites. There is thus much scope for investigation into how these properties are affected by the particle size and loading fraction of algae fillers, as well as the effect of grafting of PLA on such composites.

Blends or composites of PLA with several other natural fibres have also been studied. Nanocomposites of PLA and cellulose were prepared via solution casting, with enhanced thermal and mechanical properties as compared to the native PLA (Petersson et al., 2007). PLA was used to reinforce composites of natural fibres such as bamboo, kenaf and coir (Yusoff et al., 2016), while the fire retardancy characteristics of PLA were improved by the addition of banana fibres and nanoclay (Sajna et al., 2017), as well as alkali-treated pineapple leaf fibres (Siakeng et al., 2020), and via melt blending with melamine polyphosphate and cloisite 30B (Guo et al., 2017). Composites of PLA with cotton were also prepared and showed improved load-bearing strength for application in furniture or automotive materials (Battegazzore et al., 2019).

Composites of PLA with various natural products have been studied for food packaging, including with chitosan and basil olive oil for improved tensile and barrier properties with antioxidant effect (Salmas et al., 2021), with talc as nucleating agent for improved thermal resistance (Li & Huneault, 2007), with clay nanofillers for improved permeability as well as UV resistance (Sanchez-Garcia & Lagaron, 2010), and with grape oil to improve light stability (Nikvarz et al., 2020). PLA composites with clay-lignin nanofillers were prepared to improve toughness and barrier properties (Yang et al., 2020), and composites of PLA with chitin and cellulose nanofibers were synthesized for improved mechanical and thermal properties (Rizal et al., 2021). Nanocomposites of PLA and cellulose were prepared via solution casting, with enhanced thermal and mechanical properties compared to the native PLA (Petersson et al., 2007).

Blending of PLA with other biodegradable polymers is another strategy to optimise the properties and costs of the PLA matrix. PLA-based composites often suffer from reduced toughness (Nomai & Jarukumjorn, 2014). In response, blending of PLA with different biodegradable polymers has been investigated, including blending with polycaprolactone (PCL) (Vieira et al., 2011, Simoe et al., 2009, Noroozi et al., 2012), PHB (Armentano et al., 2015), poly(butylene succinate) (PBS) (Park & Im 2002a, 2002b, Xu & Huang, 2012) and polybutylene adipate terephthalate (PBAT) (Teamsinsungvon & Ruksakulpiwat, 2012, Jiang et al., 2006, Deng et al., 2018).

Among other reinforcements, chitosan is emerging as a suitable filler for the production of biocomposites, especially in areas requiring antimicrobial properties. Chitosan is a cationic polysaccharide, an abundant, inexpensive and renewable polymer derived from chitin by deacetylation in the presence of an alkali. However, chitosan cannot be converted into plastic by using processing techniques such as extrusion, injection moulding and thermoforming because it is not thermoplastic (Dabertrand et al., 2022). This is attributed to the strong intermolecular bonds that prevent melting, flow and deformation from occurring (Matet, 2014). Nonetheless, chitosan can be utilised as reinforcement in polymer composites by thermomechanical processes.

Many efforts have been made to incorporate chitosan as a bioactive material in polymers such as PLA. Several scholars have developed chitoson-PLA biodegradable composites by solution mixing and film casting methods (Li et al., 2004, Grande & Carvalho, 2011, Chen et al., 2005, Sebastien et al., 2006). Bonilla et al. (2013) studied the effects of particle size and content of chitosan on the physicochemical and antimicrobial properties of PLA-chitosan composite films synthesized via a melt method. They found that the incorporation of chitosan particles led to less rigid films with no effect on the thermal properties of PLA. The films produced demonstrated significant antimicrobial activity against total aerobic and coliform microorganisms, with greater antimicrobial activity in films where the particle size of chitosan had been reduced. As with other fillers, sufficient dispersion of chitosan particles throughout the PLA matrix is important. It was shown that agglomeration of chitosan particles negatively affected the compatibility of a PLA-chitosan composite prepared via melt processing (Correlo et al., 2005, Matet. 2014).

Composites of PLA with hydroxyapatite (HA) are often studied for use in biomedical applications. HA is known to stimulate osteogenesis and, thus, bone regeneration and enhances the flexibility of PLA, while PLA in turn enhances the physical and mechanical properties of HA (DeStefano et al., 2020). An amorphous poly(D,L-lactic acid) matrix was reinforced with HA particles, as well as with semi-crystalline poly(L-lactic acid) fibres for application in bone repair. The inclusion of both fillers resulted in improved thermomechanical properties (Nazhat et al., 2001). A composite of PLA and HA was prepared via 3D printing for the possible use as a bioresourable implant for bone regeneration (Amnael Orozco-Díaz et al., 2020). The in vitro biocompatibility of the composites was maintained after 3D printing, indicating that the process is a viable option for manufacturing implantable devices.

1.3.2 Polyhydroxyalkanoates (PHA) and Their Composites

PHA compounds are classified according to the length of the repeat unit (3-hydroxyalkanoate). Poly(3-hydroxybutyrate) (PHB) was the first PHA discovered in 1926 by Maurice Lemoigne (1926). Commercially, the PHAs P3HB, P(3HB-co-3HV) and P(3HB-co-3HH) are the most widely available (Nakajima et al., 2017). PHAs can be soft and ductile, or brittle, depending on the strain of bacteria, the feedstock type or the fermentation conditions used (Naser et al., 2021).

Commercial suppliers of PHA raw materials and products include Danimer Scientific, Biomer (Germany), Biocycle (Brazil), Kaneka (Japan), TianAn (China) and Tepha (USA).

PHA is produced naturally by bacteria, in hundreds of both gram-positive as well as gram-negative types (Song et al., 2018). It can also be produced via modified plants or through chemical reactions (Madison & Huisman, 1999). During the fermentation cycle, the cells are allowed to proliferate under optimal growth conditions. This stage is followed by the accumulation of PHA polymer by limiting the nutrient supply and, thus, the growth of the organisms. As PHA is used as energy storage medium, this stage is crucial for accumulation of PHA. These two stages are repeated for a few cycles before the cells are harvested and isolated. PHA is recovered by lysing the cells, followed by processing into powder or pellets (Madison & Huisman, 1999).

Optimal PHA production is dependent on many factors, including the strain of bacteria utilised, the type of carbon source, and processing conditions such as pH, temperature and fermentation time. Fermentation can be performed in batch, fed-batch or continuous modes (Bedade et al., 2021). PHA processing remains more challenging and expensive than that of conventional plastics due to the variations in production rates and molecular weights of PHAs produced, as well as due to challenges in developing cost-effective downstream processing and manufacturing processes (Chen, 2009, Chen & Jiang, 2017). Bedade and colleagues published a comprehensive review on the production of PHA to gain more in-depth knowledge on the subject (Bedade et al., 2021).

Synthesis conditions such as the carbon source and the strain or strains of bacteria are also varied in order to alter the properties such as crystallinity, mechanical properties and amphilicity of the final product (Li et al., 2017, Saito & Doi, 1994). The use of synthetic biology has been applied to alter certain bacteria to produce diverse PHA structures (Chen et al., 2016, Li et al., 2017). This approach has been helped by the rise of genetic and molecular studies enabling the characterization of various genes in PHA-producing as well as PHA-degrading microorganisms to elucidate the multitude of enzymatic pathways utilised in the processes (Madison & Huisman, 1999). Post-production chemical modification can also be utilized to reduce the hydrophobicity of the PHAs via the insertion of hydrophilic sections, or PHAs can be blended with other PHAs, other biodegradable polymers or grafted or copolymerized (Hazer & Steinbüchel, 2007).

The addition of modified MMT to poly(hydroxybutyrate-co-hydroxyvaleate), for example, improved the thermal stability of the composite (Corrêa et al., 2012), while a biocomposite of PHA with nanoclay was prepared without compromising mechanical and thermal properties (Garcia-Quiles et al., 2019). Composites with bacterial cellulose nanowhiskers were also prepared (Martinez-Sanz et al., 2014).

PHAs display excellent biocompatibility and, as such, are well suited to biomedical applications (Wang et al., 2013). PHAs, however, lack functional groups and, as such, are thus often incompatible with drugs (Chen & Wu, 2005), while their high hydrophobicity makes them unsuitable for many biomedical applications (Li et al., 2016). An exception is PHBV, a PHA with a high degree of crystallinity and resistance to UV radiation, along with being biocompatible and biodegradable (Naser et al., 2021). PHB and PHBV are, however, suitable for packaging applications due

TABLE 1.1
Applications of Polyhydroxyalkanoates

PHA	Application	References
PHA	Bone tissue engineering	Lim et al. (2017)
Poly(3-hydroxybutyrate-co-4-hydroxybutyrate)	Accelerated wound healing	Shishatskaya et al. (2016)
PHBV	Drug delivery	Vilos et al. (2013)
PHBV (poly(3-hydroxybutyrate-co-3-hydroxyvalerate))	Bone regeneration (with HA)	Öner and Ilhan (2016)
Poly(3-hydroxyalkanoate)	Tissue engineering (electrospun scaffolds)	Grande et al. (2017)
Poly(3-hydroxybutyrate-co-3-hydroxyhexanoate)	With cotton, for improved load-bearing material	Battegazzore et al. (2019)
PHB	Coated onto cellulose cardboard for packaging	Cyras et al. (2009)
PHBV	With clay nanofillers for improved barrier properties (gas and water permeability as well as UV resistance) for packaging	Sanchez-Garcia and Lagaron (2010)

to their favourable water permeability characteristics and hydrophobic nature (Naser et al., 2021).

PHAs have been applied in various applications, as listed in Table 1.1.

1.3.3 Thermoplastic Starch and Composites of TPS

Native starch is highly hydrophilic and thus loses mechanical strength when exposed to moisture. It is also brittle and lacking in mechanical strength (Diyana et al., 2021). The incorporation of a plasticizer is required to convert the starch to a thermoplastic material, as well as the application of heat and shear force. Plasticizing agents such as glycerol, glucose, sorbitol or compounds with nitrogen such as urea are incorporated into the starch during conventional plastic processing technologies such as extrusion or injection moulding (Montilla-Buitrago et al., 2021). The semi-crystalline nature of the starch is altered via plasticisation into an amorphous polymer-type structure. Once plasticized, the TPS is able to flow like a conventional plastic. The final physical characteristics of the produced TPS are dependent on the proportion and type of plasticizer added (Sanyang et al., 2015, Abera et al., 2020) as well as the conditions and shear force under which processing took place.

TPS products have good oxygen barrier properties but are still more susceptible to moisture than other materials. Changes in the relative humidity can thus affect TPS properties. TPS also lacks in mechanical strength compared to conventional plastics (Martinez Villadiego et al., 2022). TPS is often blended with PLA to reduce costs inherent to native PLA and to improve the mechanical properties of TPS (Reis et al., 2018, Schwach et al., 2008, Volpe et al., 2018). The hydrophobicity of PLA, in contrast to the hydrophilicity of TS, results in incompatibility between the two compounds. (Martinez Villadiego et al., 2022). Compatibilizers are thus often added to improve blending and thus also final product characteristics (Xiong et al., 2013, Noivoil & Yoksan, 2020).

TPS is often reinforced with filler compounds to improve mechanical and barrier properties. Cassava starch films were reinforced with lignocellulose nanofibers from cassava bagasse with improved mechanical and barrier properties (Travalini et al., 2019), while fillers of cassava and ahipa starch (Florencia et al., 2020) or sugarcane bagasse fibres (dos Santos et al., 2018) incorporated into TPs from corn starch resulted in improved mechanical properties compared to the native TPS films. The use of fried sunflower oil, a waste product from fast food production, was utilised in combination with glycerol as a natural plasticizer for processing TS from corn starch via twin screw extrusion (Volpe et al., 2018). It was found that when a mixture of oil and glycerol was utilised, the thermal stability and mechanical properties of the material were improved (Volpe et al., 2018).

Barrier properties of TPS films for food packaging applications can be improved via several different fillers. TPS films from corn starch were mixed with microalgae via melt mixing, and the addition of specific algae resulted in improved water and oxygen permeability (Fabra et al., 2018). The mechanical, thermal and barrier properties of tapioca thermoplastic starch films were improved by the addition of microcrystalline cellulose for food packaging applications (Othman et al., 2019).

Foams of cassava starch reinforced with grape stalks were prepared via thermal expansion, and it was observed that these structures would make suitable packaging material for food with low moisture content (Engel et al., 2019). Trays for food packaging were prepared from PLA and TS via flat extrusion and thermopressing and coated with beeswax in order to reduce water permeability (Reis et al., 2018).

1.3.4 Cellulose Ester, Regenerated Cellulose, Wood and Other Natural Materials

Nanocellulose is applied as cellulose nanocrystals (CNCs), microfibrillated cellulose or bacterial cellulose (BC). The type of nanocellulose formed depends on the sources of cellulose and the processing conditions used. Nanostructures with specific dimensions and functions are formed, thus presenting combinations of the inherent properties of cellulose with features specific to nanoscale materials (Klemm et al., 2011). Recent reviews into nanocellulose and its applications are available for those interested (Vanderfleet & Cranston, 2020, Eichhorn et al., 2022).

CNCs are generally synthesised via acid hydrolysis of cellulose sources (Vanderfleet & Cranston, 2020). Cellulose nanofibers (CNFs) are produced via mechanical processing of wood pulp via processes such as high-pressure homogenisation, cryocrushing, grinding and high-intensity ultrasonication (Abdul Khalil et al., 2014). Pretreatment can be applied to reduce the high energy requirements of the mechanical processes as well as improve the extent of nanofibrillation of the process (Isogai, 2013). Both CNCs and CNFs have very good mechanical properties and are thus often used as reinforcement for other compounds, although proper dispersion of the reinforcement is critical to an even improvement in mechanical strength (Wang et al., 2020). CNFs also display large surface areas, are easily functionalized and transparent with low thermal expansion, making them suitable for applications such as water treatment and energy storage (Klemm et al., 2018).

CNCs are amphiphilic with good optical properties and can be produced with consistency (Reid et al., 2017). The average size distribution of produced CNCs is also narrower than those achieved for CNFs and BC. Even though acids used for hydrolysis during CNC production are not environmentally friendly per se, they can be recycled, and sugars that are produced in waste can be utilised to produce biofuel (Klemm et al., 2018).

BC, as the name suggests, is produced by bacteria, most typically from the *Gluconacetobacter* genus, in the form of hydrogels. Drying of the gel is performed via air and press drying, freeze drying or dewatering via solvent exchange, and the chosen method impacts significantly on the properties of the dried product such as porosity, fibre aggregation, brittles or changes to the 3D network of the BCs (Klemm et al., 2018). In general, BC is highly crystalline with good mechanical properties (Klemm et al., 2011). It has high purity and is biocompatible, thus making it very well suited to biomedical applications. Due to the biological synthesis of BC, it can be altered or manipulated via synthetic biology without the need for further post-treatment. CNC is also known to be biocompatible and thus is suitable for targeted drug delivery and cell culturing (Roman et al., 2009, Dugan et al., 2010).

Cellulose nanocomposites are prepared via solvent-casting, melt processing techniques (batch or continuous), with foams prepared via compression or injection moulding and fibres prepared via spinning techniques (Oksman et al., 2016). Nanocellulose is used in various applications, such as energy storage and electronic applications (Sabo et al., 2016, Du et al., 2017), packaging, biomedical applications (Eichhorn et al., 2022) and water treatment (Carpenter et al., 2015). NC is widely studied for biomedical applications due to its inherent biocompatibility, biodegradability, favourable mechanical properties and non-toxicity (Lin & Dufresne, 2014, Tortorella et al., 2020). Nanocellulose is especially well suited as additive or matrix material in drug delivery due to its ability to modulate release and due to good mechanical properties (Eichhorn et al., 2022). Pure nanocellulose is however not bioactive and must generally be functionalized to produce scaffolds for tissue engineering with sufficient bioactivity (Hua et al., 2016). Nanocellulose is also utilised in water treatment by acting either as the active adsorbent (in the case of charged or functionalized materials, or due to inherent porosity and large specific surface area) or as a support structure for other active agents (Carpenter et al., 2015).

A summary of the use of nanocellulose or composites of nanocellulose with other materials from largely renewable sources in various biomedical applications is given in Table 1.2.

Commercially available biomedical products based on nanocellulose include Membracel® by Vuelopharma for wound dressing (Vuelopharma, 2023), TuniCell® for cell culturing and tissue engineering (Ocean TuniCell, 2023), Xcell® wound dressing and Gengiflex® membranes for dental applications, amongst others.

A composite of CNF with carbonated HA was studied for the pH-independent removal of metals, phosphates and nitrates from water (Hokkanen et al., 2014), while carboxymethylated CNFs from wood pulp were used for the successful adsorption of

TABLE 1.2
Biomedical Applications of Nanocellulose

Type of Nanocellulose	Application	Effects/Improvements	References
CNFs and CNCs (electrospun scaffolds)	Tissue engineering scaffolds	Improved mechanical and thermal properties Enhanced cell proliferation	He et al. (2014)
CNFs from wood (coated onto commercial mesh)	Cell culture	Promoted skin cell proliferation	Pajorova et al. (2020)
CNFs (bioactive 3D printed scaffolds with alginate and single-walled carbon nanotubes)	Cell culture	Enhanced cell differentiation	Bordoni et al. (2020)
CNC (as reinforcement in collagen hydrogel)	Tissue engineering	Enhanced delivery of mesenchymal stem cells for cartilage healing	Zhang et al. (2020)
BC (modified with silk fibroin proteins to create sponge-type scaffold)	Tissue engineering	Induced cell attachment and proliferation	Oliveira Barud et al. (2015)
CNFs (aerogels with drug-filled protein-coated nanoparticles)	Drug delivery	Controlled drug release	Valo et al. (2013)
CNC (hydrogel with starch)	Drug delivery	Sustained oral delivery of Vit B12	Mauricio et al. (2015)
CNC	Drug delivery	Topical delivery of hydroquinone	Taheri and Mohammadi (2015)
CNC	Drug delivery	Enhanced release of ampicillin	Poonguzhali et al. (2018)
BC and plant CNF combined with high amylose starch and pectin	Drug delivery	Improved controlled release	Meneguin et al. (2017)
BC	Drug delivery	Controlled release of albumin	Muller et al. (2013)
BC with sodium alginate	Drug delivery	Modulated ibuprofen release	Shi et al. (2014)
BC (film with calcium carbonate and seaweed extract)	Drug delivery	Controlled and sustained drug release	Cacicedo et al. (2015)

Cu(II) (Wang et al., 2019). BC, CNFs and CNCs were used to make nanopapers via thermopressing and the characteristics measured indicated that the nanopapers can be used as ultrafiltration membranes (Mautner et al., 2015). Unmodified CNFs were prepared from waste from coffee processing with a porous structure and could sufficiently remove dye from water (elAchaby et al., 2019). Nanocellulose-based films for packaging applications have also been studied with the addition of compounds for antimicrobial bioactive packaging (Cozzolino et al., 2013) or with PLA additive for improved barrier properties and hydrophobicity as compared to the native nanocellulose (Urbina et al., 2016).

Even though true commercial applications of nanocellulose are not yet widely available, the rise in the amount of patents filed in the last decade or so (Charreau et al., 2020) indicates that this could soon change.

The use of waste materials from renewable sources in other applications also contributes to the goal of increasing sustainability of human activity. The use of short sugar cane bagasse fibres for additives to cement for construction was investigated (Hernandez-Olivares et al., 2020), and it was found that the reinforced cement showed good physical and mechanical properties, even at relatively high bagasse fractions. Aerogels prepared from waste newspaper were prepared and showed good physical properties such as low density and hydrophobicity and sorption capacity for oils and solvents for potential use as pollution absorbents (Han et al., 2016).

1.4 FUTURE PERSPECTIVES

Currently the biodegradable polymer market is still extremely small when compared to the conventional fossil fuel-based polymer market. This is largely due to the differences in costs between conventional and generally much more expensive biodegradable polymers. As a result, current and upcoming research and industry efforts will largely focus on improving the material properties as well as cost-effectiveness of biopolymer production to improve competitiveness with the conventional plastics market. The increasing use of agricultural residues and abundant naturally occurring materials such as algae as fillers can improve the material properties of PLA and thermoplastic starches at reduced costs and minimal energy requirements. The increasing use of inexpensive agricultural waste as fermentation feedstock will reduce PHA prices, as well as the development of more cost-effective downstream processing, while more efficient production can be obtained via genetically engineering bacterial strains (Bedade et al., 2021).

1.5 CONCLUSION

Biodegradable polymers have been shown to be viable alternatives to conventional plastics in a wide range of applications. The development of biodegradable polymers derived from non-food lignocellulose biomass is an especially promising alternative, and the production efficiency of microbially synthesized PHAs can still be much improved.

The level of biodegradation of bioplastic products can be tailored and controlled to suit specific applications and disposal requirements. Options available to consumers for the responsible waste disposal of biodegradable polymers are increasing, including industrial and home composting, soil biodegradation, anaerobic digestion, and reuse and recyclability. It is, however, important that the uptake of biopolymers be accompanied by increased public awareness of the correct avenues for disposal of products to ensure efficient biodegradation and minimal environmental impact.

Ultimately, the uptake of biopolymers will be greatly affected by increased global awareness of the damage done to the natural environment and, ultimately, humans themselves as a result of careless and unfettered use of non-biodegradable plastics from finite sources. This increased awareness will also accelerate the implementation of more regulations on the usage of non-biodegradable plastics. Regulatory pressure in turn will further advance efforts to develop more cost-effective and efficient biopolymers.

REFERENCES

Abdul Khalil, H. P. S., Davoudpour, Y., Islam, M. N., Mustapha, A., Sudesh, K., Dungani, R., & Jawaid, M., 2014. Production and modification of nanofibrillated cellulose using various mechanical processes: A review. *Carbohydrate Polymers*, 99, 649–665. https://doi.org/10.1016/J.CARBPOL.2013.08.069.

Abera, G., Woldeyes, B., Demash, H. D., & Miyake, G., 2020. The effect of plasticizers on thermoplastic starch films developed from the indigenous Ethiopian tuber crop Anchote (Cocciniaabyssinica) starch. *International Journal of Biological Macromolecules*, 155, 581–587. https://doi.org/10.1016/J.IJBIOMAC.2020.03.218.

Ahmed, J., & Varshney, S. K., 2011. Polylactides-chemistry, properties and green packaging technology: A review. *International Journal of Food Properties*, 14, 37–58.

Alexandre, M., & Dubois, P., 2000. Polymer layered silicate nanocomposites: Preparation, properties and uses of a new class of materials. *Materials Science and Engineering*, 28, 1–163.

Allred, R. E., & Busselle, L. D., 2000. Tertiary recycling of automotive plastics and composites. *Journal of Thermoplastic Composite Materials*, 13, 92.

Almeida, D., 2011. Life cycle engineering approach to analyse the performance of biodegradable injection moulding plastics. Master, Instituto Superior TecnicoUniversidadeTecnica de Lisboa.

Amnael Orozco-Díaz, C., Moorehead, R., Reilly, G. C., Gilchrist, F., & Miller, C., 2020. Characterization of a composite polylactic acid-hydroxyapatite 3D-printing filament for bone-regeneration. *Biomedical Physics & Engineering Express*, 6(2), 025007. https://doi.org/10.1088/2057-1976/AB73F8.

Arakawa, K., Madaa, T., Parka, S., & Todoa, M., 2006. Tensile fracture behaviour of a biodegradable polymer, poly (lactic acid). *Polymer Testing*, 25, 628–634.

Armentano, I., Fortunati, E., Burgos, N., Dominici, F., Luzi, F., Flori, S., Jimenz, A., Yoon, K., Ahn, J., Kang, S., & Kenny, J. M., 2015. Processing and characterization of plasticized PLA/PHB blends for biodegradable multiphase system. *Express Polymer Letter*, 9, 583–596.

Auras, R., Harte, B., & Selke, S., 2004. An overview of polylactides as packaging materials. *Macromolecular Bioscience*, 4(9), 835–864. https://doi.org/10.1002/MABI.200400043.

Auras, R. A., Harte, B., Selke, S., & Hernandez, R., 2003. Mechanical, physical and barrier properties of poly(lactide) films. *Journal of Plastic Film and Sheeting*, 19, 123–135.

Auras, R. A., Lim, L. T., Selke, S. E., & Tsuji, H., 2011. *Poly (lactic acid): synthesis, structure, properties, processing and applications*, John Wiley & Sons, Hoboken, NJ.

Auras, R. A., Singh, S. P., & Singh, J. J., 2005. Evaluation of oriented poly(lactide) polymers vs existing PET and oriented PS for fresh food service containers. *Packaging Technology and Science*, 18, 207–216.

Baiardo, M., Frisoni, G., Scandola, M., Rimelen, M., Lips, D., Ruffieux, K., & Wintermantel, E., 2003. Thermal and mechanical properties of plasticized poly(L-lactic acid). *Journal of Applied Polymer Science*, 90, 1731–1738.

Baimark, Y., & Srisa-Ard, M., 2012. Preparation of drug-loaded microspheres of linear and star-shaped poly(D,L-lactide)s and their drug release behaviors. *Journal of Applied Polymer Science*, 124(5), 3871–3878. https://doi.org/10.1002/APP.35473.

Barghini, A., Ivanova, V. I., Imam, S. H., & Chiellini, E., 2010. Poly-(Ɛ-caprolactone)(PCL) and poly (hydroxyl-butyrate)(PHB) blends containing seaweed fibers: Morphology and thermal-mechanical properties. *Journal of Polymer Science Part A: Polymer Chemistry*, 48, 5282–5288.

Barnes, D. K. A., Galgani, F., Thompson, R. C., & Barlaz, M., 2009. Accumulation and fragmentation of plastic debris in global environments. *Philosophical Transactions of the Royal Society*, 364, 1985–1998.

Battegazzore, D., Abt, T., Maspoch, M. L., & Frache, A., 2019. Multilayer cotton fabric bio-composites based on PLA and PHB copolymer for industrial load carrying applications. *Composites Part B: Engineering*, 163, 761–768. https://doi.org/10.1016/J.COMPOSITESB.2019.01.057.

Bedade, D. K., Edson, C. B., & Gross, R. A., 2021. Emergent approaches to efficient and sustainable polyhydroxyalkanoate production. *Molecules*, 26(11), 3463. https://doi.org/10.3390/MOLECULES26113463.

Bohdan, K., 2022. Estimating global marine surface microplastic abundance: Systematic literature review. *Science of the Total Environment*, 832, 155064. https://doi.org/10.1016/J.SCITOTENV.2022.155064.

Bonilla, J., Fortunati, E., Vargas, M., Charalt, A., & Kenny, J. M., 2013. Effects of chitosan on the physicochemical and antimicrobial properties of PLA films. *Journal of Food Engineering*, 119, 236–243.

Bordoni, M., Karabulut, E., Kuzmenko, V., Fantini, V., Pansarasa, O., Cereda, C., & Gatenholm, P., 2020. 3D printed conductive nanocellulose scaffolds for the differentiation of human neuroblastoma cells. *Cells*, 9(3), 682. https://doi.org/10.3390/CELLS9030682.

Browne, M. A., Dissanayake, A., Galloway, T. S., Lowe, D. M., & Thompson, R. C. 2008. Ingested microscopic plastic translocates to the circulatory system of the mussel, *Mytilus edulis* (L.). *Environmental Science and Technology*, 42(13), 5026–5031. https://doi.org/10.1021/ES800249A/SUPPL_FILE/ES800249A-FILE002.PDF.

Bulota, M., & Budtova, T., 2015. PLA/algae composites: Morphology and mechanical properties. *Composites: Part A*, 73, 109–115.

Cacicedo, M. L., Cesca, K., Bosio, V. E., Porto, L. M., & Castro, G. R., 2015. Self-assembly of carrageenin-CaCO3 hybrid microparticles on bacterial cellulose films for doxorubicin sustained delivery. *Journal of Applied Biomedicine*, 13(3), 239–248. https://doi.org/10.1016/J.JAB.2015.03.004.

Carpenter, A. W., de Lannoy, C. F., & Wiesner, M. R., 2015. Cellulose nanomaterials in water treatment technologies. *Environmental Science and Technology*, 49(9), 5277–5287. https://doi.org/10.1021/ES506351R/ASSET/IMAGES/MEDIUM/ES-2014-06351R_0006.GIF.

Castro-Aguirre, E., Iñiguez-Franco, F., Samsudin, H., Fang, X., & Auras, R., 2016. Poly(lactic acid)-Mass production, processing, industrial applications, and end of life. *Advanced Drug Delivery Reviews*, 107, 333–366. https://doi.org/10.1016/J.ADDR.2016.03.010.

Charreau, H., Cavallo, E., & Foresti, M. L., 2020. Patents involving nanocellulose: Analysis of their evolution since 2010. *Carbohydrate Polymers*, 237, 116039. https://doi.org/10.1016/J.CARBPOL.2020.116039.

Chen, C., Dong, L., & Cheung, M. K., 2005. Preparation and characterization of biodegradable poly(L-lactice)/chitosan blends. *European Polymer Journal*, 41, 958–966.

Chen, G. Q., 2009. A microbial polyhydroxyalkanoates (PHA) based bio- and materials industry. *Chemical Society Reviews*, 38(8), 2434–2446. https://doi.org/10.1039/B812677C.

Chen, G. Q., & Jiang, X. R., 2017. Engineering bacteria for enhanced polyhydroxyalkanoates (PHA) biosynthesis. *Synthetic and Systems Biotechnology*, 2(3), 192–197. https://doi.org/10.1016/J.SYNBIO.2017.09.001.

Chen, G. Q., Jiang, X. R., & Guo, Y., 2016. Synthetic biology of microbes synthesizing polyhydroxyalkanoates (PHA). *Synthetic and Systems Biotechnology*, 1(4), 236–242. https://doi.org/10.1016/J.SYNBIO.2016.09.006.

Chen, G. Q., & Wu, Q., 2005. The application of polyhydroxyalkanoates as tissue engineering materials. *Biomaterials*, 26(33), 6565–6578. https://doi.org/10.1016/J.BIOMATERIALS.2005.04.036.

Chen, R., Curran, S. J., Curran, J. M., & Hunt, J. A., 2006. The use of poly(l-lactide) and RGD modified microspheres as cell carriers in a flow intermittency bioreactor for tissue engineering cartilage. *Biomaterials*, 27(25), 4453–4460. https://doi.org/10.1016/J.BIOMATERIALS.2006.04.011.

Convery, F., McDonnell, S., & Ferreira, S., 2007. The most popular tax in Europe? Lessons from the Irish plastic bags levy. *Environmental and Resource Economics*, 38, 1–11.

Corrêa, M. C. S., Branciforti, M. C., Pollet, E., Agnelli, J. A. M., Nascente, P. A. P., & Avérous, L., 2012. Elaboration and characterization of nano-biocomposites based on plasticized poly(hydroxybutyrate-co-hydroxyvalerate) with organo-modified montmorillonite. *Journal of Polymers and the Environment*, 20(2), 283–290. https://doi.org/10.1007/S10924-011-0379-0/FIGURES/7.

Correlo, V. M., Boesel, L. F., Bhattacharya, M., Mano, J. F., Neves, N. M., & Reis, R. L., 2005. Properties of melt processed chitosan and aliphatic polyester blends. *Materials Science and Engineering*, 403, 57–68.

Cozzolino, C. A., Nilsson, F., Iotti, M., Sacchi, B., Piga, A., & Farris, S., 2013. Exploiting the nano-sized features of microfibrillated cellulose (MFC) for the development of controlled-release packaging. *Colloids and Surfaces B: Biointerfaces*, 110, 208–216. https://doi.org/10.1016/J.COLSURFB.2013.04.046.

Cyras, V. P., Soledad, C. M., & Analía, V., 2009. Biocomposites based on renewable resource: Acetylated and non acetylated cellulose cardboard coated with polyhydroxybutyrate. *Polymer*, 50(26), 6274–6280. https://doi.org/10.1016/J.POLYMER.2009.10.065.

Dabertrand, M., Audonnet, F., & De Baynast, H., 2022. Chitosan as reinforcement for biopolymers—A mini review. *Polymer Science: Peer Review Journal*, 3, 1–4.

de Souza Machado, A. A., Kloas, W., Zarfl, C., Hempel, S., & Rillig, M. C., 2018. Microplastics as an emerging threat to terrestrial ecosystems. *Global Change Biology*, 24(4), 1405–1416. https://doi.org/10.1111/GCB.14020.

Deng, Y., Yu, C., Wongwiwattana, P., & Thomas, N. L., 2018. Optimising ductility of poly(lactic acid)/poly(butylene adipate-co-terephthalate) blends through co-continuous phase morphology. *Journal of Polymers and the Environment*, 26, 3802–3816.

DeStefano, V., Khan, S., & Tabada, A., 2020. Applications of PLA in modern medicine. *Engineered Regeneration*, 1, 76–87. https://doi.org/10.1016/J.ENGREG.2020.08.002.

Dimonie, M., & Rapa, M., 2010. Biodegradable blends based on PHB and wood fiber. *University Politehnica of Bucharest Scientific Bulletin*, 72, 3–10.

Diyana, Z. N., Jumaidin, R., Selamat, M. Z., Ghazali, I., Julmohammad, N., Huda, N., & Ilyas, R. A., 2021. Physical properties of thermoplastic starch derived from natural resources and its blends: A review. *Polymers*, 13(9), 5–20. https://doi.org/10.3390/POLYM13091396.

dos Santos, B. H., de Souza Do Prado, K., Jacinto, A. A., & da Silva Spinacé, M. A., 2018. Influence of sugarcane bagasse fiber size on biodegradable composites of thermoplastic starch. *Journal of Renewable Materials*, 6(2), 176. https://doi.org/10.7569/JRM.2018.634101.

Du, X., Zhang, Z., Liu, W., & Deng, Y., 2017. Nanocellulose-based conductive materials and their emerging applications in energy devices—A review. *Nano Energy*, 35, 299–320. https://doi.org/10.1016/J.NANOEN.2017.04.001.

Dugan, J. M., Gough, J. E., & Eichhorn, S. J., 2010. Directing the morphology and differentiation of skeletal muscle cells using oriented cellulose nanowhiskers. *Biomacromolecules*, 11(9), 2498–2504. https://doi.org/10.1021/BM100684K/ASSET/IMAGES/LARGE/BM-2010-00684K_0006.JPEG.

Eichhorn, S. J., Etale, A., Wang, J., Berglund, L. A., Li, Y., Cai, Y., Chen, C., Cranston, E. D., Johns, M. A., Fang, Z., Li, G., Hu, L., Khandelwal, M., Lee, K. Y., Oksman, K., Pinitsoontorn, S., Quero, F., Sebastian, A., Titirici, M. M., ... Frka-Petesic, B., 2022. Current international research into cellulose as a functional nanomaterial for advanced applications. *Journal of Materials Science*, 57(10), 5697–5767. https://doi.org/10.1007/S10853-022-06903-8.

elAchaby, M., Ruesgas-Ramón, M., Fayoud, N. E. H., Figueroa-Espinoza, M. C., Trabadelo, V., Draoui, K., & ben Youcef, H., 2019. Bio-sourced porous cellulose microfibrils from coffee pulp for wastewater treatment. *Cellulose*, 26(6), 3873–3889. https://doi.org/10.1007/S10570-019-02344-W/TABLES/2.

Engel, J. B., Ambrosi, A., & Tessaro, I. C., 2019. Development of biodegradable starch-based foams incorporated with grape stalks for food packaging. *Carbohydrate Polymers*, 225, 115234. https://doi.org/10.1016/J.CARBPOL.2019.115234.

Everaert, G., van Cauwenberghe, L., de Rijcke, M., Koelmans, A. A., Mees, J., Vandegehuchte, M., & Janssen, C. R., 2018. Risk assessment of microplastics in the ocean: Modelling approach and first conclusions. *Environmental Pollution*, 242, 1930–1938. https://doi.org/10.1016/J.ENVPOL.2018.07.069.

Fabra, M. J., Martínez-Sanz, M., Gómez-Mascaraque, L. G., Gavara, R., & López-Rubio, A., 2018. Structural and physicochemical characterization of thermoplastic corn starch films containing microalgae. *Carbohydrate Polymers*, 186, 184–191. https://doi.org/10.1016/J.CARBPOL.2018.01.039.

Farah, S., Anderson, D. G., & Langer, R., 2016. Physical and mechanical properties of PLA, and their functions in widespread applications—A comprehensive review. *Advanced Drug Delivery Reviews*, 107, 367–392. https://doi.org/10.1016/J.ADDR.2016.06.012.

Florencia, V., López, O. V., & García, M. A., 2020. Exploitation of by-products from cassava and ahipa starch extraction as filler of thermoplastic corn starch. *Composites Part B: Engineering*, 182, 107653. https://doi.org/10.1016/J.COMPOSITESB.2019.107653.

Fossi, M. C., Coppola, D., Baini, M., Giannetti, M., Guerranti, C., Marsili, L., Panti, C., de Sabata, E., & Clò, S., 2014. Large filter feeding marine organisms as indicators of microplastic in the pelagic environment: The case studies of the Mediterranean basking shark (Cetorhinus maximus) and fin whale (*Balaenoptera physalus*). *Marine Environmental Research*, 100, 17–24. https://doi.org/10.1016/J.MARENVRES.2014.02.002.

Garcia-Quiles, L., Cuello, Á. F., & Castell, P., 2019. Sustainable materials with enhanced mechanical properties based on industrial polyhydroxyalkanoates reinforced with organomodified sepiolite and montmorillonite. *Polymers*, 11(4), 696. https://doi.org/10.3390/polym11040696.

Gasperi, J., Wright, S. L., Dris, R., Collard, F., Mandin, C., Guerrouache, M., Langlois, V., Kelly, F. J., & Tassin, B., 2018. Microplastics in air: Are we breathing it in? *Current Opinion in Environmental Science & Health*, 1, 1–5. https://doi.org/10.1016/J.COESH.2017.10.002.

Grande, D., Ramier, J., Versace, D. L., Renard, E., & Langlois, V., 2017. Design of functionalized biodegradable PHA-based electrospun scaffolds meant for tissue engineering applications. *New Biotechnology*, 37, 129–137. https://doi.org/10.1016/J.NBT.2016.05.006.

Grande, R., & Carvalho, A. J. F., 2011. Compatible ternary blends of chitosan/poly(vinyl alcohol)/poly(lactic acid) produced by oil-in-water emulsion processing. *Biomacromolecules*, 12, 907–914.

Gregory, M. R., 2009. Environmental implications of plastic debris in marine settingsentanglement, ingestion, smothering, hangers-on, hitch-hiking and alien invasions. *Philosophical Transactions of the Royal Society B: Biological Sciences*, 364(1526), 2013–2025. https://doi.org/10.1098/RSTB.2008.0265.

Guo, Y., Chang, C. C., Cuiffo, M. A., Xue, Y., Zuo, X., Pack, S., Zhang, L., He, S., Weil, E., & Rafailovich, M. H., 2017. Engineering flame retardant biodegradable polymer nanocomposites and their application in 3D printing. *Polymer Degradation and Stability*, 137, 205–215. https://doi.org/10.1016/J.POLYMDEGRADSTAB.2017.01.019.

Hamad, K., Kaseem, M., Ayyoob, M., Joo, J., & Deri, F., 2018. Polylactic acid blends: The future of green, light and tough. *Progree in Polymer Science*, 85, 83–127.

Han, S., Sun, Q., Zheng, H., Li, J., & Jin, C., 2016. Green and facile fabrication of carbon aerogels from cellulose-based waste newspaper for solving organic pollution. *Carbohydrate Polymers*, 136, 95–100. https://doi.org/10.1016/J.CARBPOL.2015.09.024.

Hazer, B., & Steinbüchel, A., 2007. Increased diversification of polyhydroxyalkanoates by modification reactions for industrial and medical applications. *Applied Microbiology and Biotechnology*, 74(1), 1–12. https://doi.org/10.1007/S00253-006-0732-8/TABLES/3.

He, X., Xiao, Q., Lu, C., Wang, Y., Zhang, X., Zhao, J., Zhang, W., Zhang, X., & Deng, Y., 2014. Uniaxially aligned electrospun all-cellulose nanocomposite nanofibers reinforced with cellulose nanocrystals: Scaffold for tissue engineering. *Biomacromolecules*, 15(2), 618–627. https://doi.org/10.1021/BM401656A/SUPPL_FILE/BM401656A_SI_001.PDF.

Hernández-Olivares, F., Elizabeth Medina-Alvarado, R., Burneo-Valdivieso, X. E., & Rodrigo Zúñiga-Suárez, A., 2020. Short sugarcane bagasse fibers cementitious composites for building construction. *Construction and Building Materials*, 247, 118451. https://doi.org/10.1016/J.CONBUILDMAT.2020.118451.

Hokkanen, S., Repo, E., Westholm, L. J., Lou, S., Sainio, T., & Sillanpää, M., 2014. Adsorption of Ni2+, Cd2+, PO43− and NO3− from aqueous solutions by nanostructured microfibrillated cellulose modified with carbonated hydroxyapatite. *Chemical Engineering Journal*, 252, 64–74. https://doi.org/10.1016/J.CEJ.2014.04.101.

Hua, K., Rocha, I., Zhang, P., Gustafsson, S., Ning, Y., Strømme, M., Mihranyan, A., & Ferraz, N., 2016. Transition from bioinert to bioactive material by tailoring the biological cell response to carboxylated nanocellulose. *Biomacromolecules*, 17(3), 1224–1233.

Huang, D., Chen, H., Shen, M., Tao, J., Chen, S., Yin, L., Zhou, W., Wang, X., Xiao, R., & Li, R., 2022. Recent advances on the transport of microplastics/nanoplastics in abiotic and biotic compartments. *Journal of Hazardous Materials*, 438, 129515. https://doi.org/10.1016/J.JHAZMAT.2022.129515.

Huang, L., Wu, Q., Wang, Q., & Wolcott, M., 2020. Interfacial crystals morphology modification in cellulose fiber/polypropylene composite by mechanochemical method. *Composites Part A: Applied Science and Manufacturing*, 130, 105765.

Isogai, A., 2013. Wood nanocelluloses: Fundamentals and applications as new bio-based nanomaterials. *Journal of Wood Science*, 59(6), 449–459. https://doi.org/10.1007/S10086-013-1365-Z/FIGURES/5.

Ivar Do Sul, J. A., & Costa, M. F., 2014. The present and future of microplastic pollution in the marine environment. *Environmental Pollution*, 185, 352–364. https://doi.org/10.1016/J.ENVPOL.2013.10.036.

Jacobsen, S., Fritz, H. G., Degée, P., Dubois, P., & Jérôme, R., 2000. New developments on the ring opening polymerisation of polylactide. *Industrial Crops and Products*, 11(2–3), 265–275. https://doi.org/10.1016/S0926-6690(99)00053-9.

Jiang, L., Wolcott, M. P., & Zhang, J., 2006. Study of biodegradable polylactide/poly(butylene adipate-co-terephthalate) blends, *Biomacromolecules*, 7, 199–207.

Klemm, D., Cranston, E. D., Fischer, D., Gama, M., Kedzior, S. A., Kralisch, D., Kramer, F., Kondo, T., Lindström, T., Nietzsche, S., Petzold-Welcke, K., & Rauchfuß, F., 2018. Nanocellulose as a natural source for groundbreaking applications in materials science: Today's state. *Materials Today*, 21(7), 720–748. https://doi.org/10.1016/J.MATTOD.2018.02.001.

Klemm, D., Kramer, F., Moritz, S., Lindström, T., Ankerfors, M., Gray, D., & Dorris, A., 2011. Nanocelluloses: A new family of nature-based materials. *AngewandteChemie International Edition*, 50(24), 5438–5466. https://doi.org/10.1002/ANIE.201001273.

Kontrick, A. V., 2018. Microplastics and human health: Our great future to think about now. *Journal of Medical Toxicology*, 14(2), 117–119. https://doi.org/10.1007/S13181-018-0661-9.

Kuciel, S., Mazur, K., & Hebda, M., 2020. The influence of wood and basalt fibers on mechanical, thermal and hydrothermal properties of PLA composite. *Journal of Polymer and Environmental*, 28, 1204–1215.

La Mantia, F. P., & Morreale, M., 2011. Green composites: A brief review. *Composites Part A: Applied Science and Manufacturing*, 42(6), 579–588. https://doi.org/10.1016/J.COMPOSITESA.2011.01.017.

Lannace, S., Nocilla, G., & Nicolais, L., 1999. Biocomposites based on sea algae fibers and biodegradable thermoplastic matrices. *Applied Polymer Science*, 73, 583–592.

Lau, W. W. Y., Shiran, Y., Bailey, R. M., Cook, E., Stuchtey, M. R., Koskella, J., Velis, C. A., Godfrey, L., Boucher, J., Murphy, M. B., Thompson, R. C., Jankowska, E., Castillo, A. C., Pilditch, T. D., Dixon, B., Koerselman, L., Kosior, E., Favoino, E., Gutberlet, J., … Palardy, J. E., 2020. Evaluating scenarios toward zero plastic pollution. *Science*, 369(6509), 1455–1461. https://doi.org/10.1126/science.aba9475.

Lee, H., & Kim, D. S., 2009. Preparation and physical properties of wood/polypropylene/clay nanocomposites. *Applied Polymer Science*, 111, 2769–2776.

Leiggener, C. S., Curtis, R., Müller, A. A., Pfluger, D., Gogolewski, S., & Rahn, B. A., 2006. Influence of copolymer composition of polylactide implants on cranial bone regeneration. *Biomaterials*, 27(2), 202–207. https://doi.org/10.1016/J.BIOMATERIALS.2005.05.068.

Lemoigne, M., 1926. De l'acide 3-oxybutyrique. *Bulletin de la Société de Chimie Biologique*, 8, 770.

Li, H., & Huneault, M. A., 2007. Effect of nucleation and plasticization on the crystallization of poly(lactic acid). *Polymer*, 48(23), 6855–6866. https://doi.org/10.1016/J.POLYMER.2007.09.020.

Li, L., Ding, S., & Zhou, C., 2004. Preparation and degradation of PLA/chitosan composite materials. *Journal of Applied Polymer Science*, 91, 274–277.

Li, Z., Yang, J., & Loh, X. J., 2016. Polyhydroxyalkanoates: Opening doors for a sustainable future. *NPG Asia Materials*, 8(4), e265–e265. https://doi.org/10.1038/am.2016.48.

Li, Z. J., Qiao, K., Che, X. M., & Stephanopoulos, G., 2017. Metabolic engineering of *Escherichia coli* for the synthesis of the quadripolymer poly(glycolate-co-lactate-co-3-hydroxybutyrate-co-4-hydroxybutyrate) from glucose. *Metabolic Engineering*, 44, 38–44. https://doi.org/10.1016/J.YMBEN.2017.09.003.

Lim, J., You, M., Li, J., & Li, Z., 2017. Emerging bone tissue engineering via polyhydroxyalkanoate (PHA)-based scaffolds. *Materials Science and Engineering: C*, 79, 917–929. https://doi.org/10.1016/J.MSEC.2017.05.132.

Lin, N., & Dufresne, A., 2014. Nanocellulose in biomedicine: Current status and future prospect. *European Polymer Journal*, 59, 302–325. https://doi.org/10.1016/J.EURPOLYMJ.2014.07.025.

Ljungberg, N., & Wesslen, B., 2003. Tributyl citrate oligomers as plasticizer for poly (lactic acid): Thermos-mechanical film properties and aging. *Polymer*, 44, 7679–7688.

Llorens, E., Calderón, S., del Valle, L. J., & Puiggalí, J., 2015. Polybiguanide (PHMB) loaded in PLA scaffolds displaying high hydrophobic, biocompatibility and antibacterial properties. *Materials Science and Engineering: C*, 50, 74–84. https://doi.org/10.1016/J.MSEC.2015.01.100.

Luckachan, G. E., & Pillai, C. H. S., 2011. Biodegradable polymers—A review on recent trends and emerging perspectives. *Journal of Polymer and Environment*, 19, 637–676.

Madison, L. L., & Huisman, G. W., 1999. Metabolic engineering of poly(3-hydroxyalkanoates): From DNA to plastic. *Microbiology and Molecular Biology Reviews*, 63(1), 21–53. https://doi.org/10.1128/MMBR.63.1.21-53.1999.

Maji, R., Dey, N. S., Satapathy, B. S., Mukherjee, B., & Mondal, S., 2015. Preparation and characterization of tamoxifen citrate loaded nanoparticles for breast cancer therapy. *International Journal of Nanomedicine*, 9(1), 3107–3118. https://doi.org/10.2147/IJN.S63535.

Martinez Villadiego, K., Judith Arias Tapia, M., Useche, J., & Escobar Macías, D., 2022. Thermoplastic Starch (TPS)/polylactic acid (PLA) blending methodologies: A review. *Journal of Polymers and the Environment*, 30, 75–91. https://doi.org/10.1007/s10924-021-02207-1.

Martinez-Sanz, M., Villano, M., Oliveira, C., Albuquerque, M. G. E., Majone, M., Reis, M., Lopez-Rubio, A., & Lagaron, J. M., 2014. Characterization of polyhydroxyalkanoates synthesized from microbial mixed cultures and of their nanobiocomposites with bacterial cellulose nanowhiskers. *New Biotechnology*, 31(4), 364–376. https://doi.org/10.1016/J.NBT.2013.06.003.

Matet, M., 2014. Ecole Polytechnique de montreal, Montréal, Canada. PhD Thesis.

Mauricio, M. R., da Costa, P. G., Haraguchi, S. K., Guilherme, M. R., Muniz, E. C., & Rubira, A. F., 2015. Synthesis of a microhydrogel composite from cellulose nanowhiskers and starch for drug delivery. *Carbohydrate Polymers*, 115, 715–722. https://doi.org/10.1016/J.CARBPOL.2014.07.063.

Mautner, A., Lee, K. Y., Tammelin, T., Mathew, A. P., Nedoma, A. J., Li, K., & Bismarck, A., 2015. Cellulose nanopapers as tight aqueous ultra-filtration membranes. *Reactive and Functional Polymers*, 86, 209–214. https://doi.org/10.1016/J.REACTFUNCTPOLYM.2014.09.014.

Mazur, K. E., Borucka, A., Kaczor, P., Gadek, S., Bogucki, R., Mirzewinski, D., & Kucie, S., 2022. Mechanical, thermal and microstructural characteristic of 3D printed ploylactide composites with natural fibers: Wood, bamboo and cork. *Journal of Polymers and the Environment*, 30, 2341–2354.

Mehta, N., Cunningham, E., Roy, D., Cathcart, A., Dempster, M., Berry, E., & Smyth, B. M., 2021. Exploring perceptions of environmental professionals, plastic processors, students and consumers of bio-based plastics: Informing the development of the sector. *Sustainable Production and Consumption*, 26, 574–587.

Meneguin, A. B., Ferreira Cury, B. S., dos Santos, A. M., Franco, D. F., Barud, H. S., & da Silva Filho, E. C., 2017. Resistant starch/pectin free-standing films reinforced with nanocellulose intended for colonic methotrexate release. *Carbohydrate Polymers*, 157, 1013–1023. https://doi.org/10.1016/J.CARBPOL.2016.10.062.

Mihai, M., Huneault, M. A., & Favis, B. D., 2010. Rheology and extrusion foaming of chain-branched poly(lactic acid). *Polymer Engineering & Science*, 50(3), 629–642. https://doi.org/10.1002/PEN.21561.

Molnar, K., Moczo, J., Murariu, M., Dubois, P., & Pukanszky, B., 2009. Factors affecting the properties of PLA/CaSO4 composites: Homogeneity and interactions. *Express Polymer Letter*, 3, 49–61.

Montilla-Buitrago, C. E., Gómez-López, R. A., Solanilla-Duque, J. F., Serna-Cock, L., & Villada-Castillo, H. S., 2021. Effect of plasticizers on properties, retrogradation, and processing of extrusion-obtained thermoplastic starch: A review. *Starch—Stärke*, 73(9–10), 2100060. https://doi.org/10.1002/STAR.202100060.

Moshood, T. D., Nawanir, G., Mahmud, F., Mohamad, F., Ahmad, M. H., & AbdulGhani, A., 2022. Biodegradable plastic applications towards sustainability: A recent innovations in the green product. *Cleaner Engineering and Technology*, 6, 100404–100417.

Müller, A., Ni, Z., Hessler, N., Wesarg, F., Müller, F. A., Kralisch, D., & Fischer, D., 2013. The biopolymer bacterial nanocellulose as drug delivery system: Investigation of drug loading and release using the model protein albumin. *Journal of Pharmaceutical Sciences*, 102(2), 579–592. https://doi.org/10.1002/JPS.23385.

Nakajima, H., Dijkstra, P., & Loos, K., 2017. The recent developments in biobased polymers toward general and engineering applications: Polymers that are upgraded from biodegradable polymers, analogous to petroleum-derived polymers, and newly developed. *Polymers*, 9(10), 523. https://doi.org/10.3390/polym9100523.

Naser, A. Z., Deiab, I., & Darras, B. M., 2021. Poly(lactic acid) (PLA) and polyhydroxyalkanoates (PHAs), green alternatives to petroleum-based plastics: A review. *RSC Advances*, 11(28), 17151–17196. https://doi.org/10.1039/D1RA02390J.

Nazhat, S. N., Kellomäki, M., TörmäLä, P., Tanner, K. E., & Bonfield, W., 2001. Dynamic mechanical characterization of biodegradable composites of hydroxyapatite and polylactides. *Journal of Biomedical Materials Research*, 58(4), 335–343.

Ng, E. L., Huerta Lwanga, E., Eldridge, S. M., Johnston, P., Hu, H. W., Geissen, V., & Chen, D., 2018. An overview of microplastic and nanoplastic pollution in agroecosystems. *Science of the Total Environment*, 627, 1377–1388. https://doi.org/10.1016/J.SCITOTENV.2018.01.341.

Nikvarz, N., Khayati, G. R., & Sharafi, S., 2020. Preparation of UV absorbent films using polylactic acid and grape syrup for food packaging application. *Materials Letters*, 276, 128187. https://doi.org/10.1016/J.MATLET.2020.128187.

Noivoil, N., & Yoksan, R., 2020. Oligo(lactic acid)-grafted starch: A compatibilizer for poly(lactic acid)/thermoplastic starch blend. *International Journal of Biological Macromolecules*, 160, 506–517. https://doi.org/10.1016/J.IJBIOMAC.2020.05.178.

Nomai, J., & Jarukumjorn, K., 2014. Effect of maleic anhydride grafted poly(lactic acid) on properties of sawdust/poly(lactic acid) composites toughened with poly(butylene adipate-co-terephthalate). *Advanced Materials Research*, 970, 74–78.

Noroozi, N., Schafer, L. L., & Hatzikiriakos, S. G., 2012. Thermorheological properties of poly (Ɛ-caprolactone)/polylactide blends. *Polymer Engineering Science*, 52, 2348–2359.

Ocean TuniCell, 2023. https://oceantunicell.com/product/

Oksman, K., Aitomäki, Y., Mathew, A. P., Siqueira, G., Zhou, Q., Butylina, S., Tanpichai, S., Zhou, X., & Hooshmand, S., 2016. Review of the recent developments in cellulose nanocomposite processing. *Composites Part A: Applied Science and Manufacturing*, 83, 2–18. https://doi.org/10.1016/J.COMPOSITESA.2015.10.041.

Okubo, K., Fujii, T., & Thostenson, E. T., 2009. Multi-scale hybrid biocomposite: Processing and mechanical characterization of bamboo fiber reinforced PLA with microfibrillated cellulose. *Composites Part A: Applied Science and Manufacturing*, 40, 469–475.

Oliveira Barud, H. G., Barud, H. D. S., Cavicchioli, M., do Amaral, T. S., de Oliveira Junior, O. B., Santos, D. M., de Oliveira Almeida Petersen, A. L., Celes, F., Borges, V. M., de Oliveira, C. I., de Oliveira, P. F., Furtado, R. A., Tavares, D. C., & Ribeiro, S. J. L., 2015. Preparation and characterization of a bacterial cellulose/silk fibroin sponge scaffold for tissue regeneration. *Carbohydrate Polymers*, 128, 41–51. https://doi.org/10.1016/J.CARBPOL.2015.04.007.

Öner, M., & Ilhan, B., 2016. Fabrication of poly(3-hydroxybutyrate-co-3-hydroxyvalerate) biocomposites with reinforcement by hydroxyapatite using extrusion processing. *Materials Science and Engineering: C*, 65, 19–26. https://doi.org/10.1016/J.MSEC.2016.04.024.

Othman, S. H., Majid, N. A., Tawakkal, I. S. M. A., Basha, R. K., Nordin, N., & Shapi'i, R. A., 2019. Tapioca starch films reinforced with microcrystalline cellulose for potential food packaging application. *Food Science and Technology*, 39(3), 605–612. https://doi.org/10.1590/FST.36017.

Pajorova, J., Skogberg, A., Hadraba, D., Broz, A., Travnickova, M., Zikmundova, M., Honkanen, M., Hannula, M., Lahtinen, P., Tomkova, M., Bacakova, L., & Kallio, P., 2020. Cellulose mesh with charged nanocellulose coatings as a promising carrier of skin and stem cells for regenerative applications. *Biomacromolecules*, 21(12), 4857–4870. https://doi.org/10.1021/ACS.BIOMAC.0C01097/SUPPL_FILE/BM0C01097_SI_002.AVI.

Pan, D., Su, F., Liu, C., & Guo, Z., 2020. Research progress for plastic waste management and manufacture of value-added products. *Advanced Composites and Hybrid Materials*, 3, 443–461.

Panichsombat, K., Panbangpong, W., Poompiew, N., & Potiyaraj, P., 2019. Biodegradable fibers from poly (lactic acid)/poly (butylene succinate) blends. *IOP Conference Series: Materials Science and Engineering*, 600(1), 012004. https://doi.org/10.1088/1757-899X/600/1/012004.

Park, J. W., & Im, S. S., 2002a. Morphological changes during heating in poly(L-lactic acid)/Poly(butylene succinate) blend systems as studied by synchroton X-ray scattering. *Journal of Polymer Science, Part B: Polymer Physics*, 40, 1931–1939.

Park, J. W., & Im, S. S., 2002b. Phase behaviour and morphology in blends of poly(L-lactic acid) and poly(butylene succinate). *Journal of Applied Polymer Science*, 86, 647–655.

Perego, G., Cella, G. D., & Bastioli, C., 1996. Effect of molecular weight and crystallinity on poly(lactic acid) mechanical properties. *Journal of Applied Polymer Science*, 59(1), 37–43. https://doi.org/10.1002/(SICI)1097-4628(19960103)59:1<37::AID-APP6>3.0.CO;2-N.

Petersen, K., Nielsen, P. V., & Olsen, M. B., 2001. Physical and mechanical properties of biobased materials - starch, polylactate and polyhydroxybutyrate. *Starch/Staerke*, 53, 356–361.

Petersson, L., Kvien, I., & Oksman, K., 2007. Structure and thermal properties of poly(lactic acid)/cellulose whiskers nanocomposite materials. *Composites Science and Technology*, 67(11–12), 2535–2544. https://doi.org/10.1016/J.COMPSCITECH.2006.12.012.

Poonguzhali, R., Khaleel Basha, S., & Sugantha Kumari, V., 2018. Synthesis of alginate/nanocellulose bionanocomposite for in vitro delivery of ampicillin. *Polymer Bulletin*, 75(9), 4165–4173. https://doi.org/10.1007/S00289-017-2253-2/FIGURES/5.

Qiang, T., Yu, D., & Gao, H., 2012. Wood flour/polylactide biocomposites toughened with polyhydroxyalkanoates. *Journal of Applied Polymer Science*, 124, 1831–1839.

Reid, M. S., Villalobos, M., & Cranston, E. D., 2017. Benchmarking cellulose nanocrystals: From the laboratory to industrial production. *Langmuir*, 33(7), 1583–1598. https://doi.org/10.1021/ACS.LANGMUIR.6B03765/ASSET/IMAGES/LARGE/LA-2016-037659_0003.JPEG.

Reis, M. O., Olivato, J. B., Bilck, A. P., Zanela, J., Grossmann, M. V. E., & Yamashita, F., 2018. Biodegradable trays of thermoplastic starch/poly (lactic acid) coated with beeswax. *Industrial Crops and Products*, 112, 481–487. https://doi.org/10.1016/J.INDCROP.2017.12.045.

Rizal, S., Olaiya, F. G., Saharudin, N. I., Abdullah, C. K., Olaiya, N. G., Mohamad Haafiz, M. K., Yahya, E. B., Sabaruddin, F. A., Ikramullah, & Abdul Khalil, H. P. S., 2021. Isolation of textile waste cellulose nanofibrillated fibre reinforced in polylactic acid-chitin biodegradable composite for green packaging application. *Polymers*, 13(3), 325. https://doi.org/10.3390/POLYM13030325.

Roman, M., Dong, S., Hirani, A., & Lee, Y. W., 2009. Cellulose nanocrystals for drug delivery. *ACS Symposium Series*, 1017, 81–91. https://doi.org/10.1021/BK-2009-1017.CH004.

Rosenboom, J. G., Langer, R., & Traverso, G., 2022. Bioplastics for a circular economy. *Nature Reviews Materials*, 7, 117–137.

Sabo, R., Yermakov, A., Law, C. T., & Elhajjar, R., 2016. Nanocellulose-enabled electronics, energy harvesting devices, smart materials and sensors: A review. *Journal of Renewable Materials*, 4(5), 297–312. https://doi.org/10.7569/JRM.2016.634114.

Saito, Y., & Doi, Y., 1994. Microbial synthesis and properties of poly(3-hydroxybutyrate-co-4-hydroxybutyrate) in Comamonasacidovorans. *International Journal of Biological Macromolecules*, 16(2), 99–104. https://doi.org/10.1016/0141-8130(94)90022-1.

Sajna, V. P., Mohanty, S., & Nayak, S. K., 2017. A study on thermal degradation kinetics and flammability properties of poly(lactic acid)/banana fiber/nanoclay hybrid bionanocomposites. *Polymer Composites*, 38(10), 2067–2079. https://doi.org/10.1002/PC.23779.

Salmas, C. E., Giannakas, A. E., Baikousi, M., Leontiou, A., Siasou, Z., & Karakassides, M. A., 2021. Development of poly(L-lactic acid)/chitosan/basil oil active packaging films via a melt-extrusion process using novel chitosan/basil oil blends. *Processes*, 9(1), 88. https://doi.org/10.3390/PR9010088.

Sanchez-Garcia, M. D., & Lagaron, J. M., 2010. Novel clay-based nanobiocomposites of biopolyesters with synergistic barrier to UV light, gas, and vapour. *Journal of Applied Polymer Science*, 118(1), 188–199. https://doi.org/10.1002/APP.31986.

Sanyang, M. L., Sapuan, S. M., Jawaid, M., Ishak, M. R., & Sahari, J., 2015. Effect of plasticizer type and concentration on dynamic mechanical properties of sugar palm starch-based films. *International Journal of Polymer Analysis and Characterization*, 20(7), 627–636. https://doi.org/10.1080/1023666X.2015.1054107.

Schwach, E., Six, J. L., & Avérous, L., 2008. Biodegradable blends based on starch and poly(lactic acid): Comparison of different strategies and estimate of compatibilization. *Journal of Polymers and the Environment*, 16(4), 286–297. https://doi.org/10.1007/S10924-008-0107-6/FIGURES/17.

Sebastien, F., Stephane, G., Copinet, A., & Coma, V., 2006. Novel biodegradable films made from chitosan and poly(lactic acid) with antifungal properties against mycotoxinogen strains. *Carbohydrate Polymers*, 65, 185–193.

Shi, X., Zheng, Y., Wang, G., Lin, Q., & Fan, J., 2014. pH- and electro-response characteristics of bacterial cellulose nanofiber/sodium alginate hybrid hydrogels for dual controlled drug delivery. *RSC Advances*, 4(87), 47056–47065. https://doi.org/10.1039/C4RA09640A.

Shishatskaya, E. I., Nikolaeva, E. D., Vinogradova, O. N., & Volova, T. G., 2016. Experimental wound dressings of degradable PHA for skin defect repair. *Journal of Materials Science: Materials in Medicine*, 27(11), 1–16. https://doi.org/10.1007/S10856-016-5776-4/TABLES/2.

Siakeng, R., Jawaid, M., Asim, M., Saba, N., Sanjay, M. R., Siengchin, S., & Fouad, H., 2020. Alkali treated coir/pineapple leaf fibres reinforced PLA hybrid composites: Evaluation of mechanical, morphological, thermal and physical properties. *eXPRESS Polymer Letters*, 14(8), 717–730. https://doi.org/10.3144/expresspolymlett.2020.59.

Sim, K. J., Han, S. O., & Seo, Y. B., 2009. Dynamic mechanical and thermal properties of red algae fiber reinforced Poly(lactic acid) biocomposites. *Macromolecular Research*, 18, 489–495.

Simoe, C. L., Viana, J. C., & Cunha, A. M., 2009. Mechanical properties of poly (Ɛ-caprolactone) and poly(lactic acid) blends. *Journal of Applied Polymer Science*, 112, 345–352.

Solomon Ogunola, O., Onada, O. A., & Falaye, A. E., 2018. Mitigation measures to avert the impacts of plastics and microplastics in the marine environment (a review). *Environmental Science and Pollution Research*, 25, 9293–9310. https://doi.org/10.1007/s11356-018-1499-z.

Song, R., Murphy, M., Li, C., Ting, K., Soo, C., & Zheng, Z., 2018. Current development of biodegradable polymeric materials for biomedical applications. *Drug Design, Development and Therapy*, 12, 3117–3145. https://doi.org/https://doi.org/10.2147/DDDT.S165440.

Spindola Vilela, C. L., Damasceno, T. L., Thomas, T., & Peixoto, R. S., 2022. Global qualitative and quantitative distribution of micropollutants in the deep sea. *Environmental Pollution*, 307, 119414. https://doi.org/10.1016/J.ENVPOL.2022.119414.

Sun, J., Shen, J., Chen, S., Cooper, M. A., Fu, H., Wu, D., & Yang, Z., 2018. Nanofiller reinforced biodegradable PLA/PHA composites: Current status and future trends. *Polymers*, 10(5), 505. https://doi.org/10.3390/POLYM10050505.

Taheri, A., & Mohammadi, M., 2015. The use of cellulose nanocrystals for potential application in topical delivery of hydroquinone. *Chemical Biology & Drug Design*, 86(1), 102–106. https://doi.org/10.1111/CBDD.12466.

Teamsinsungvon, A., & Ruksakulpiwat, Y., 2012. Poly (lactic acid)/poly (butylene adipate-co-terephthalate) blend and its composite: Effect of maleic anhydride grafted poly (lactic acid) as a compatibilizer. *Advanced Materials Research*, 410, 51–54.

Tortorella, S., Buratti, V. V., Maturi, M., Sambri, L., Franchini, M. C., & Locatelli, E., 2020. Surface-modified nanocellulose for application in biomedical engineering and nanomedicine: A review. *International Journal of Nanomedicine*, 15, 9909–9937. https://doi.org/10.2147/IJN.S266103.

Travalini, A. P., Lamsal, B., Magalhães, W. L. E., & Demiate, I. M., 2019. Cassava starch films reinforced with lignocellulose nanofibers from cassava bagasse. *International Journal of Biological Macromolecules*, 139, 1151–1161. https://doi.org/10.1016/J.IJBIOMAC.2019.08.115.

Tyler, B., Gullotti, D., Mangraviti, A., Utsuki, T., & Brem, H., 2016. Polylactic acid (PLA) controlled delivery carriers for biomedical applications. *Advanced Drug Delivery Reviews*, 107, 163–175. https://doi.org/10.1016/J.ADDR.2016.06.018.

United Nations Environment Programme., 2024. Global Waste Management Outlook 2024: Beyond an age of waste – Turning rubbish into resource, Nairobi. https://www.unep.org/resources/global-waste-management-outlook-2024.

Urbina, L., Algar, I., García-Astrain, C., Gabilondo, N., González, A., Corcuera, M., Eceiza, A., & Retegi, A., 2016. Biodegradable composites with improved barrier properties and transparency from the impregnation of PLA to bacterial cellulose membranes. *Journal of Applied Polymer Science*, 133(28), 43669. https://doi.org/10.1002/APP.43669.

Valo, H., Arola, S., Laaksonen, P., Torkkeli, M., Peltonen, L., Linder, M. B., Serimaa, R., Kuga, S., Hirvonen, J., & Laaksonen, T., 2013. Drug release from nanoparticles embedded in four different nanofibrillar cellulose aerogels. *European Journal of Pharmaceutical Sciences*, 50(1), 69–77. https://doi.org/10.1016/J.EJPS.2013.02.023.

Vanderfleet, O. M., & Cranston, E. D., 2020. Production routes to tailor the performance of cellulose nanocrystals. *Nature Reviews Materials*, 6(2), 124–144. https://doi.org/10.1038/s41578-020-00239-y.

Vieira, A., Vieira, J., Ferra, J., Magalhaes, F., Guedes, R., & Marques, A., 2011. Mechanical study of PLA-PCL fibers during in vitro degradation. *Journal of the Mechanical Behaviour of Materials*, 4, 451–460.

Vilos, C., Morales, F. A., Solar, P. A., Herrera, N. S., Gonzalez-Nilo, F. D., Aguayo, D. A., Mendoza, H. L., Comer, J., Bravo, M. L., Gonzalez, P. A., Kato, S., Cuello, M. A., Alonso, C., Bravo, E. J., Bustamante, E. I., Owen, G. I., & Velasquez, L. A., 2013. Paclitaxel-PHBV nanoparticles and their toxicity to endometrial and primary ovarian cancer cells. *Biomaterials*, 34(16), 4098–4108. https://doi.org/10.1016/J.BIOMATERIALS.2013.02.034.

Volpe, V., de Feo, G., de Marco, I., & Pantani, R., 2018. Use of sunflower seed fried oil as an ecofriendly plasticizer for starch and application of this thermoplastic starch as a filler for PLA. *Industrial Crops and Products*, 122, 545–552. https://doi.org/10.1016/J.INDCROP.2018.06.014.

Vuelo Pharma (2023). https://www.vuelopharma.com/en/membracel-en/

Wang, J., Liu, M., Duan, C., Sun, J., & Xu, Y., 2019. Preparation and characterization of cellulose-based adsorbent and its application in heavy metal ions removal. *Carbohydrate Polymers*, 206, 837–843. https://doi.org/10.1016/J.CARBPOL.2018.11.059.

Wang, L., Du, J., Cao, D., & Wang, Y., 2013. Recent advances and the application of poly(3-hydroxybutyrate-co-3-hydroxyvalerate) as tissue engineering materials. *Journal of Macromolecular Science, Part A*, 50(8), 885–893. https://doi.org/10.1080/10601325.2013.802540.

Wang, L., Gardner, D. J., Wang, J., Yang, Y., Tekinalp, H. L., Li, K., Zhao, X., Neivandt, D. J., Han, Y., Ozcan, S., & Anderson, J., 2020. Towards industrial-scale production of cellulose nanocomposites using melt processing: A critical review on structure-processing-property relationships. *Composites Part B: Engineering*, 201, 108297. https://doi.org/10.1016/J.COMPOSITESB.2020.108297.

Widiastuti, I., 2014. Mechanical performance of a PLA-based biodegradable plastic for liquid packaging application. Doctor of Philosophy, Swinburne University of Technology.

Wright, S. L., Rowe, D., Thompson, R. C., & Galloway, T. S., 2013. Microplastic ingestion decreases energy reserves in marine worms. *Current Biology*, 23(23), R1031–R1033. https://doi.org/10.1016/J.CUB.2013.10.068.

Xiong, Z., Yang, Y., Feng, J., Zhang, X., Zhang, C., Tang, Z., & Zhu, J., 2013. Preparation and characterization of poly(lactic acid)/starch composites toughened with epoxidized soybean oil. *Carbohydrate Polymers*, 92(1), 810–816. https://doi.org/10.1016/J.CARBPOL.2012.09.007.

Xu, L. Q., & Huang, H. X., 2012. Relaxation behaviour of poly(lactic acid)/poly(butylene succinate) blend and a new method for calculating its interfacial tension. *Journal of Applied Polymer Science*, 125, E272–E277.

Yaguchi, Y., Takeuchi, K., Waragai, T., & Tateno, T., 2020. Durability evaluation of an additive manufactured biodegradable composite with continuous natural fiber in various conditions reproducing usage environment. *International Journal of Automotive Technology*, 14, 959–965.

Yang, W., Weng, Y., Puglia, D., Qi, G., Dong, W., Kenny, J. M., & Ma, P., 2020. Poly(lactic acid)/lignin films with enhanced toughness and anti-oxidation performance for active food packaging. *International Journal of Biological Macromolecules*, 144, 102–110. https://doi.org/10.1016/J.IJBIOMAC.2019.12.085.

Yusoff, R. B., Takagi, H., & Nakagaito, A. N., 2016. Tensile and flexural properties of polylactic acid-based hybrid green composites reinforced by kenaf, bamboo and coir fibers. *Industrial Crops and Products*, 94, 562–573. https://doi.org/10.1016/J.INDCROP.2016.09.017.

Zhang, S., Huang, D., Lin, H., Xiao, Y., & Zhang, X., 2020. Cellulose nanocrystal reinforced collagen-based nanocomposite hydrogel with self-healing and stress-relaxation properties for cell delivery. *Biomacromolecules*, 21(6), 2400–2408. https://doi.org/10.1021/ACS.BIOMAC.0C00345/ASSET/IMAGES/LARGE/BM0C00345_0004.JPEG.

Zhu, J., & Wang, C., 2020. Biodegradable plastics: Green hope or greenwashing? *Marine Pollution Bulletin*, 161, 111774.

2 Bioplastics, Biodegradable Polymers and Biocomposites

An Overview

Mandla V. Khumalo and Sudhakar Muniyasamy

2.1 INTRODUCTION

The end-of-life option (reuse, recyclability, and renewability) of post-consumer plastic materials plays an important role in the development of sustainable polymer products. It was estimated that 60% of the world's plastics production is derived from petroleum source non-biodegradable plastics, and it is mainly used for single and short-use disposable packaging and agricultural applications. World production and consumption of plastics is at 320 million tons/year, and it is still expanding at a 12% growth rate per year. However, those recalcitrant polymeric materials and their extensive use have led to their accumulation in the environment, which poses a serious environmental problem due to the low degradability and pollution in aquatic and terrestrial habitats (Sudhakar et al., 2008; Muniyasamy, Reddy, et al., 2013; Corti, Sudhakar, and Chiellini, 2012; Sudhakar et al., 2007; Mierzwa-Hersztek, Gondek, and Kopeć, 2019). In that regard, much attention has been given to more environmentally friendly solutions, leading to the support of biodegradable materials as an alternative to non-biodegradable polymeric materials.

Particularly, renewable agricultural and biomass feedstock has shown much promise for use in eco-efficient packaging to replace petroleum feedstock without competing with food crops (Abdelwahab et al., 2012). However, as compared to thermoplastic synthetic polymers, biopolymers present problems when processed with traditional technologies and show inferior performances in terms of functional and structural properties (Mensitieri et al., 2011). However, the fate of environmentally degradable polymeric materials and plastics must be their conversion by microorganisms into final elemental products such as carbon dioxide, water, and new microbial biomass (i.e., mineralisation). Hence, it is necessary to evaluate their biodegradability in natural environments to meet various commercial and environmental needs for their sustainable growth. Searching and studying new innovative materials that could be used as economic and environmentally friendly packages was an important purpose (Weeg-Aerssens, 1998).

Therefore, many attempts have been made to reduce plastic waste. Several physical and chemical degradation methods, such as UV treatment, physical

DOI: 10.1201/9781003304142-2

stress, oxidants, methanolysis, ammonolysis, and hydrolysis, have been developed (Hauenstein, Agarwal, and Greiner, 2016; Gewert, Plassmann, and MacLeod, 2015; Kamini and Iefuji, 2001). Those processes usually require elevated temperatures and generally produce toxic substances. However, the extent of plastic biodegradability depends on its physical and chemical properties (Abareshi, Zebarjad, and Goharshadi, 2009). Photodegradation could affect the plastic films in two ways. First, it could cause random main chain scission either via Norrish I or Norrish II mechanisms. Generally, biodegradation of synthetic polymers involves enzymatic and chemical degradation by living microorganisms. The primary mechanism for the enzymatic degradation of polymers takes place by hydrolysis and oxidation. Most synthetic polymers cannot be degraded by microorganisms. To increase their biodegradability, non-biodegradable polymers were associated with natural biopolymers such as starch, cellulose, lignin, and dextrin. Adding those additives to synthetic polymers increases the polymer chain oxidation reaction. The microbes in turn release nonspecific oxidative enzymes that could attack the synthetic polymers. This accelerates degradation of the synthetic polymers by diffusion of oxygen, moisture, and enzymes into the porous polymer matrix (El-Rehim et al., 2004; Wool et al., 2000; Steller and Meissner, 1998). The investigations of biodegradation processes of polyolefin-starch blends have been done for many years. The results obtained within the last three decades were presented in books, review articles and many other publications.

Polyolefins have significantly obtained a main position in the packaging industry because of their low cost, lightweight, required properties, and low-energy consumption during their processing (Guilbert, Gontard, and Cuq, 1995; Craver and Carraher, 2000). Among those polymers, low-density polyethylene (LDPE), which is hard to degrade in the environment, is one of the fastest-growing commercial thermoplastic materials. However, the continuous use of polyethylene (PE) plastics in different applications led to the growing problem of environmental pollution. To overcome the problem, production of degradable and biodegradable polyolefins was necessary (Chiellini and Solaro, 1996; Kubowicz and Booth, 2017). The degradation of polymers might proceed by one or more mechanisms, including biodegradation, chemical degradation, photo, and thermal oxidation, depending on the polymer environment and desired application. The combination of different environmental factors, such as sunlight, heat, oxygen, humidity, and microorganisms, has synergistic effects on the degradation rate of starch-filled (LDPE) (Albertsson and Karlsson, 1993). Natural polymers were susceptible to microbial attack, and derived from agricultural products (such as starch, proteins, cellulose, and plant oils), they were the major resource for developing renewable and biodegradable industrial applications due to increased environmental concern and diminished petrochemical resources (Raquez et al., 2013; Zhang et al., 2010; Le Corre, Bras, and Dufresne, 2010).

Native starch is the term used to describe starch in the form in which it occurs in plants such as potatoes, wheat, cassava, rice, and maize. In plants, starch occurs in the form of granules. The granules vary in shape, size, and relative proportions of amylose and amylopectin depending on the source of the starch. Starch was therefore described by its plant source as corn starch, potato starch, tapioca starch, etc. (Souza and Andrade, 2001). After its extraction from plants, starch occurred as a flour-like white powder insoluble in cold water. The powder consisted of microscopic granules with diameters ranging from 2 to 100 μm, depending on the botanic origin, and with a density of 1.5 (Buléon et al., 1998). Amylose is defined as a linear molecule of glucose units linked by (1–4) α-D-glycoside bonds, slightly branched by (1–6) α-linkages. Amylopectin is highly branched polymer consisting of relatively short branches of α-D-(1–4) glycopyranose that are interlinked by α-D-(1–6)-glycosidic linkages approximately every 22 glucose units (Dufresne, 2006).

Starch, which is a mixture of amylose and amylopectin, is a renewable and fully biodegradable polymer that is readily available in high purity and at low cost. It has drawn a lot of attention in the preparation of biodegradable plastic. Starch could be thermoplastically processed without chemical modifications, providing adequate moisture. However, the mechanical properties of starch were rather poor (Rodriguez-Gonzalez, Ramsay, and Favis, 2003). Blending starch with PE or any other polyester improves the mechanical properties of starch and makes the blends suitable for biodegradation under compost conditions. Another possibility for obtaining biodegradable packages was the blending of traditional, synthetic polymers with biopolymers (based on starch or cellulose). In case of such compositions, only the natural organic component undergoes decomposition to carbon dioxide and water (in air conditions) or to water and methane (in oxygen-free circumstances), whereas the synthetic component only breaks up into small pieces and dissipates (Halim Hamid, 2000).

Biodegradation (i.e., biotic degradation) is a chemical degradation of materials (i.e., polymers) deferred by the action of microorganisms such as bacteria, fungi, and algae. The most common definition of a biodegradable polymer is "a degradable polymer wherein the primary degradation mechanism was through the action of metabolism by microorganisms." Biodegradation is considered as a type of degradation that involved biological activity and major mechanism of loss for most chemicals released into the environment (Gautam, Bassi, and Yanful, 2007). The most important organisms in biodegradation are fungi, bacteria, and algae (Eubeler, Bernhard, and Knepper, 2010). In contrast to the abiotic process, biotic breakdown starts at the surface of the material. The first step depolymerises the molecule mainly by extracellular enzymes. When broken-down parts were small enough for uptake by microorganisms' mineralisation occurred within the organisms in the second step. Microorganisms were involved in the degradation and deterioration of both synthetic and natural polymers (Gu, 2003; Albertsson, Andersson, and Karlsson, 1987).

During degradation, exoenzymes from microorganisms break down complex polymers, yielding short chains or smaller molecules, for example, oligomers, dimers, and monomers, were smaller enough (water soluble) to pass the semi-permeable outer bacterial membranes and then to be utilised as carbon and energy sources. Biodegradability of polymers is affected by two main factors, namely, exposure conditions and characteristic features of polymers. Exposure conditions could further be categorised as abiotic and biotic factors. The main chain scission from photodegradation reduced the average molecular weight, which provides greater accessibility to the polymer chain by moisture and microorganisms (Göpferich, 1996). In the case of aliphatic aromatic polyesters, photodegradation could result in both main chain scission and cross-linking (Fukuda and Osawa, 1991; Schnabel, 1982).

PE, PP, and polyethylene terephthalate (PET) degrade through the mechanisms of photo-, thermal, and biodegradation. The three polymers degrade at different rates and different pathways. Under normal conditions, photo- and thermal degradations were similar. For PP, photodegradation results in sharper peaks in the bands representing ketones, esters, acids, etc., on their infrared spectrum. The same was true for PP, which was more resistant to photodegradation. The photo-oxidation of PET involves the formation of hydroperoxide species through oxidation of the CH_2 groups adjacent to the ester linkages, and the hydroperoxide species involved in the formation of photoproducts through several pathways (Fotopoulou and Karapanagioti, 2017; Khabbaz and Albertsson, 2000; Vargha et al., 2016; Cornell, Kaplan, and Rogers, 1984). The different microorganisms that are responsible for the degradation of different groups of plastics are listed in their article, and bacterium could constantly synthesis all the enzyme required for degradation or else could activate enzyme synthesis as necessary to metabolize when needed or is thermodynamically favourable (Shimao, 2001; Devi et al., 2016). Microbial enzymes induce the rate of biodegradation of plastics very effectively without causing any harm to the environment. Target markets for biodegradable polymers include packaging materials like trash bags, loose-fill foam, food containers, film wrapping, laminated paper, hygiene products like diaper back sheets and cotton swabs, consumer goods like fast-food tableware and containers, egg cartons, and toys, and agricultural tools like mulch films and planters, etc.

The aim of this review is to improve mechanisms for technology transfer and the exchange of knowledge about biodegradable plastics with industrial users. The outcome improvement will gain access to scientific findings, the current knowledge and its edition will be implemented in manufacturing of biodegradable polymers and plastics. The implementation of sustainable practices will help minimise the impact on the environment and conserve resources for future generations. Industrial progress in packaging technology in future appears to be moving forward with newer breed of biomaterials.

2.2 CLASSIFICATION OF BIOBASED POLYMERS

Biobased polymers can be produced from various renewable resources (Thakur et al., 2016). In the current era, the first generation of biobased polymers is mainly obtained from the agricultural sectors that are rich in carbohydrate feedstocks such as grain, sugar beet, sugarcane, and other crops. Since these sources are mainly used for food and feed purposes, this has shifted focus toward the development of biobased polymers from non-food renewable resources such as lignocellulose, agricultural waste, and food waste and is referred to as second-generation biobased materials (Thakur and Thakur, 2014; Thakur et, al., 2014 (1); Thakur et, al., 2014 (2); Thakur et, al., 2016). Biobased polymers produced from renewable resources can be classified into three categories, as the principal ways as shown in Figure 2.1: (i) polymers extracted from plants and animals (e.g., protein and carbohydrates); (ii) polymers synthesized from biobased monomers by fermentation processes followed

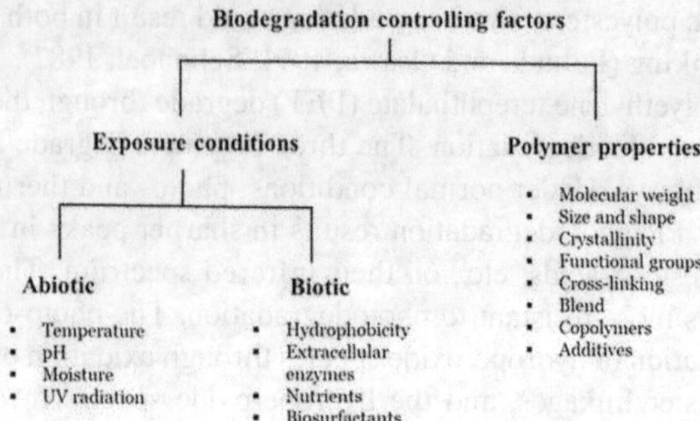

FIGURE 2.1 Factors affecting biodegradation (Muniyasamy and Maya, 2017).

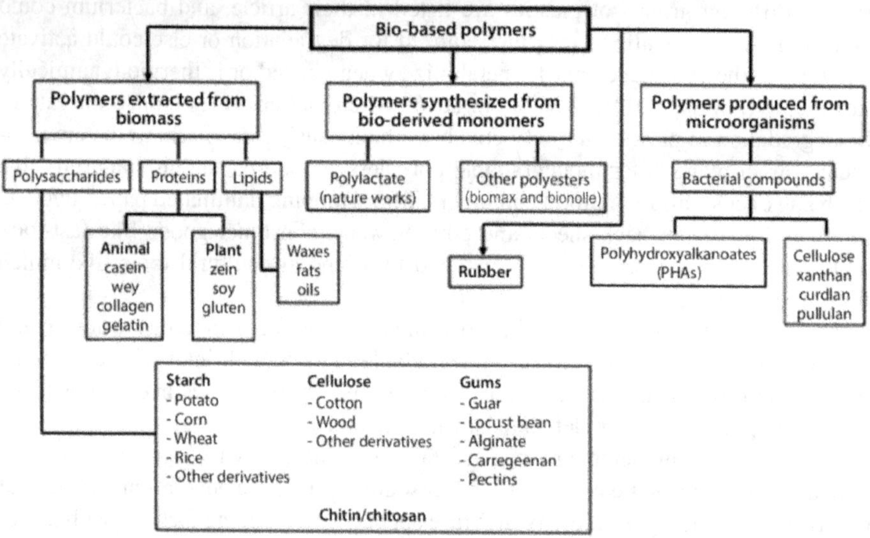

FIGURE 2.2 Classification of bio-based polymers from renewable resources (Briassoulis and Giannoulis, 2018).

by polymerisation or condensation of monomers product (e.g., polylactic acid (PLA), polybutylene succinate (PBS), and biopolyethylene (PE)); and (iii) polymers produced from microorganisms, that is, biochemical synthesis of polymer in the microbial cells (e.g., polyhydroxyalkanoates (PHA)) (Figure 2.2).

2.2.1 Extraction of Biopolymers from Biomass

Natural biopolymers can be directly extracted from biomass, such as from proteins and carbohydrates. Natural fibres such as cotton, hemp, and wood are also good examples. However, the native form of biopolymers such as starch and cellulose cannot be used as plastics and need to be modified chemically, thermally, and mechanically to obtain any technological usefulness. Starch is the most widely used biopolymer from agricultural biomass for the development of thermoplastic starch (TPS), which uses thermal processing with the addition of plasticisers (Bastioli, 2005). The starch-based biodegradable plastics are mainly used for applications in food packaging, disposable bags, mulching films, etc. These starch-based plastic materials are potentially biodegradable in the natural environment (soil, compost, marine, and anaerobic digesters). Cellulose is another natural biopolymer derived from the cell walls of plants, and it is the most abundant biopolymer available on the earth. Cellulose-based plastics can be made from the extracted cellulose using chemical hydrolysis and modifications (Faruk et al., 2012; Reddy et al., 2013; Thakur et al., 2014; Thakur, Thakur et al., 2014 (2)).

The extracted cellulose can be obtained either by micro or nano size level during the hydrolysis processing. The obtained micro- and nanocellulose possess thin transparent films and have high thermal stability properties and strength. In addition, cellulose can be chemically modified to form thermoplastic cellulose by using plasticisers, which include cellulose acetate, propionate, and butyrate. Some plastic industries are producing cellulose acetate by blending it with petrochemical polyesters for packaging films, twist wraps, tape, and envelope window applications. The end-of-life options of these cellulose-based polymers are biodegradable in compost (industrial and home composting), soil, water, and solid marine environments. Similarly, the recycling of cellulose acetate with cardboard paper is also possible.

Proteins are another well-known renewable resource-based biopolymer derived from plant and animal origin, offering many advantages, including their ready availability, unique structure, renewability, natural resource origin, and degradability (Verbeek and van den Berg, 2010; Song et al., 2011). These advantages make proteins an ideal candidate for exploring the possibility of their usage in developing biodegradable materials. Some examples of proteins of animal origin are whey, casein, collagen, and keratin, and some examples of proteins of plant origin are gluten, soy, pea, and potato. Proteins are random chain copolymers of amino acids, and their sidechains are mainly suitable for chemical modification and materials engineering of tailored properties for packaging film applications. Both hydrophilic (water and polyols) and lipidic (oils, waxes, and fatty acids) compounds are used as plasticisers for protein-based biodegradable plastics. Generally, two types of methods are used to prepare protein plastics: wet and dry. In the wet method, protein and plasticiser are solubilised in a solvent, leading to the formation of proteins upon evaporation of the solvent (Reddy, Mohanty, and Misra, 2012; Reddy, Mohanty, and Misra, 2010). The dry method involves the use of hot pressing to form thermoplastic materials from various sources of proteins, such as soy proteins, corn proteins, wheat gluten, myofibrillar proteins from fish, animal proteins, and pea proteins.

Chitin is another biopolymer abundantly found in invertebrate animals such as insects and crustaceans. Chitin is mainly obtained from crabs and shrimps by removing the protein and dissolving the calcium carbonate (Suh and Matthew, 2000; Kumar, 2000). Chitin has a structure like cellulose and the degradation of chitin undergoes chitinase (Muzzarelli, 1997). Chitin-based materials are mainly used in biomedical applications such as wound dressing and absorbable sutures.

2.2.2 Biobased Polymers Derived from Biobased Monomer

With the current technologies, biobased polymers can also be produced from chemical synthesis and polymerisation of biobased monomer obtained from bacterial fermentation. Recent studies have shown that some biobased biodegradable polymers derived from renewable resources have comparable mechanical properties and cost of production compared to non-biodegradable conventional plastics (Reddy et al., 2013). The most important biobased polymer is PLA, which falls under the thermoplastic polyester group derived from renewable resources such as corn starch, tapioca roots, chips or starch, and sugarcane. The currently available bioplastics PLA made from renewable resources (agricultural feedstock) offer an attractive alternative to fossil fuel-based petroleum-derived polymers, which significantly reduce the carbon footprint, reduce the emission of GHG and the benefit of compostable, recyclable, and energy recovery. Despite these advantages, some limitations of PLA, such as brittle behaviours and challenging processability are restricting their broader applications like common commodity polymers (PE, PP, and polystyrene (PS)). PLA can be processed in melt extrusion, injection moulding, film, and sheet casting. PLA is the second highest volume consumed of biobased polymers in the world (Keshavarz and Roy, 2010; Mooney, 2009; Oksman, Skrifvars, and Selin, 2003; Philp, Ritchie, and Guy, 2013; Song et al., 2009; Stevens and Goldstein, 2002). Thus, it is not surprising that PLA-based products are widely used in various industrial sectors, including automotive, pharmaceutical, medical, packaging and consumer products. Another great example is biobased polyethylene (bio-PE), produced by using sugarcane renewable feedstock. Then again, bio-PE is not biodegradable. Some studies reported that since sugarcane contributes to the food supply at a very small level, the unused land for growing sugarcane can be used for production bio-PE that will help speed up the process as equivalent for fossil fuel-based PE. Moreover, the biobased properties, applications and recycling are identical to fossil fuel-based PE.

2.2.3 Biobased Polymers from Bacterial Synthesis

Biobased polymers polyhydroxyalkanoate (PHA) are synthesized and intercellularly accumulated as energy and carbon storage source of most bacteria under unfavourable growth conditions, such as limitation of oxygen, nitrogen, phosphorus, or magnesium in the presence of an excess supply of carbon source. PHA possesses valuable thermoplastic properties and its high resistance to water. The PHA physical properties are like conventional non-biodegradable polypropylene (PP). Furthermore, PHAs are biodegradable and compostable under both aerobic and anaerobic conditions. By using its copolymer, a wide range of biobased polymers with different mechanical and thermal properties can be produced for automotive and packaging applications. Some examples of PHA copolymers are polyhydroxybutyrate (PHB), polyhydroxybutyratevalerate (PHBV), and poly-3-hydroxybutyrate-co-3-hydroxyhexanoate (PHBH) (Chen, 2009; Povolo et al., 2013). The main disadvantages of microbial PHAs are the production capacity is relatively low, associated with high cost and energy, which results in a negative contribution on LCA. Researchers also reported the blend of PHB with PLA is a promising technique, resulting in a biodegradable plastic with advanced properties. By using PLA, which is a biobased polymer from plant resources, and is currently available in large quantities with a lower cost value, constitutes an environmental-friendly biobased polymer and composite that can be obtained for various durable and biodegradable product applications.

Over the past years, several bioplastic products have emerged. They were sometimes labelled "biobased," "biodegradable," and/or "compostable." Bioplastics, according to the Society of the Plastics Industry (SPI) Bioplastics Council, were defined as plastic that is biodegradable, has biobased content, or both. A biobased plastic made from renewable resources instead of fossil fuels. Examples of renewable carbon resources include corn, potatoes, rice, soy, sugarcane, wheat, and vegetable oil. A biobased plastic can be partly or entirely biobased. Note that just because a plastic product is biobased does not necessarily mean the product is biodegradable or compostable.

For biodegradable bioplastics, products are typically intended for short-life applications such as single-use packaging, food waste collection bags, or food service ware (e.g., utensils, cups, plates) and were promoted for composting at the end of life. Compostable products have the potential to streamline the collection of food scraps and yard trimmings for composting, helping to divert it from disposal in landfills and incinerators (SPI Bioplastics Council, 2013).

2.3 BIODEGRADABLE AND COMPOSTABLE POLYMERIC MATERIALS

Compostable plastic is biodegradable in a composting environment, yielding H_2O, CO_2, biomass, and inorganic compounds. The biodegradation during composting should be at a rate like other known compostable materials and should not leave visual or toxic residue. Compostable plastic is "a plastic that undergoes biological degradation during composting to yield carbon dioxide, water, inorganic compounds and biomass at a rate consistent with other known compostable materials and leaves not visually distinguishable or toxic residues." The standard guideline definitions have been given to establish and identify biodegradable polymeric materials or products. Based on these guidelines, American Society for Testing and Materials (ASTM) standards have provided various test methods to evaluate the potential biodegradability of polymeric materials and products in various disposal environmental conditions such as soil, compost, marine water, and sewage. In the case of compostable plastics, the polymeric organic carbon must be converted into CO_2, water, and cell biomass within 180 days (Muniyasamy, Anstey, et al., 2013).

Because of the dependence of biodegradation on both polymer structure and environmental factors, a wide range of tests are required to fully quantify the biodegradability of a given plastic (Krzan et al., 2006; Kroschwitz et al., 1999). To ensure a global standard of plastic biodegradability, standardised testing procedures have been defined by international organisations, such as the International Organization for Standardization (ISO) and ASTM. These organisations define and regulate number of standard tests that are used to properly assess biodegradable polymers and encompass a variety of environmental factors. The definition of these biodegradability tests was established by the ISO (Stoica et al., 2022; Brodhagen et al., 2017; Voeller, 2010). For biodegradable plastics, these factors include humidity, temperature, pH level, oxygen presence, and, most importantly, the presence of microorganisms (both in compost and soil). Each factor affects the rate of degradation and changing them could have a major effect on the result of the experiment; as a result, these factors must be closely monitored for accurate testing. Through the addition of natural fibre to the biodegradable polymer, the polymer properties and biodegradability of the resulting composites can be changed in comparison to the base polymer. Just as with pure polymer materials, the true biodegradability of polymer composites (in terms of rate of degradation and toxicity for safe disposal in environmental conditions) must be evaluated using specific standardised testing methods. These tests cover a broad scope of conditions to accommodate all types of plastics and the various environmental conditions they could be exposed to as plastic waste. Thus, with continuous development and innovation in the biopolymer industry, these standard test methods for evaluating the biodegradability of polymeric materials are essential to the future of the plastics industry and will be necessary to reduce the global issue of plastic pollution.

Biodegradation is defined as a process in which materials undergo chemical changes due to the action of enzymes that are secreted by living organisms (bacteria, fungi, and algae). The process of biodegradation is comprised of two phases, including the initial phase (primary biodegradation) and the secondary phase (ultimate biodegradation). During the initial phase, a polymer undergoes significant weight loss, reduction in molecular weight and fragmentation, and is degraded into soluble low molecular weight compounds. The second phase of biodegradation is the ultimate conversion of primary degraded low molecular weight compounds into CO_2, water, and cell biomass (in aerobic conditions) and CH_4, CO_2, and cell biomass (in anaerobic conditions). The primary and ultimate biodegradation must occur at specific times with specific rates under biological environmental conditions (soil, compost, aqueous, and anaerobic digestate to prevent the accumulation of plastics in the environment). The ASTM and the International Organization for Standardization (ISO) define:

1. **Degradation** is "an irreversible process leading to a significant change of the structure of a material, typically characterised by a loss of properties (e.g., integrity, molecular weight, structure, or mechanical strength) and/or fragmentation. Degradation is affected by environmental conditions and proceeds over a period comprising one or more steps."
2. **Biodegradable plastic** is "a degradable plastic in which the degradation results from the action of naturally occurring microorganisms such as bacteria, fungi and algae."
3. **Compostable plastic** is "a plastic that undergoes biological degradation during composting to yield carbon dioxide, water, inorganic compounds and biomass at a rate consistent with other known compostable materials and leaves not visually distinguishable or toxic residues."

The standard guideline definitions have been given to establish and identify biodegradable polymeric materials or products. Based on these guidelines, ASTM standards have provided various test methods to evaluate the potential biodegradability of polymeric materials and products in various disposal environmental conditions such as soil, compost, marine water, and sewage. In the case of compostable plastics, the polymeric organic carbon must be converted into CO_2, water, and cell biomass within 180days.

As defined by established standards, polymer biodegradation undergoes two phases: primary degradation and mineralisation (ultimate biodegradation). The primary degradation is significantly associated with biodeterioration in physical properties, such as discoloration, embrittlement, and fragmentation. This primary degradation can occur when microorganisms starving conditions secrete extracellular enzymes to break down the longer chains into smaller molecular or oligomers. This disintegration process is sometimes referred to as primary degradation. The second phase is the ultimate biodegradation (bioassimilation) conversion of polymer low molecular fragments/compounds into CO_2, water, production of new microbial cell biomass (aerobic conditions), CH_4, CO_2, and production of new microbial cell biomass in the case of anaerobic conditions.

The process of polymer degradation relies on a combination of various environmental abiotic and biotic factors and mechanisms. Most common environmental abiotic factors include heat, UV radiation, oxygen, water, mechanical stress, and pH, and greatly influence the polymer disintegration/fragmentation (primary degradation) followed by microbial assimilation (mineralisation) of primary degraded low-molecular-weight compounds influenced by biotic factors such as bacteria, fungi, and algae. The primary and ultimate biodegradation must occur at a specific time with a specific rate for the exclusion of plastic waste in the environment. Apart from that, the biodegradation of plastics can be affected by various physical and chemical properties of the plastic materials and their products, for example, lack of active functional group of polymers, high hydrophobic nature of materials, high molecular weight of polymers, physical form (films, pellets, powder or fibres), distribution of crystalline and amorphous regions, structure of the polymer (linear chain or branching or cross-linking), chemical composition of the polymer (blends, additives, UV stabilisers, antioxidants, etc.), microorganisms in the form of mixed culture present in environment, properties of microorganisms including their ability to produce biosurfactant, or other exo polysaccharides and hydrophobic nature of the bacterial cell wall. In the case of aerobic biodegradation conditions, when oxygen is present, plastic carbon is oxidised by the action of microorganisms (bacteria and fungi), leading to the conversion of carbon to CO_2, H_2O, production of new microbial biomass, and some of the carbon undigested and remain as residual sample or in metabolites:

$$C_{sample} + O_2 \rightarrow CO_2 + H_2O + C_{biomass} + C_{residual} \tag{2.1}$$

Equation 2.1: Mineralisation process of polymeric carbon (C) under aerobic conditions. Under anaerobic biodegradation conditions, when oxygen is not present, microorganisms digest plastic carbon and release end products such as CH_4 and CO_2, residual sample or metabolites and biomass. The anaerobic environmental conditions typically occur in some natural environments such as landfill, anaerobic digester, and deep sea:

$$C_{sample} \rightarrow CH_4 + CO_2 + C_{biomass} + C_{residual} \tag{2.2}$$

Equation 2.2: Mineralisation process of polymeric carbon under anaerobic conditions. Clearly, from these equations, the true biodegradation (mineralisation) is mainly based on the production of CO_2 and CH_4 from plastic carbon. However, other parameters such as weight loss, changes in chemical groups, mechanical strength, colour changes, and microbial growth on polymer surfaces are also primary effects, which can only be established for partial or incomplete biodegradation. Another important criterion for the biodegradation of plastics is in the environmental conditions in which plastic biodegradation takes place. The rate and degree of biodegradation may vary from one environment to another, such as landfill, home and industrial composting, freshwater, marine environments, wastewater, and an anaerobic digester. In each environment, the various factors, including humidity, oxygen availability, temperature, type of microbiology (bacteria and/or fungi), microbial density, and salt concentrations, play a major role in the rate and degree of biodegradation.

Therefore, some biodegradable plastic may degrade in one environment, with limitations in biodegradability in another environment. For example, PLA is a compostable polymer but not biodegradable in soil and aqueous environments. Therefore, it is necessary to define the biodegradation claims of polymeric materials in terms of the specific environments where biodegradation can take place. Biodegradable plastics are not a solution to the problem of litter and the "plastic soup." On the other hand, biodegradability can be a useful characteristic for specific marine and soil-related applications like fishing lines and mulch films. When lost at sea or when only partly recovered from the land, marine or soil biodegradable plastics would at least result in a lower risk of harmful consequences than if they fail to break down at all. As a certified claim like OK biodegradable MARINE or OK biodegradable SOIL could stimulate consumers to leave a certified product behind in the environment, a clear distinction should be made between certification of the claim and authorisation to communicate about this certification.

2.3.1 Biopolymers

The term biopolymer generally understood as an organic polymer produced naturally by living organisms. Major advantage of biopolymers are fully capable of biodegradation at speedy rates, they can break down simple molecules found in the environment, such as carbon dioxide, water, or methane, under the enzymatic environment within short period of time (Armentano et al., 2013; Alshehrei, 2017). Biopolymers produced from various natural resources, such as starch, cellulose, chitosan, and various proteins from plant and animal origins, have been considered as attractive alternatives for non-biodegradable petroleum-based plastic packaging materials, since they were abundant, renewable, inexpensive, environmentally friendly, biodegradable, and biocompatible (Luckachan and Pillai, 2011; Sorrentino, Gorrasi, and Vittoria, 2007; Siracusa, 2019).

The main synthetic biopolymers include polylactide or PLA, poly(ϵ-caprolactone) (PCL), poly(butylene succinate) (PBS), poly(butylene-adipate-coterephthalate) (PBAT) and poly(hydroxyalkanoates) (PHAs). The most used biopolymers in food packaging are PLA, PHAs, TPS, cellulose, and proteins because they are abundantly available, odourless, and non-toxic. For biopolymers to be considered a suitable food packaging material, they should meet certain requirements such as protecting food quality from contamination, non-toxic when contacting food, forming a good barrier, and having adequate mechanical properties.

In addition, biopolymer-based packaging materials have some beneficial properties as packaging materials improve food quality and extend the shelf life through minimised microbial growth in the product. Furthermore, biopolymer films were excellent vehicles for incorporating a wide variety of additives, such as antioxidants, antifungal agents, antimicrobials, colours, and other nutrients (Appendini and Hotchkiss, 2002). However, there were some limitations to the commercial use of biopolymer films due to their poor mechanical properties and high sensitivity to moisture (Cabedo et al., 2006). Biobased non-biodegradable polymers possess characteristics like those of fossil-based polymers (e.g., biobased and fossil-based polyethylene, PE). The major difference in this category is the source of the monomer. On the other hand, biobased biodegradable polymers consist of synthetic and naturally occurring biopolymers, such as cellulose and starch.

Saha, Zatloukal, and Saha (2003) explained ways to reach the goal via the modification of polymer with starch, cellulose, protein and other water-soluble materials. An investigation on the modification of metallocene-based linear-low-density polyethylene (mLLDPE) with HP was reported. It was interesting to note that HP easily blends with mLLDPE, but like other biopolymers, it also reduces the original mechanical properties of the mLLDPE.

However, not all biodegradable polymers are biobased, as others can be obtained from fossil fuels. For example, biobased poly(ethylene terephthalate) (PET) is derived from renewable resources but is not biodegradable, whereas poly(ϵ-caprolactone) (PCL) is petroleum-based, but it is 100% biodegradable. Biopolymers also find applications in the electrical/electronic and automobile sectors. Electrical/electronic accessory makers use environmentally friendly/green materials as marketing vehicles, which could benefit from specialised bioplastic films suitable for such applications (Mtibe et al., 2021).

2.3.1.1 Sugar-Based Biopolymers

The term "biobased" means that the material or product is (partly) derived from biomass (plants). Biomass used for bioplastics stems from, for example, corn, sugarcane, or cellulose. Plants, animals, microorganisms, and agricultural wastes are examples of natural biological sources of biopolymers. Biopolymers from plant sources, such as rice, maize (Awad et al., 2014), wheat (Karmanov et al., 2021), sorghum (Tanamool et al., 2013), yams (Gutiérrez et al., 2014), cassava (Perotti, Auras, and Constantino, 2013), potatoes (Borah, Das, and Badwaik, 2017), banana (Redondo-Gómez et al., 2020), tapioca (Cabral et al., 2018), corn (Lauer, Tennyson, and Smith, 2021), cotton (Aguilar et al., 2019), and barley (Yu, 2007), can be produced chemically from monomeric components, such as oils, sugars, and amino acids. Polysaccharides are biopolymers composed of monosaccharides (sugars) connected by glycosidic linkages generated during condensation. The linking of monosaccharides into chains results in radically varying lengths, ranging from two monosaccharides to polysaccharides with thousands of sugars. Polysaccharides serve a variety of essential functions in the biology of living processes (Gangalla et al., 2021), and many have found crucial commercial uses. They are split into two classes based on their role in nature: structural and storage polysaccharides. Guar gum is a polygalactomannan derived from Cyamopsis tetragonolobus endosperm, a leguminous plant. It possesses a backbone of (1-4)-linked beta-D-mannopyranosyl units that contain a single alpha-D-galactopyranosyl team joined every second on the average central chain unit by (1-6) linkage. Furthermore, guar gum has recently been used to produce biodegradable food packaging films and as a wall composition for flavor encapsulation. Paper, textiles, ceramics, mining, cosmetics, paint, explosives, and pharmaceuticals are just a few sectors that use guar gum and its numerous chemical derivatives (Baranwal et al., 2022).

Biopolymers based on sugar can be produced by blowing, vacuum forming injection, and extrusion. Lactic acid polymers (polylactides) are created from milk sugar (lactose) which is extracted from maize, potatoes, wheat, and sugar beet. Polylactides, manufactured by methods like vacuum forming, blowing and injection moulding, are resistant to water (Fanun, 2014). Poly (L-lactide) is a slow-degrading polymer compared to polyglycolide (PGA). L-lactide (LPLA) homopolymer is a semicrystalline polymer with high tensile strength exhibition, low elongation, and a high modulus consequently which make them suitable in orthopaedics fixation and sutures that have wild applications like load bearing (Maurus and Kaeding, 2004).

Sugar-based biopolymers possess numerous indispensable advantages, including high biocompatibility and biodegradability, low systemic toxicity, availability of different molecular weights (MWs) and particle sizes, versatile physicochemical properties, and facile chemical modifications. As such, sugar-based biopolymers have been extensively studied for the preparation of a variety of targeted imaging agents or imaging guidance for drug delivery systems (Goodarzi et al., 2013). They are glucose polymers produced by bacteria from sucrose or by chemical synthesis and, therefore, only available as exogenous agents. They are available in multiple MWs ranging from 1 kD to 2 MD (diameters from 1 to 54 nm, respectively) (Armstrong et al., 2004; Dreher et al., 2006). Dextrans have been used clinically for more than six decades for plasma volume expansion, peripheral flow promotion, and antithrombolytic treatment with an excellent safety profile (Dubick and Wade, 1994; Thorén et al., 1980). Dextran70 (dextran with MW ~70 kD) is on the World Health Organization model list of essential medicines.

Lignocellulose biomass is a plant material derived from common plants such as softwoods, hardwoods, and grasses. Lignocellulosic biomass is mainly comprised of three polymers, namely lignin, cellulose, and hemicelluloses (Lee, Padmanabhan, and Whang, 1997). Other constituents that make up a smaller part of the biomass are pectin, protein, wax, and extractives (Kumar et al., 2009). Cellulose forms microfibrils which are held together by hemicellulose and lignin through a combination of hydrogen and covalent bonds.

2.3.1.2 Starch-Based Biopolymers

Starch-based plastics are complex blends of starch with compostable polyesters (such as PLA, PBAT, PBS, PCL and PHAs) and additives like plasticisers and compatibilisers. Blending of starch with polyesters is necessary to improve water resistance, processing properties and mechanical properties. Commonly, the starch content of starch-based plastics is lower than 50% and the overall biobased content of these materials depends on the nature of the other components. Starch-based plastics are used in applications where biodegradability is an advantage like agricultural products (mulching films), service ware, garbage bags for green waste and carrier bags. The advantage is these products can be composted together with organic waste. Starch blends are generally compostable, and various materials (depending on the other blend components) are also biodegradable in soil, anaerobic digestion installations, fresh water, or marine environment.

Starch is an inexpensive and renewable resource; it is biodegradable and biocompatible. Its small granule size makes starch a good particulate filler in many polymer blending systems. Native starch is not processable like a thermoplastic, which can be ascribed to the strong intermolecular and intramolecular hydrogen bonds in starch. However, in the presence of water and plasticisers, starch can be converted into TPS (Shogren, 1992) with conventional equipment and conditions used for processing synthetic polymers. The main disadvantages of TPS are its pronounced hydrophilic character, the fast degradation rate and, in some cases, unsatisfactory mechanical properties, particularly in wet environment (Wang et al., 2007).

Moreover, the use of starch in the plastics industry could reduce dependence on synthetic polymers. Starch is a widely used material for making biodegradable plastics, but pure starch-based films possess low mechanical properties (Sharmin et al., 2012) and are the most important polysaccharides; they are most abundant in nature and relatively inexpensive. Natural starch exists in granular form and, as such, is used as filler in polymers but is also processed with classical plastic processed technologies such as extrusion, foaming and film blowing after thermoplasticisation. The main limitation of starch is its hydrophilic nature, which limits its use in high-moisture environments (Mensitieri et al., 2011; Lu, Xiao, and Xu, 2009).

Polysaccharides are good candidates for renewable nanofillers because they have partly crystalline structures conferred interesting properties. Recent reviews have published on cellulose presents (Azizi Samir, Alloin, and Dufresne, 2005; Siqueira, Bras, and Dufresne, 2008; Dubief, Samain, and Dufresne, 1999; Grossman and Nwabunma, 2013), starch nanoparticles would presented in detail with a complete overview of their preparation, characterisation, and applications, and they could be modified by esterification, etherification, and oxidation. Due to their abundance, high strength and stiffness, low weight and biodegradability, nanoscale polysaccharide materials serve as promising candidates for the preparation of bionanocomposites. A broad range of applications of these nanoparticles exists, even if many unknowns remain. Polysaccharide nanocrystals (cellulose nanocrystal (CNC), starch nanocrystal (SNC), and chitin nanowhiskers (ChW)) have unique properties such as high specific area and sustainability, making them ideal candidates for reinforcing various polymeric matrices.

TPS could be considered as a new class of inexpensive, green polymers could returned to its natural state without any pollution after use. TPS materials often based on a combination of starch, glycerol, and water. The limitations of TPS in various applications were its sensitivity to moisture and retrogradation processes. Therefore, more studies on starch (native and modified), plasticisers, and nanoparticles focusing on reduced water absorption and decreased retrogradation of material to avoid decreased mechanical strength or stiffness during storage in the development of TPSs are required. TPS is a renewable material and could be incorporated into soil as an organic fertiliser (Bastioli, 2001).

Zhang et al. (2008) studied aliphatic amide diol and glycerol used as a novel mixed plasticiser for corn starch to prepare TPS. The mixed plasticiser proves to be a good plasticiser for corn starch. Compared with glycerol, the mixed plasticiser formed stronger and more stable hydrogen bonds with starch, as detected by FTIR spectra and DMTA. Shi et al. (2007) studied a soft TPS with an improved ageing-resistant property prepared by the melt blending method for a biodegradable biomaterial. The glycerol content varies from 30 to 60 wt%. FTIR results show the shifting speed of the peak of hydroxyl group stretches fell as glycerol content increased. The glycerol content had no obvious effect on the mechanical properties when it was high enough. The higher content of glycerol possesses an obviously inhibitory effect on ageing.

Tena-Salcido et al. (2008) show TPS mixes with LDPE and different melt flow indexes in a one-step extrusion process to produce LDPE/TPS blends varied from 32% to 62% by weight of TPS. Morphology of LDPE/TPS blends is evidently affected by both TPS concentration and LDPE viscosity. The two blends extended bacterial biodegradation was 39% and 22%, respectively, which is less than 60% of the available TPS, as demonstrated by hydrolytic degradation. Tensile properties are affected by biodegradation due to the formation of holes on the lateral surfaces of tensile specimens. Majid, Ismail, and Taib (2009) explained the effects of PE-g-MA on tensile properties, morphology, and water absorption of LDPE/Thermoplastic sago starch (TPSS) blends studied. The mixture melt-compounded used heated two roll-mills at 150°C for 10 minutes. TPSS and LDPE were incompatible and a compatibiliser, PE-g-MA, needed to improve their compatibility. Tensile properties of the LDPE/TPSS blends improve with the addition of PE-g-MA. SEM micrographs of tensile surface fracture showed better TPSS particle dispersion in the blends with PE-g-MA.

St-Pierre et al. (1997) processed and characterised TPS/PE (LDPE and LLDPE) blends, and used glycerol as plasticiser. It was found that the modulus decreases as expected with addition of TPS in the TPS/LDPE blends. It also found the blends containing 22% TPS in LDPE and 39% TPS in LLDPE maintain high elongation even at those high loading and morphology, showing the incompatibility in TPS/PE blends. Bikiaris and his co-workers prepared TPS/LDPE blends; two different PE-g-MA copolymers contained 0.4 and 0.8 mol% anhydride groups used as reactive compatibilisers. It was found that compatibilised blends have only a slightly lower biodegradation rate compared to uncompatibilised blends (Bikiaris et al., 1998).

LDPE/plasticised starch blends varying in starch content are processed using conventional extrusion, injection moulding, and film-blowing techniques. PE-g-MA acts as compatibiliser. The effect of TPS on the mechanical properties of LDPE/TPS materials is found to depend not only on starch content and developed morphology, but on employed shaping process as well. TPS contains 20 wt% glycerol, which, under the experimental conditions used, could act as a reinforced agent in injection-moulded materials but not in films. LDPE/TPS injection-moulded products containing up to 50 wt% TPS could prepared without any problem. In the case of blown films, a 50 wt% TPS content appears to cause processing problems (Matzinos et al., 2001).

Wang, Yu, and Yu (2004) prepared compatible TPS/LLDPE blends by one-step reactive extrusion in a single-screw extruder, and MA was added to improve compatibility between TPS and LLDPE. It was found that blends with MA have better mechanical, morphological, and thermal properties than the blends without MA. In addition to MA, compatibility between TPS and LLDPE in blends truly improved. The author further explained that TPS/PE blends found from these studies, PE-g-MA, act as compatibiliser between TPS and PE. The thermal stability of blends increases with the addition of MA and DCP.

Jang et al. (2001) studied the mechanical properties and morphology of modified TPS/ HDPE blends, HDPE chemically modified to enhance compatibility with TPS. It found the compatibility of TPS/HDPE improved by increasing HDPE-g-glycidyl methacrylate content. In another work, Sailaja and his colleagues used HDPE-g-MA as compatibiliser for TPS/HDPE blends. It was also found that the mechanical properties of the blends improved by the addition of HDPE-g-MA. The authors further studied (Sailaja, Reddy, and Chanda, 2001; Sailaja and Chanda, 2000) the effect of epoxy-functionalised compatibiliser on mechanical properties of TPS/LDPE blends, and poly(ethylene co-glycidyl methacrylate) (PEGMA) was used as compatibiliser. It was found that the mechanical properties of the TPS/LDPE improve when PEGMA is used as compatibiliser. They further used poly(ethylene co-vinyl alcohol) (EVOH) as compatibiliser in TPS/LDPE blends. The blends of LDPE and TS compatibilised with EVOH copolymer moulded into dumbbell specimens studied their mechanical properties. While the impact strength of the uncompatibilised blends decreases with increases in TS loading, compatibilised blends show a significant improvement in impact strength, nearly reaching the value obtained for neat LDPE. The modulus and tensile strength values also increased with the addition of EVOH.

Huang, Yu, and Ma (2005) elaborated on the Ethanolamine-Plasticised Thermoplastic Starch prepared using novel plasticiser ethanolamine. Mixtures of urea and ethanolamine are used as plasticisers for preparing TPS in a single-screw extruder. FTIR and DSC analysis seconded the results. The mixture of urea and ethanolamine could form more stable and strong hydrogen bonds with hydroxyl groups of starch molecules than conventional plasticiser and glycerol. Ethanolamine is a good solvent for urea, and both could exist in molecular form in TPS. The TPS plasticised by urea (15 wt/wt%) and ethanolamine (15 wt/wt%) showed better thermal stability and mechanical properties (Ma, Yu, and Wan, 2006).

Blends TPS and chitosan (TPC) were obtained by melt extrusion. Results showed it was possible to successfully produce corn starch–chitosan blends by extrusion with high dispersion and distribution degree of the TPC phase in TPS as observed by SEM micrographs showing blends with homogeneous surface, and cryofractured samples displayed no agglomeration of chitosan within a completely restructured starch matrix. These blends had good thermal stability, and the addition of chitosan produced more thermally stable films (Mendes et al., 2016). Photodegradable polymers are produced by introducing chromophores intentionally into the polymer structure by copolymerisation (e.g., ethylene and carbon monoxide) or by mixing photosensitive additives (ketone base compounds) with the polymer (Al-Malaika, 1993).

Méndez-Hernández et al. (2011) explained thin films of HDPE and HDPE/TPS blends exposed to UV irradiation to analyse the effect of the presence of TPS on photodegradation of HDPE. Evaluation of oxidation rate and molecular weight of HDPE showed that there was almost no effect of TPS on photodegradation mechanisms of HDPE. Conversely, the presence of TPS had important influence on mechanical performance of HDPE/TPS blends due to increment of ΔH_m during ageing. Increment of ΔH_m was related to retrogradation process of starch molecules during accelerated ageing. The results evidenced the important effect of starch retrogradation on the performance of HDPE/TPS blends during ageing at high-energy and high-relative-humidity environment.

2.3.1.3 Cellulose-Based Biopolymer

Cellulose is the principal component of cell walls in all higher forms of plant life at varying percentages. It is therefore the most common organic compound and the most common polysaccharide (multi-sugar). A cellulose fibre consists of bundles of single-cellulose fibres, which have diameters of 25–30 μm. This single-cellulose fibre is made up of bundles of microfibres, which have diameters of 0.1–1 μm. Nanofibres, which have a diameter in the range of 10–70 nm and lengths of thousands of nanometres, are the constituent of microfibres (Wang, 2010).

Cellulose is another natural biopolymer derived from the cell walls of plants, and it is the most abundant biopolymer available on the earth. Cellulose-based plastics can be made from the extracted cellulose using chemical hydrolysis and modifications (Faruk et al., 2012; Reddy et al., 2013; Thakur, Thakur, and Gupta, 2014; Thakur et al., 2014). The extracted cellulose can be obtained either by micro or nanosize level during the hydrolysis processing. The obtained micro- and nanocellulose possess thin transparent films and have high thermal stability properties and strength. In addition, cellulose can be chemically modified to form thermoplastic cellulose by using plasticisers, which include cellulose acetate, propionate, and butyrate. Some plastic industries are producing cellulose acetate by blending it with petrochemical polyesters for packaging films, and twist wraps, tape, and envelope window applications. The end-of-life options of these cellulose-based polymers are biodegradable in compost (industrial and home composting), soil, water, and solid marine environments. Similarly, the recycling of cellulose acetate with cardboard paper is also possible.

The cellulose microfibre has been defined as a fibre consisting of continuous cellulose chains with negligible lignin and hemicelluloses content and having a diameter of 0.1–1 μm, with a minimum corresponding length of 2–20 μm (Chakraborty et al., 2007). Traditionally, cellulose nanofibre has been defined as purely crystalline

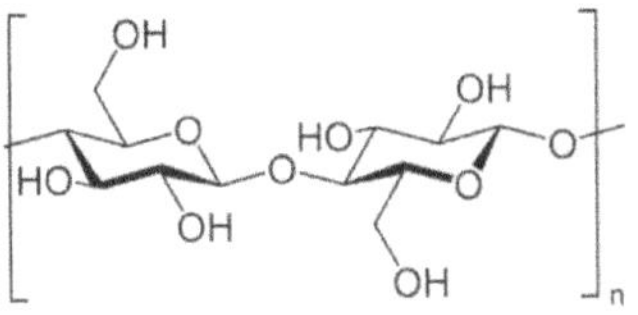

FIGURE 2.3 Structure of cellulose.

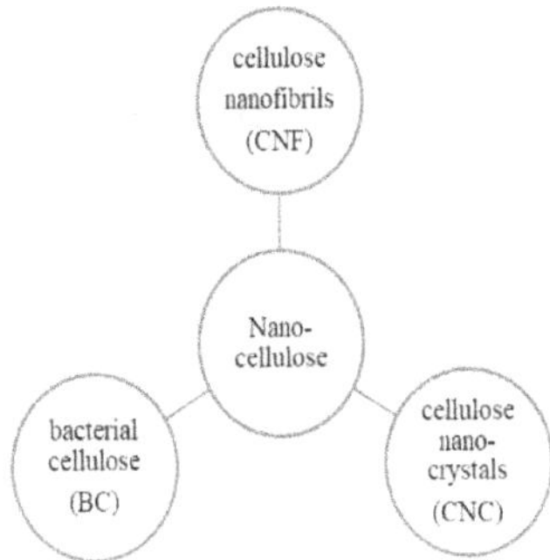

FIGURE 2.4 Different types of nanocelluloses classification of nanocellulose.

cellulose chains having diameters within the range of 5–40 nm with lengths of a few microns (Youssef et al., 2015) (Figure 2.3).

Nanocellulose is the most natural material that is currently used in many studies and is extracted from cellulose. Nanocellulose has attracted the interest of researchers due to its physicochemical properties such as biocompatibility, availability, non-toxicity, hydrophilicity, low cost, environmental friendliness and it is easy to handle. These properties allow the application of nanocellulose in several fields (Prado and Spinacé, 2019; Prado, Gonzales, and Spinace, 2019) such as automotive, structural components, and packaging. Nanocellulose can be categorised into three types, as shown in Figure 2.4.

A classification of nanocellulose material is given in Table 2.1. Nanocellulose nomenclature has not been used in a completely uniform manner in the past. Islam et al. have used the terms micro fibrillated cellulose (MFC), NCC, and BNC, the latter standing for bacterial nanocellulose. According to them, "the name micro fibrillated cellulose (MFC) was coined by the original investigators and is widely used in the scientific and commercial literature, whereas NCC and BNC seem simple and descriptive. Maybe, over time, the nomenclature 'nano fibrillated cellulose' will prevail, and the nanocellulose terminology will become more consistent".

Based on the methods that used, like enzymatic, mechanical, and chemical, the cellulose chain reduced in size in diameter, length, or both. According to the article nanocellulose materials must have at least one side within the range of nano. Therefore, nanocrystal cellulose (CNC), bacterial cellulose, and nanofibres cellulose (NFC) are the three types of nanocellulose (Figure 2.5).

TABLE 2.1
The Family of Nanocellulose Materials (Islam et al., 2014)

Type of Nanocellulose	Synonyms	Typical Sources	Formation and Average Size
Microfibrillated Cellulose (MFC)	Microfibrillated cellulose, nanofibrils and microfibrils, nanofibrillated cellulose	Wood, sugar beet, potato tuber, hemp, flax	Delamination of wood pulp by mechanical pressure before and/or after chemical or enzymatic treatment diameter: 5–60 nm length: several micrometres
Nanocrystalline Cellulose (NCC)	Cellulose nanocrystals, crystallites, whiskers, rod like cellulose microcrystals	Wood, cotton, hemp, flax, wheat straw, mulberry bark, ramie, Avicel, tunicin, cellulose from algae and bacteria	Acid hydrolysis of cellulose from many sources' diameter: 5–70 nm length: 100–250 nm (from plant celluloses);100 nm to several micrometres (from celluloses of tunicates, algae, bacteria)
Bacterial Nanocellulose (BNC)	Bacterialcellulose, microbialcellulose, biocellulose	Low molecular weight sugars and alcohols	Bacterial synthesisdiameter:20–100 nm; different types of nanofibre networks

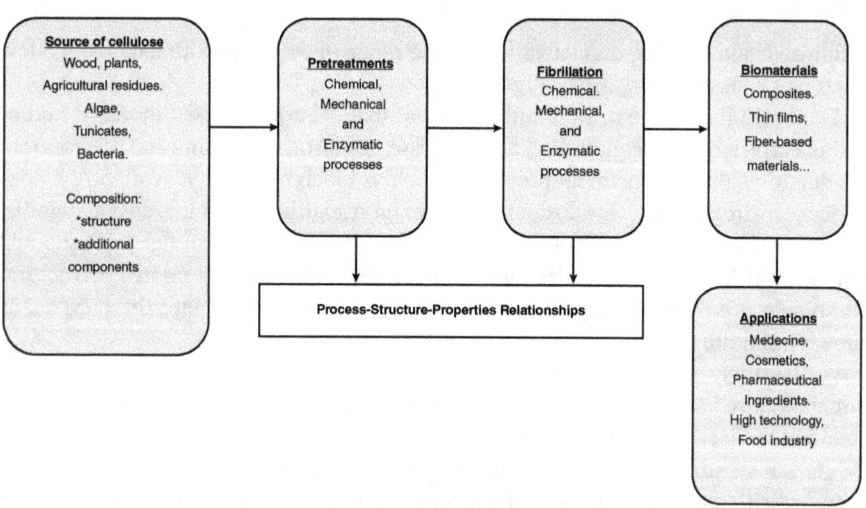

FIGURE 2.5 Flow chart depicting materials and steps during the processing of nanocellulose.

2.4 BIOPOLYMERS BASED ON SYNTHETIC MATERIALS

Three groups of bioplastics are biobased, non-biodegradable plastics, Fossil-based biodegradable plastics, and plastics that are both biobased and non-biodegradable biobased biodegradable. Most modern plastics are synthesized from fossil petrochemicals such as natural gas or oil. However, such plastics are resistant to biodegradation

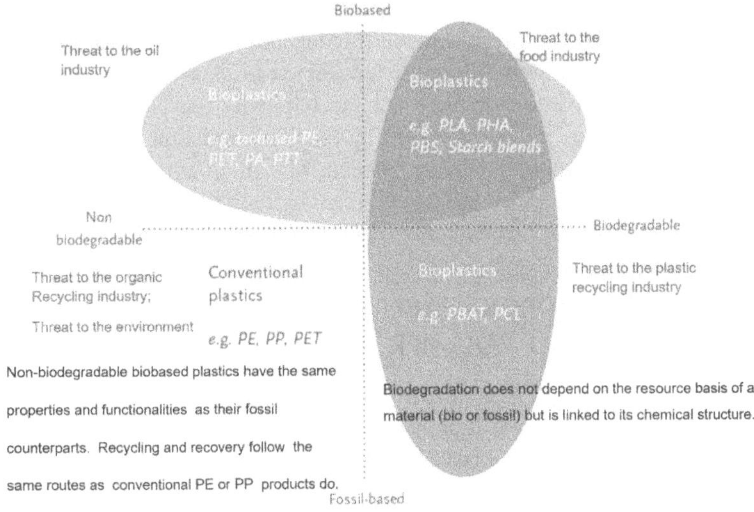

FIGURE 2.6 Bio-based non-biodegradable vs petro-based non-biodegradable.

and, therefore, their widespread use Led to the accumulation of a huge amount of plastic waste. According to a 2016 plastic industry report, global plastic manufacture has expanded by 8.6% each year since 1950, from 1.5 million tonnes to more than 330 million tonnes. Non-decomposed plastic enters our ecosystem, pollutes the environment, reduces soil fertility, and leads to the death of millions of animals (Figure 2.6).

2.5 BIOBASED NON-BIODEGRADABLE VS PETRO-BASED NON-BIODEGRADABLE POLYMERS

"Biobased" is defined in European standard EN 16575 as "derived from biomasses." Therefore, a biobased product is a product wholly or partly derived from biomass. Biomass is material of biological origin, excluding material embedded in geological formations and/or fossilised (EN 16575, 2014). Examples are paper and wood, but also plastics such as PLA whose building blocks are produced from sugars. Biobased materials that are not biodegradable include bio-PE, bio-PP and bio-PET. These materials can be recycled. It is the polymerisation process of the biobased material that enables it to perform like conventional plastic PE, PP or PET. The significant difference is in the origin of the feedstock used to make the biobased material.

Recent statistics show that the world produces ≈381 million tons of polymeric waste annually, and out of this, 50% originates from single-use polymers (Alberghini et al., 2023; Mtibe et al., 2021). Considering the sustainable usage of plastic products, recycling and biodegradation play an essential role, particularly the end-of-life options. Thermoplastic polymers are melt-processable and can be streamlined for recycling purpose.

Conventional polymeric materials are made up of petroleum-based resources and are non-biodegradable. As a result, those that are not recycled end up in landfills, unnecessarily and prematurely filling the landfill sites.

Many Fossils fuel–based polymers are mechanically recycled on an industrial scale. However, there are concerns regarding the contamination of fossil fuel-based recycling streams by biopolymers. The presence of a small amount of moisture during processing can trigger hydrolysis of polymer chains, which leads to a reduction in molecular weight and poor mechanical properties (Mtibe et al., 2021).

2.6 BIOBASED BIODEGRADABLE VS PETRO-BASED BIODEGRADABLE POLYMER

Petrochemicals or fossil plastics are made of fossil feedstocks like petroleum and natural gas which have taken millions of years to be formed. Nowadays, about 7% of all petroleum is converted into plastics (European Bioplastics. 2013). (. Examples of fossil-based plastics are polyethylene (PE), polypropylene (PP), PET, PS and others. Whereas presently these materials are predominantly made from fossil feedstock, they could also be produced from biomass and would then be biobased.

The main applications of biobased and biodegradable plastics are currently in (food) packaging, food service ware, (shopping) bags, fibres/nonwovens and agricultural applications. Biobased drop-in plastics such as bio-PE and bio-PET are identical to fossil-based counterparts and can be used in the same applications. The three most used biobased plastics with unique properties are PLA, Starch-based plastics, and Cellophane. As for fossil-based plastics, careful selection of a biobased and/or biodegradable packaging material is necessary to ensure that a packed product has the required shelf life. Some characteristics of plastic can be a disadvantage in one application and an advantage in another; the low water vapour barrier of the biobased plastic PLA is a disadvantage for a water bottle but an advantage in the (breathable) packaging of vegetables and fruit. Further, biobased and/or biodegradable plastics must comply with the same regulations with respect to food safety as fossil-based plastics, and many biobased plastics have certificates to prove that they can be used in food-contact applications.

The most suitable end-of-life solution depends on the type of biobased and biodegradable plastic, the volume on the market, the application and the available collection and processing infrastructure. Apart from PET bottle recycles, all post-consumer fossil plastic recycles are not 100% pure and cannot be used in food packaging products due to food safety regulations. Biodegradable plastic packaging, which is not mechanically recycled, and which does not add benefits in the composting route, can be incinerated, thus allowing energy recovery.

The environmental impact of biobased plastics and fossil-based plastics are in different impact categories. Substitution of fossil-based plastics by biobased plastics generally leads to lower non-renewable energy use and greenhouse gas (GHG) emissions. The GHG emission reduction, however, may be negatively influenced by direct and/or indirect land-use change. The GHG emission reduction that is reached by biobased plastics is generally significantly larger than that by biofuels. For the categories related to agriculture, such as eutrophication and acidification, biobased

TABLE 2.2
Properties of Biodegradable Plastics vs Synthetic Plastics (Wani, Shaikh, and Sayyed, 2015)

Parameters	PHB	Polypropylene (PP)
Source	Microbial	Fossil fuel (Petrochemicals)
Melting point Tm [°C]	171–182	171–186
Crystallinity [%]	65–80	65–70
MW ($\times 10^{-5}$)	1–8	2.2–7
Biodegradability	Good degraded within six months	Degraded require >500 years
Environmental aspect	eco-friendly	Not eco-friendly
Density [g/cm^3]	1.23–1.25	0.905–0.94
Tensile strength [MPa]	40	39
Solvent resistance	Poor	Good

plastics generally have a higher impact than fossil plastics. Because there are large differences in impact both in the family of biobased plastics as well as in the family of fossil plastics no absolute rule can be given (Van den Oever et al., 2017; Molenveld and Bos, 2017).

A variety of biodegradable polyesters has been developed in the eighties of the previous century. These biodegradable polyesters are predominantly based on fossil fuels. These materials are mainly used in combination with biobased polymers like starch and PLA. The most frequently used biodegradable polyesters are PBAT, PCL and PBS(A). Partly biobased PBS is commercially available. Furthermore, some biobased and biodegradable plastics, such as PLA, have barrier properties that can help food to stay fresh longer. For instance, the shelf life of lettuce packed in PLA can be extended by two days. It is well known that approximately 30% of the total food production is wasted; part of this occurs in the supermarket due to the passing of the sell-by date, and part of it occurs at home.

Biodegradable polymers posse's thermoplastic characteristics and resemble synthetic polymers to a larger extent. Table 2.2 illustrates the properties of biodegradable polymers v/s synthetic plastics.

Vinyl polymers, such as poly(ethylene), poly(propylene), and poly(vinyl chloride), with carbon-only backbones, are normally not susceptible to hydrolysis or biodegradation. An exception is poly(vinyl alcohol) or PVOH, which is biodegradable because of its high hydrolysability. The enzymatic oxidation of hydroxy groups (–OH) forms carbonyl groups (C––O) in the polymer backbone, and hydrolysis of two carbonyl groups (–CO–CH_2–CO–) causes polymer chain cleavage, which leads to a decrease in molecular weight (Kale et al., 2007).

2.7 DEGRADABLE PLASTICS

Degradable plastic as "a plastic designed to undergo a significant change in its structure under specific environmental conditions." The degradation process of polymeric materials can be typically characterised by a loss of properties (e.g., integrity, molecular weight, structure, or mechanical strength) and/or fragmentation. Biodegradable plastic is "a degradable plastic in which the degradation results from the action of naturally occurring microorganisms such as bacteria, fungi and algae."

Bioplastics are defined as materials that are biobased, biodegradable or both; they can provide excellent biodegradability and can be used to help alleviate environmental problems. All plastics can be divided into two main categories, thermosetting and thermoplastics, according to their reactions to heating. Thermoplastics are plastics that can be melted and moulded by heating, and when cooled, they take the shape in which they were formed. If the formed thermoplastic is reheated, it softens and melts again. Known thermoplastics are polyacrylates, polyesters, polyolefins, polyamides, etc. At present, many scientists and engineers worldwide are very interested in bioplastics because of their vulnerability and water exposure, lack of good relation, and low melting point relative to plastic petroleum. However, the development of bioplastics is hampered by higher costs of production compared to traditional plastics.

Several have been found to degrade PET, which is the most widely used synthetic polymer in the world (Yoshida et al., 2016), with almost 70 million tons produced worldwide each year for use in textiles and packaging (Markit, 2018). PET is composed of repeating units of ethylene glycol and aromatic terephthalic acid, which increases the rigidity of the polymer chain and decreases its susceptibility to biodegradation (Tokiwa et al., 2009).

As defined by established standards, polymer biodegradation undergoes two phases: primary degradation and mineralisation (ultimate biodegradation). The primary degradation is significantly associated with biodeterioration in physical properties, such as discoloration, embrittlement, and fragmentation. This primary degradation can occur when microorganisms' starving conditions secrete extracellular enzymes to break down the longer chains into smaller molecular or oligomers. This disintegration process is sometimes referred to as primary degradation. The second phase is the ultimate biodegradation (bioassimilation) conversion of polymer low molecular fragments/compounds into CO_2, water, production of new microbial cell biomass (aerobic conditions), CH_4, CO_2, and production of new microbial cell biomass in the case of anaerobic conditions (Figure 2.7).

Polymer degradation, in general, is influenced by both abiotic and biotic factors, whereby the term "abiotic" describes parameters like mechanical stress, light or temperature, and "biotic" the involvement of naturally occurring microorganisms like bacteria, fungi, and algae (Kyrikou and Briassoulis, 2007). Traditional polymeric materials are not easily degraded because they are resistant to microbial attack; they accumulate in the environment and represent a significant source of environmental pollution. Packaging materials with biodegradable plastic is more expensive than traditional petroleum-based plastic. Packaging materials based on these natural materials might be a solution to help control environmental pollution and resolve other problems posed by non-degradable synthetic polymers. Biodegradation and composting techniques have been accepted worldwide as one of the most promising technologies to determine the biodegradability of polymeric materials.

In that regard, much attention has been given to more environmentally friendly solutions, leading to the support of biodegradable materials as an alternative to

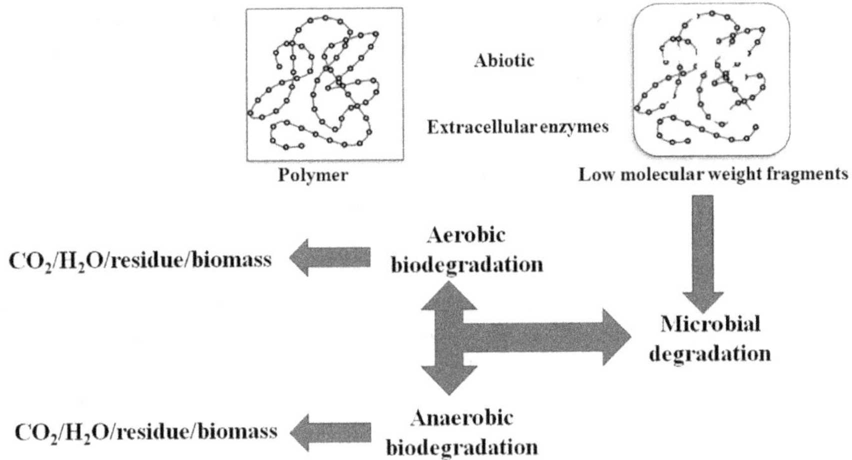

FIGURE 2.7 Diagrammatic representation of the chemistry of biodegradation (Luckachan and Pillai, 2011).

non-biodegradable polymeric materials. Reengineering polymer new composite materials from synthetic and biobased polymers have gained attention from industries, consumers, and governments as a potential solution for dependence on petroleum because any carbon sourced from biomass could be recycled into bio-resource. Particularly, renewable agricultural and biomass feedstock has shown much promise for use in eco-efficient packaging to replace petroleum feedstock without competing with food crops (Muniyasamy, Reddy, et al., 2013; Muniyasamy et al., 2008).

However, as compared to thermoplastic synthetic polymers, biopolymers present problems when processed with traditional technologies and show inferior performances in terms of functional and structural properties. Although the fate of environmentally degradable polymeric materials and plastics must be their conversion by microorganisms into final elemental products such as carbon dioxide, water, and new microbial biomass (i.e., mineralisation), it is necessary to evaluate their biodegradability in natural environments to meet various commercial and environmental needs for their sustainable growth.

Polyolefins has significantly obtained a main position in packaging industry because of their low cost, lightweight, required properties, and low energy consumption during their processing. Among those polymers, LDPE, which was hard to degrade in environment, is one of the fastest-growing commercial thermoplastic materials. However, the continuous use of polyethylene (PE) plastics in different applications led to the growing problem of environmental pollution. To overcome the problem, production of degradable and biodegradable polyolefins (Kubowicz and Booth, 2017).

The degradation of polymers might proceed by one or more mechanisms, including biodegradation, chemical degradation, and photo and thermal oxidation, depending on the polymer environment and desired application. The combination of different environmental factors such as sunlight, heat, oxygen, humidity, and microorganisms have synergistic effects on the degradation rate of starch-filled low-density polyethylene (LDPE). Natural polymers are susceptible to microbial attack, and they are often derived from agricultural products (such as starch, proteins, cellulose, and plant oils). They are the major resources for the development of many renewable and biodegradable polymers for industrial applications due to increased environmental concerns and diminished petrochemical resources (Le Corre, Bras, and Dufresne, 2010) (Table 2.3).

Many of the biodegradation studies of biobased polymers have focused on the microbial enzymes and molecular size level, which is related to different molecular weights, copolymer content, thickness, size particles, etc., and identifying the influence on the degradation and biodegradation behaviours under natural environmental conditions. Biodegradation studies of biobased polyesters such as PLA, PHA, TPS, PBAT, PBS(A), and PCL under controlled composting conditions in accordance with ISO and ASTM biodegradation test methods, identify that biobased polymer blended with starch have great influence on the improving mechanical strength, elongation properties, and increased biodegradation rate (Iovino et al., 2008; Jayasekara, Dissanayake, and Jayakody, 2022; Signori, Pelagaggi, et al., 2012) (Figure 2.8). The research by Jayasekara et al. (2003) shows a higher biodegradation result of blends of starch and PBSA under industrial composting conditions that reached 90% biodegradation within 45 days.

Kijchavengkul et al. (2008) and Kijchavengkul and Auras (2008) studied biodegradation of PLA and PLA/PCL blend under controlled composting, according to ISO 14855 and the biodegradability results of both samples reached 100% relative biodegradation within 90 days. In studies by Iovino et al. (2008) and Richards et al. (2008), commercial-grade PLA was studied in compost with high-humidity conditions at a temperature of 58–60°C, the PLA reached 61% biodegradation within 45 days and complete biodegradation after 80 days. Interestingly, Muniyasamy, Anstey, et al. (2013) studied the petroleum-based biodegradable polyester PBAT and its biocomposites containing 20% and 30% biofuel co-products dried distiller grain solubles (DDGS) under controlled composting conditions, and the biodegradation results of all PBAT biocomposites reached 60%–65% biodegradation after 45 days with positive trend reaching 90% biodegradation in 180 days.

The microorganisms reported for the biodegradation of the polyethylene include fungi (*Aspergillus niger, Aspergillus flavus, Aspergillus oryzae, Chaetomium globusum, Penicillium funiculosum, Pullularia pullulan*), bacteria (*Pseudomonas aeruginosa, Bacillus cereus, Coryneformesbacterium, Bacillus* sp., *Mycobacterium, Nocardia, Corynebacterium, Candida*, and *Pseudomonas*) and Actinomycetales (*Streptomycetaceae*). Their activity on the polymer was studied by growth tests on solid agar medium for a definite period. The changes in molecular weight, structure, crystallinity, density, weight loss, mechanical, optical, or dielectric properties were also measured (Weiland, Daro, and David, 1995; Volke-Sepúlveda et al., 2002; Yamada-Onodera et al., 2001; Orhan and Büyükgüngör, 2000; Saha, Zatloukal, and Saha, 2003; Pinchuk, Von Plessen, and Kreibig, 2004).

Composting is a natural process by which organic material is decomposed into a soil-like substance, called humus, a soil conditioner. Decomposition is mainly performed by microorganisms (mesophilic and thermophilic), including bacteria, fungi,

TABLE 2.3
Overview of Biodegradation Results of Biodegradable Polymer in Different Environmental Conditions (Muniyasamy and John, 2017)

Polymer materials	Results	Condition	Reference
PCL	100% biodegradable after 67 days	Compost	Hoshino et al. (2007)
PBAT (Ecoflex)	81% biodegradation after 33 days	Compost	Hoshino et al. (2007)
PLA (Nature works)	100% biodegradation after 150 days	Compost	Hoshino et al. (2007)
PLA/PCL blend	100% biodegradation after 76 days for both bio-based polymers	Compost	Kijchavengkul and Auras (2008)
PHB	90% biodegradation after 81 days	Compost	Hoshino et al. (2007)
PBAT/PLA blend (Ecovio)	100% biodegradation after 81 days	Compost	Touchaleaume et al. (2016)
PLA (Bioflex)	>90% biodegradation after 45 days	Compost	Signori et al. (2012)
PBSA/starch blend	>90% biodegradation after 50 days	Compost	Jayasekara et al. (2003)
PLA	61% biodegradation after 45 days and increasing with a slow biodegradation phase	Compost	Iovino et al. (2008)
TPS	60% biodegradation after 45 days and increasing with a slow biodegradation phase	Compost	Iovino et al. (2008)
PHA	64% biodegradation after 45 days and increasing	Compost	Richards et al. (2008)
PBAT (Ecoflex)	60% biodegradation after 45 days and increasing with a slow biodegradation phase	Compost	Muniyasamy, Reddy, et al. (2013)
PLA	65% biodegradation after 63 days and increasing with a slow biodegradation phase	Compost	Kale et al. (2007)
PLA PLA natural fibre composite	100% biodegradation for PLA after 120 days and 100% biodegradation of PLA natural fibre composite within 60 days	Compost	Pradhan, Misra, et al. (2010)
PHB and PHBV	81% biodegradation after 60 days for both biopolymer	Compost	Richards et al. (2008)
PBAT (Ecoflex)	70% after 25 days	Soil	Muniyasamy (2011)
PBS (Bionelle)	100% biodegradation after 110 days	Soil	Arcos-Hernandez et al. (2012)
PBSA	70% biodegradation after 368 days	Soil	Ratto et al. (1999)
PCL/starch blend (Mater-bi)	90% biodegradation after 84 days	Soil	Feuilloley et al. (2005)
PCL/starch blend (Mater-bi)	80% biodegradation after 117 days	Freshwater	Feuilloley et al. (2005)
PLA	90% biodegradation after 110 days at 55°C	Anaerobic digestion	Yagi et al. (2009b)
PCL	90% biodegradation after 60 days	Anaerobic digestion	Yagi et al. (2009a)

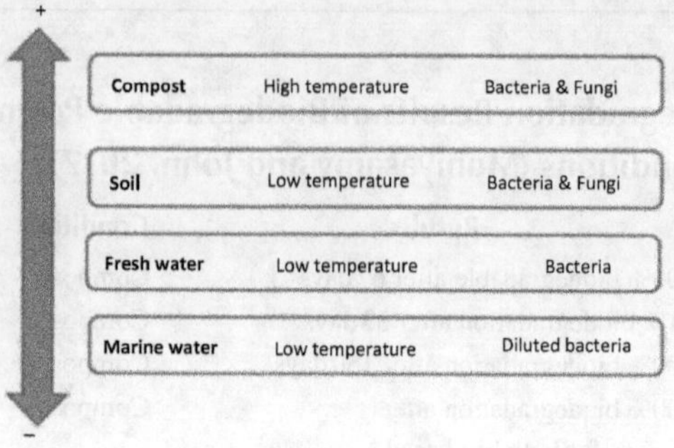

FIGURE 2.8 Relation between environment and aggressiveness of biodegradation (Muniyasamy and John, 2017).

and actinomycetes. These microorganisms use organic matter as their food source, generate CO_2, and produce humus as a product. A composting process goes through two main stages: an active composting stage and a curing period. In the first stage, the temperature rises and remains elevated, provided oxygen is available, which results in strong microbial activity. In the later stage, the temperature decreases, but the materials continue to compost at a slower rate. The compost process does not stop at a particular point; rather it continues slowly until the last remaining nutrients are consumed by the remaining microorganisms and almost all the carbon has been converted into carbon dioxide (Rynk et al., 1992). Aerobic composting takes place in the presence of oxygen; if oxygen is absent, then the process changes to anaerobic digestion. Anaerobic digestion is also a naturally occurring process of decomposition and decay, by which organic matter breaks down into simple chemical components, producing biogas and digestate (a relatively stable soil residue like compost) (Kale et al., 2007).

PLA is a 100% biobased plastic that is widely used in packaging applications. Since most PLA grades are readily industrially compostable for thicknesses of up to 2–3.2 mm (depending on the grade), PLA is very suitable for the manufacture of compostable packaging products. Specific benefits of PLA in packaging applications are its transparency, gloss, stiffness, printability, processability and excellent aroma barrier. PLA is a rigid material with (mechanical) properties that are comparable with PS and PET (Figure 2.9).

The biodegradation of biodegradable polyesters such as PLA and PBAT in compost or soil is affected by both biotic and abiotic factors of the environment, including temperature, moisture, pH, bio-surfactants, and enzymes. Their biodegradation is also affected by polymer characteristics, such as chain flexibility, crystallinity, regularity and heterogeneity, functional groups, and molecular weight.

A completely different behaviour was observed for biodegradation of the neat PBAT polymer. A microbial lag phase, starting at approximately 18 days, was observed, followed by an exponential phase after 60 days of incubation (Figure 2.10). As previously studied (Mohanty and Nayak, 2012), PBAT material undergoes biodegradation under composting conditions and reaches 80% biodegradation in

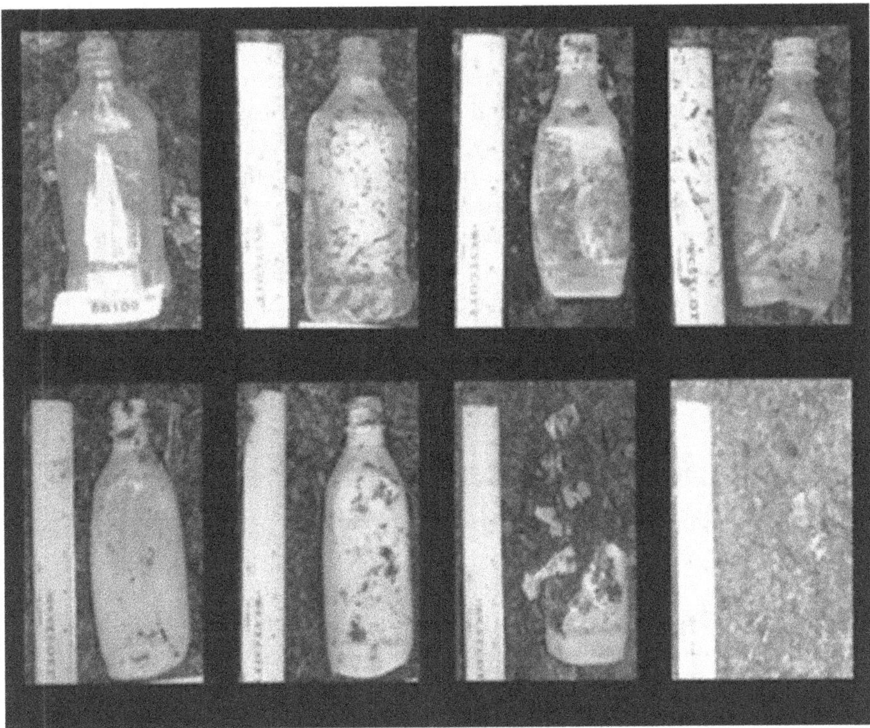

FIGURE 2.9 Biodegradation of PLA bottle under stimulated industrial composting conditions. (Figure reprinted with permission as reported by Kale et al. 2007.)

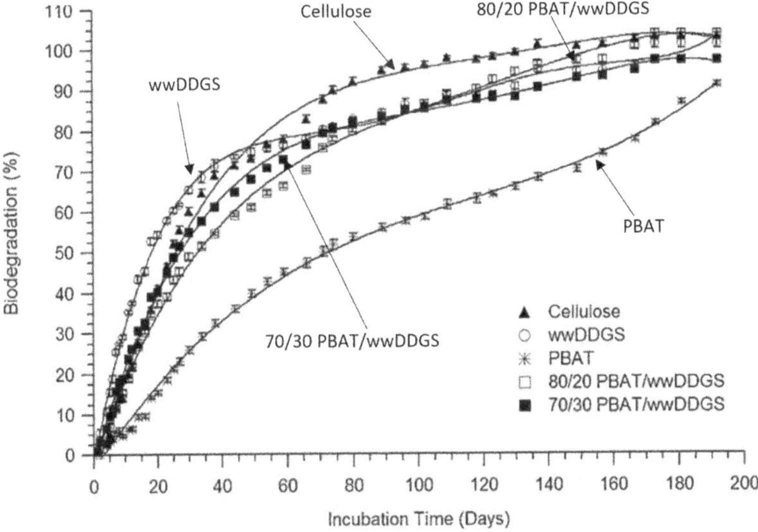

FIGURE 2.10 Biodegradation of microcrystalline cellulose (positive reference), wwDDGS, 80/20 (PBAT/wwDDGS), 70/30 (PBAT/wwDDGS) and PBAT under simulated aerobic compost (ASTM D5338) (Muniyasamy and Maya, 2017).

180 days. In our studies, similar degradation behaviour was observed, and the rate of biodegradation was about 90% within 180 days. It is noteworthy that the biodegradation behaviours of biocomposites materials containing pre-treated DDGS were 20% and 30%, very close to that of natural materials such as cellulose and DDGS. The data pertaining to the biodegradability of biocomposites clearly indicate that DDGS filler had a significant influence on the biodegradability of the PBAT matrix. During composting tests, the neat PBAT particles were observed after 10, 45 and 60 days of incubation in the test flask, and they were carefully recovered to evaluate the effect of mineralisation on the thermal properties and structural characterisation of the material. The recovered PBAT material particles were easily distinguishable and further washed with deionised water and dried in an oven at 40°C for 4 hours. In the case of the biocomposite material, recoverable particles were not distinguishable after ten days of incubation.

2.8 EFFECT OF ENVIRONMENTAL CONDITIONS

According to an investigation carried out on the biodegradation of PLA filled with corn starch (up to 60%) (Shogren et al., 2003), no degradation was observed in the PLA matrix in a soil burial test. The only weight loss that occurred in the composite was due to the degradation of the starch. In another study (Saadi et al., 2012), the PLA biodegradation test was studied and compared to reference cellulose materials in soil and compost conditions at room temperature (30°C) conditions for 200 days. The authors found that there was no PLA biodegradation in the soil or compost at room temperature conditions, while the reference cellulose materials achieved 75% biodegradation in 45 days in both conditions. When the same PLA test materials were exposed to industrial composting conditions at a temperature of 58°C, 90% biodegradation was achieved within 100 days. Therefore, it was concluded that industrial composting provides a much more suitable environment for the thermophilic bacteria and fungi that are involved in the biodegradation of PLA compared to soil burial, home compost and aqueous conditions, where mesophilic bacteria and fungi are predominant.

Researchers studied the biodegradability of poly(butylene succinate) (PBS) biocomposites with a rice husk flour (RHF) filler in soil and composting conditions (Kim et al., 2006). They observed higher biodegradability in the PBS biocomposite than in the neat PBS, and the biodegradation rate was enhanced with the addition of the RHF. The PBS/RHF biocomposites showed 18% weight loss in compost, while the PBS/RHF biocomposite had 13% weight loss in soil conditions in 80 days. The neat PBS showed a 6% weight loss in compost, while the neat PBS experienced weight loss of 3% in soil in 80 days. These comparison studies demonstrated that the RHF promoted the biodegradation of the PBS matrix in both the soil and compost environments. A comparison of the effects of environmental conditions revealed a higher rate of biodegradation in compost compared to soil due to the larger number of microbial colonies present in the compost. Biodegradation studies of aliphatic polyesters (PCL, PHBV, PBS, and PLA) with abaca fibres were investigated by Teramoto (Teramoto et al., 2004) using a soil burial test.

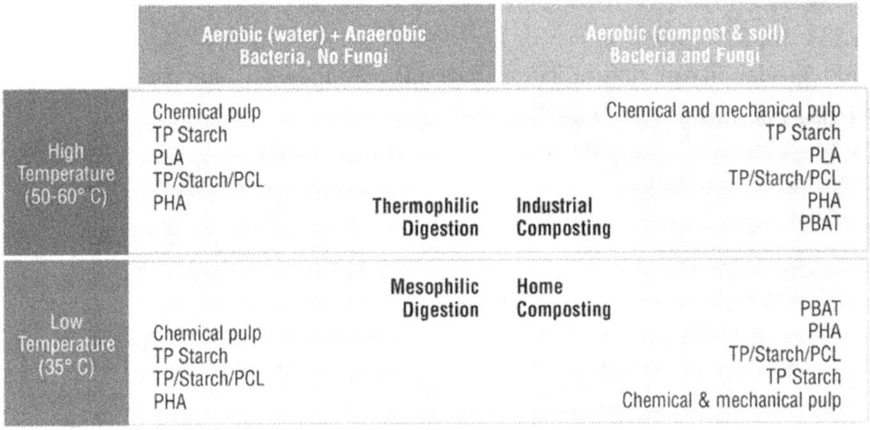

	Aerobic (water) + Anaerobic Bacteria, No Fungi		Aerobic (compost & soil) Bacteria and Fungi	
High Temperature (50-60° C)	Chemical pulp TP Starch PLA TP/Starch/PCL PHA	Thermophilic Digestion	Industrial Composting	Chemical and mechanical pulp TP Starch PLA TP/Starch/PCL PHA PBAT
Low Temperature (35° C)	Chemical pulp TP Starch TP/Starch/PCL PHA	Mesophilic Digestion	Home Composting	PBAT PHA TP/Starch/PCL TP Starch Chemical & mechanical pulp

FIGURE 2.11 Materials which are biodegradable polymers for biological waste treatment. (Figure redrawn after reference from Hermann et al. 2011.)

In this study, there was no weight loss observed for PLA in 180 days. The other composites demonstrated biodegradability, in the order of PCL>PHBV>PBS in 180 days. In another study, composites of the biodegradable polyester PBS filled with agricultural residues, such as rice husk, were exposed to biodegradation tests in an *Azospirillum brasilense* BCRC 12270 liquid culture medium. It was observed that the degree of biodegradation increased with an increasing rice husk content. To conclude, these various biodegradability studies demonstrated that a composting environment is most suitable for the disposal of bioplastic materials in a short time (within 6 months). The ultimate mineralisation of bioplastics is mainly dependent on the environmental conditions (temperature, oxygen, and active microorganisms) and the waste treatment options (composting and anaerobic digestion), which are represented in Figure 2.11.

2.9 BIOCOMPOSITES

See Table 2.4.

TABLE 2.4
Biodegradability of Natural Filler–Based Biocomposite Materials

Principal components (wt%)	Conditions for biodegradation	Results	Reference
PLA/maple wood fibre (70/30)	Compost ISO-14855	Acetyl treatment of the maple wood fibres increases their porosity, which enhances the hydrolytic biodegradation of PLA.	Way et al. (2012)

(Continued)

TABLE 2.4 (*Continued*)
Biodegradability of Natural Filler–Based Biocomposite Materials

Principal components (wt%)	Conditions for biodegradation	Results	Reference
PLA/cotton cellulose fibres (70/30)	Compost ISO-14855	Cotton cellulose fibres accelerated the biodegradability of PLA.	Funabashi and Kunioka (2005)
PLA/soy straw (70/30)	Compost (ASTM D5338)	Untreated soy straw and wheat straw promoted the biodegradability of PLA.	Pradhan, Misra, et al. (2010)
PLA/soy straw (70/30); PLA/wheat straw (70/30);PCL/distiller's dried grains with soluble (DDGS)(70/30);PCL/soy meal (70/30)	Compost (ASTM D5338)	All the biocomposites showed improved degradation compared to the neat biodegradable polymer used	Pradhan, Reddy, et al. (2010)
Distiller's dried grains with soluble (DDGS)/(aliphatic and aromatic co-polyester) PBAT (30/70)	Compost (ASTM D5338)	PBAT/DDGS biocomposite showed biodegradability similar to that of DDGS and cellulose. PBAT degradability increased with the incorporation of water washed DDGS into biocomposite system.	Muniyasamy, Reddy, et al. (2013)
Poly(ethylene sebacate)/ cellulose fibre (85/15)	Compost (ASTM D5338)	Within 30 days, the biocomposite products showed 100% biodegradability	Fernandes et al. (2011)
49% Starch /45% corn fibre/5% inorganic fillers/ 1% Arabic gum	Compost (ASTM D5338)	Approached plateau phase after 65% biodegradation in 75 days.	Chiellini et al. (2009)
45% Starch/7% poly(vinyl alcohol)/45% corn fibre/3% inorganic filler	Compost (ASTM D5338)	PVA biocomposite approached a plateau phase after 55% biodegradability in 75 days.	Chiellini et al. (2009)
80% Starch/16% poly(vinyl alcohol)/3% inorganic filler/1% Arabic gum	Compost (ASTM D5338)	Approached a plateau phase after 40% biodegradability in 75 days.	Chiellini et al. (2009)
96% Starch/3% inorganic filler/1% Arabic gum	Compost (ASTM D5338)	Approached a plateau phase after 40% biodegradability in 75 days	Chiellini et al. (2009)
49% Starch/45% corn fibre/5% inorganic fillers/1% Arabic gum	Soil (ASTM D5988)	90% biodegradability was observed within 130 days, and these materials satisfied the soil biodegradability criteria as per ASTM standard requirements.	Chiellini et al. (2009)

(*Continued*)

TABLE 2.4 (*Continued*)
Biodegradability of Natural Filler–Based Biocomposite Materials

Principal components (wt%)	Conditions for biodegradation	Results	Reference
45% Starch/7% poly(vinyl alcohol)/45% corn fibre/3% inorganic filler	Soil (ASTM D5988)	90% biodegradability observed within 130days, and these materials satisfied soil biodegradability criteria as per ASTM standard requirements.	Chiellini et al. (2009)

2.10 CONCLUSION

This chapter details the definition and classification of biobased materials, different types of biodegradable and compostable polymeric materials and a variety of biopolymers and their applications. We also emphasise biobased, biodegradable, non-degradable vs fossil-based non-biodegradable and biodegradable polymers. We also elaborated on degradable plastics, compostable polymer, oxo-biodegradable plastics, biocomposites, etc. Unmanaged plastic waste and its accumulation in nature are becoming a serious environmental issue, leaving behind carbon footprint, and contributing to GHG emissions.

To assist the background and management of plastic waste, the end-of-life treatments of plastics such as reuse, recycling, and composting (biodegradation) are important criteria for reducing the environmental burden from plastics. Currently, among different plastic waste management options, biological recycling of biobased biodegradable plastics from renewable materials is meeting solutions for plastic waste pollution.

Biobased polymeric materials are advancing in a multitude of applications, including both durable and single-use segments, due to their biodegradability, renewability, low cost, and low relative density. The high growth and global capacity production of biobased biodegradable polymers are increasing to meet the reduction of visual pollution and accumulation in nature.

The most produced biobased polymeric materials are derived from agricultural sectors, and current research is also focused on the development of biobased materials from non-food renewable lignocellulosic and agricultural and food wastes. Among biobased biodegradable plastics produced from renewable raw materials, PLA is the most important and most widely used biodegradable plastic at this moment. Even though PLA contains some challenges of thermomechanical properties for wide applications, the blending, and composites approach of PLA, in combination with other biodegradable polymers, provide advanced properties without compromising its renewability.

Unfortunately, not all the biobased polymers from renewable resources are biodegradable. The biodegradation of biobased polymer products can differ from one environment to another such as landfill, home cost, marine water, and freshwater. Also, biodegradation of polymers can be affected by the physical-chemical properties of the polymeric composition. To meet the biodegradable plastics and its scientific claims and certification, different international standards bodies such as ASTM, ISO and EN have been designed as standard and test methods for validating and labelling biodegradable plastics. Biodegradation of plastics occurs either by hydrolytic or oxidative degradation mechanisms and is influenced by various environmental aspects, biological factors and physical–chemical properties. Drastic measurements must be implemented, and the government must prioritise environmental policies and inject more funding towards a green environment.

REFERENCES

Abareshi, Maryam, Seyed Mojtaba Zebarjad, and E. K. Goharshadi. 2009. "Crystallinity behavior of MDPE-clay nanocomposites fabricated using ball milling method." *Journal of Composite Materials*, 43 (23), 2821–2830.

Abdelwahab, Mohamed A., Allison Flynn, Bor-Sen Chiou, Syed Imam, William Orts, and Emo Chiellini. 2012. "Thermal, mechanical and morphological characterization of plasticized PLA-PHB blends." *Polymer Degradation and Stability*, 97 (9), 1822–1828.

Aguilar, Nery M., F. Arteaga-Cardona, M. E. de Anda Reyes, J. J. Gervacio-Arciniega, and U. Salazar-Kuri. 2019. "Magnetic bioplastics based on isolated cellulose from cotton and sugarcane bagasse." *Materials Chemistry and Physics*, 238, 121921.

Alberghini, Leonardo, Alessandro Truant, Serena Santonicola, Giampaolo Colavita, and Valerio Giaccone. 2023. "Microplastics in fish and fishery products and risks for human health: A review." *International Journal of Environmental Research and Public Health*, 20 (1), 789.

Albertsson, Ann-Christine, Sven Ove Andersson, and Sigbritt Karlsson. 1987. "The mechanism of biodegradation of polyethylene." *Polymer Degradation and Stability*, 18 (1), 73–87.

Albertsson, Ann-Christine, and Sigbritt Karlsson. 1993. "Aspects of biodeterioration of inert and degradable polymers." *International Biodeterioration & Biodegradation*, 31 (3), 161–170.

Al-Malaika, S. 1993. *Handbook of polymer degradation*. S. Halim Hamid, Mohamed B. Amin and Ali G. Maadhah Marcel Dekker (eds.), 664 pp. New York: Elsevier.

Alshehrei, Fatimah. 2017. "Biodegradation of synthetic and natural plastic by microorganisms." *Journal of Applied & Environmental Microbiology*, 5 (1), 8–19.

Appendini, Paola, and Joseph H. Hotchkiss. 2002. "Review of antimicrobial food packaging." *Innovative Food Science & Emerging Technologies*, 3 (2), 113–126.

Arcos-Hernandez, Monica V., Bronwyn Laycock, Steven Pratt, Bogdan C. Donose, Melissa A. L. Nikolić, Paul Luckman, Alan Werker, and Paul A. Lant. 2012. "Biodegradation in a soil environment of activated sludge derived polyhydroxyalkanoate (PHBV)." *Polymer Degradation and Stability*, 97 (11), 2301–2312.

Armentano, I., N. Bitinis, E. Fortunati, S. Mattioli, N. Rescignano, Raquel Verdejo, Miguel A. López-Manchado, and José M. Kenny. 2013. "Multifunctional nanostructured PLA materials for packaging and tissue engineering." *Progress in Polymer Science*, 38 (10–11), 1720–1747.

Armstrong, Jonathan K., Rosalinda B. Wenby, Herbert J. Meiselman, and Timothy C. Fisher. 2004. "The hydrodynamic radii of macromolecules and their effect on red blood cell aggregation." *Biophysical Journal*, 87 (6), 4259–4270.

Awad, Yasser M., Sung-Chul Kim, Samy A. M. Abd El-Azeem, Kye-Hoon Kim, Kwon-Rae Kim, Kangjoo Kim, Choong Jeon, Sang Soo Lee, and Yong Sik Ok. 2014. "Veterinary antibiotics contamination in water, sediment, and soil near a swine manure composting facility." *Environmental Earth Sciences*, 71, 1433–1440.

Azizi Samir, My Ahmed Said, Fannie Alloin, and Alain Dufresne. 2005. "Review of recent research into cellulosic whiskers, their properties and their application in nanocomposite field." *Biomacromolecules*, 6 (2), 612–626.

Baranwal, Jaya, Brajesh Barse, Antonella Fais, Giovanna Lucia Delogu, and Amit Kumar. 2022. "Biopolymer: A sustainable material for food and medical applications." *Polymers*, 14 (5), 983.

Bastioli, C. 2005. "Starch-based technology." In: *Handbook of Biodegradable Polymers*, C. Bastioli, (ed.), 257-286. Rapra Technology Limited, Shawbury.

Bastioli, Catia. 2001. "Global status of the production of biobased packaging materials." *Starch-Stärke*, 53 (8), 351–355.

Bikiaris, D., J. Prinos, K. Koutsopoulos, N. Vouroutzis, E. Pavlidou, N. Frangis, and C. Panayiotou. 1998. "LDPE/plasticized starch blends containing PE-g-MA copolymer as compatibilizer." *Polymer Degradation and Stability*, 59 (1–3), 287–291.

Borah, Purba Prasad, Pulak Das, and Laxmikant S. Badwaik. 2017. "Ultrasound treated potato peel and sweet lime pomace based biopolymer film development." *Ultrasonics Sonochemistry*, 36, 11–19.

Briassoulis, Demetres, and Anastasios Giannoulis. 2018. "Evaluation of the functionality of bio-based plastic mulching films." *Polymer Testing*, 67, 99–109.

Brodhagen, Marion, Jessica R. Goldberger, Douglas G. Hayes, Debra Ann Inglis, Thomas L. Marsh, and Carol Miles. 2017. "Policy considerations for limiting unintended residual plastic in agricultural soils." *Environmental Science & Policy*, 69, 81–84.

Buléon, A., P. Colonna, V. Planchot, and S. Ball. 1998. "Starch granules: Structure and biosynthesis." *International Journal of Biological Macromolecules*, 23 (2), 85–112.

Cabedo, Lluís, José Luis Feijoo, María Pilar Villanueva, José María Lagarón, and Enrique Giménez. 2006. "Optimization of biodegradable nanocomposites based on aPLA/PCL blends for food packaging applications." *Macromolecular Symposia*, 233, 191–197.

Cabral, Horacio, Kanjiro Miyata, Kensuke Osada, and Kazunori Kataoka. 2018. "Block copolymer micelles in nanomedicine applications." *Chemical Reviews*, 118 (14), 6844–6892.

Chakraborty, Ayan, Mohini Sain, Mark Kortschot, and Sean Cutler. 2007. "Dispersion of wood microfibers in a matrix of thermoplastic starch and starch-polylactic acid blend." *Journal of Biobased Materials and Bioenergy*, 1 (1), 71–77.

Chen, Guo-Qiang. 2009. "A microbial polyhydroxyalkanoates (PHA) based bio-and materials industry." *Chemical Society Reviews*, 38 (8), 2434–2446.

Chiellini, E., P. Cinelli, V. I. Ilieva, S. H. Imam, and J. W. Lawton. 2009. "Environmentally compatible foamed articles based on potato starch, corn fiber, and poly (vinyl alcohol)." *Journal of Cellular Plastics*, 45 (1), 17–32.

Chiellini, Emo, and Roberto Solaro. 1996. "Biodegradable polymeric materials." *Advanced Materials*, 8 (4), 305–313.

Cornell, John H., Arthur M. Kaplan, and Morris R. Rogers. 1984. "Biodegradability of photo-oxidized polyalkylenes." *Journal of Applied Polymer Science*, 29 (8), 2581–2597.

Corti, Andrea, Muniyasami Sudhakar, and Emo Chiellini. 2012. "Assessment of the whole environmental degradation of oxo-biodegradable linear low density polyethylene (LLDPE) films designed for mulching applications." *Journal of Polymers and the Environment*, 20 (4), 1007–1018.

Craver, Clara, and C. Carraher. 2000. *Applied polymer science: 21st century*. Amsterdam, the Netherland: Elsevier.

Devi, Rajendran Sangeetha, Velu Rajesh Kannan, Krishnan Natarajan, Duraisamy Nivas, Kanthaiah Kannan, Sekar Chandru, and Arokiaswamy Robert Antony. 2016. "The role of microbes in plastic degradation." In *Environmental Waste Manage*, Ram Chandra (eds.), 341–370. DOI:10.1201/B19243-16

Dreher, Matthew R., Wenge Liu, Charles R. Michelich, Mark W. Dewhirst, Fan Yuan, and Ashutosh Chilkoti. 2006. "Tumor vascular permeability, accumulation, and penetration of macromolecular drug carriers." *Journal of the National Cancer Institute*, 98 (5), 335–344.

Dubick, Michael A., and Charles E. Wade. 1994. "A review of the efficacy and safety of 7.5% NaCl/6% dextran 70 in experimental animals and in humans." *Journal of Trauma and Acute Care Surgery*, 36 (3), 323–330.

Dubief, David, Eric Samain, and Alain Dufresne. 1999. "Polysaccharide microcrystals reinforced amorphous poly (β-hydroxyoctanoate) nanocomposite materials." *Macromolecules*, 32 (18), 5765–5771.

Dufresne, Alain. 2006. "Comparing the mechanical properties of high performances polymer nanocomposites from biological sources." *Journal of Nanoscience and Nanotechnology*, 6 (2), 322–330.

El-Rehim, HA Abd, El-Sayed A. Hegazy, A. M. Ali, and A. M. Rabie. 2004. "Synergistic effect of combining UV-sunlight-soil burial treatment on the biodegradation rate of LDPE/starch blends." *Journal of Photochemistry and Photobiology A: Chemistry*, 163 (3), 547–556.

Eubeler, Jan P., Marco Bernhard, and Thomas P. Knepper. 2010. "Environmental biodegradation of synthetic polymers II. Biodegradation of different polymer groups." *TrAC Trends in Analytical Chemistry*, 29 (1), 84–100.

European Bioplastics. 2013. *Bioplastics-Facts and Figures*, Vol. December 2013. Available at: http://en.european-bioplastics.org/blog/2012/10/10/pr-bioplastics-market-20121010/ (Last accessed November 2013).

Fanun, Monzer. 2014. *The role of colloidal systems in environmental protection*, 1st edn, M. Fanun (ed.). Elsevier.

Faruk, Omar, Andrzej K. Bledzki, Hans-Peter Fink, and Mohini Sain. 2012. "Biocomposites reinforced with natural fibers: 2000-2010." *Progress in Polymer Science*, 37 (11), 1552–1596.

Fernandes, Tânia F., Eliane Trovatti, Carmen SR Freire, Armando JD Silvestre, Carlos Pascoal Neto, Alessandro Gandini, and Patrizia Sadocco. 2011. "Preparation and characterization of novel biodegradable composites based on acylated cellulose fibers and poly (ethylene sebacate)." *Composites Science and Technology*, 71 (16), 1908–1913.

Feuilloley, Pierre, Guy Cesar, Ludovic Benguigui, Yves Grohens, Isabelle Pillin, Hilaire Bewa, Sandra Lefaux, and Mounia Jamal. 2005. "Degradation of polyethylene designed for agricultural purposes." *Journal of Polymers and the Environment*, 13, 349–355.

Fotopoulou, Kalliopi N., and Hrissi K. Karapanagioti. 2017. "Degradation of various plastics in the environment." In *Hazardous chemicals associated with plastics in the marine environment*, 71–92. Springer.

Fukuda, Yasushi, and Zenjiro Osawa. 1991. "Wavelength effect on the photo-degradation of polycarbonate and poly (methyl methacrylate)-Confirmation of the photo-degradation mechanism of PC/PMMA blends." *Polymer Degradation and Stability*, 34 (1–3), 75–84.

Funabashi, Masahiro, and Masao Kunioka. 2005. "Biodegradable composites of poly (lactic acid) with cellulose fibers polymerized by aluminum triflate." *Macromolecular Symposia*, 224, 309–322.

Gangalla, Ravi, Sampath Gattu, Sivasankar Palaniappan, Maqusood Ahamed, Baswaraju Macha, Raja Komuraiah Thampu, Antonella Fais, Alberto Cincotti, Gianluca Gatto, and Murali Dama. 2021. "Structural characterisation and assessment of the novel Bacillus amyloliquefaciens RK3 exopolysaccharide on the improvement of cognitive function in Alzheimer's disease mice." *Polymers*, 3 (17), 2842.

Gautam, R., A. S. Bassi, and E. K. Yanful. 2007. "A review of biodegradation of synthetic plastic and foams." *Applied Biochemistry and Biotechnology*, 141 (1), 85–108.

Gewert, Berit, Merle M. Plassmann, and Matthew MacLeod. 2015. "Pathways for degradation of plastic polymers floating in the marine environment." *Environmental Science: Processes & Impacts*, 17 (9), 1513–1521.

Goodarzi, Navid, Reyhaneh Varshochian, Golnaz Kamalinia, Fatemeh Atyabi, and Rassoul Dinarvand. 2013. "A review of polysaccharide cytotoxic drug conjugates for cancer therapy." *Carbohydrate Polymers*, 92 (2), 1280–1293.

Göpferich, Achim. 1996. "Mechanisms of polymer degradation and erosion." *Biomaterials*, 17 (2), 103–114.

Grossman, Richard F., and Domasius Nwabunma. 2013. *Biopolymer nanocomposites: Processing, properties, and applications*, D. Nwabunma, S. Thomas, R. F. Grossman, A. Dufresne, and L. A. Pothan (eds.), Vol. 8. John Wiley & Sons.

Gu, Ji-Dong. 2003. "Microbiological deterioration and degradation of synthetic polymeric materials: Recent research advances." *International Biodeterioration & Biodegradation*, 52 (2), 69–91.

Guilbert, S., N. Gontard, and B. Cuq. 1995. "Technology and applications of edible protective films." *Packaging Technology and Science*, 8 (6), 339–346.

Gutiérrez, Tomy J., Elevina Pérez, Romel Guzmán, María Soledad Tapia, and Lucia Mercedes Fama. 2014. "Physicochemical and functional properties of native and modified by crosslinking, dark-cush-cush yam (Dioscorea Trifida) and cassava (Manihot Esculenta) starch." *Journal of Polymer and Biopolymer Physics Chemistry*, 2 (1), 1–5.

Halim Hamid, S. 2000. *Handbook of polymer degradation (environmental science & pollution control series)*, S. H. Hamid, M. B. Amin, A. G. Maadhah (eds.). Marcel Dekker Inc.

Hauenstein, O., S. Agarwal, and A. Greiner. 2016. "Bio-based polycarbonate as synthetic toolbox." *Nature Communications*, 7, 11862.

Hermann, B. G., L. Debeer, B. De Wilde, K. Blok, and Martin Kumar Patel. 2011. "To compost or not to compost: Carbon and energy footprints of biodegradable materials' waste treatment." *Polymer Degradation and Stability*, 96 (6), 1159–1171.

Hoshino, Akira, Maso Tsuji, Masanori Momochi, Akiko Mizutani, Hideo Sawada, Setsuo Kohnami, Hiroki Nakagomi, Michio Ito, Hisato Saida, and Munehiro Ohnishi. 2007. "Study of the determination of the ultimate aerobic biodegradability of plastic materials under controlled composting conditions: Method by gravimetric method of evolved carbon dioxide using Microbial Oxidative Degradation Analyzer (MODA)." *Journal of Polymers and the Environment*, 15, 275–280.

Huang, Mingfu, Jiugao Yu, and Xiaofei Ma. 2005. "Ethanolamine as a novel plasticiser for thermoplastic starch." *Polymer Degradation and Stability*, 90 (3), 501–507.

Iovino, R., R. Zullo, M. A. Rao, L. Cassar, and L. Gianfreda. 2008. "Biodegradation of poly (lactic acid)/starch/coir biocomposites under controlled composting conditions." *Polymer Degradation and Stability*, 93 (1), 147–157.

Islam, Mohammad Tajul, Mohammad Mahbubul Alam, Alessia Patrucco, Alessio Montarsolo, and Marina Zoccola. 2014. "Preparation of nanocellulose: A review." *AATCC Journal of Research*, 1 (5), 17–23.

Jang, Byung Chul, Soo Young Huh, Jeong Gyu Jang, and Young Chan Bae. 2001. "Mechanical properties and morphology of the modified HDPE/starch reactive blend." *Journal of Applied Polymer Science*, 82 (13), 3313–3320.

Jayasekara, Ranjith, Simon Sheridan, Eleni Lourbakos, Henry Beh, G. B. Y. Christie, Malcolm Jenkins, Peter B. Halley, Stewart McGlashan, and Greg T. Lonergan. 2003. "Biodegradation and ecotoxicity evaluation of a bionolle and starch blend and its degradation products in compost." *International Biodeterioration & Biodegradation*, 51 (1), 77–81.

Jayasekara, Sandhya, Lakshika Dissanayake, and Lahiru N. Jayakody. 2022. "Opportunities in the microbial valorization of sugar industrial organic waste to biodegradable smart food packaging materials." *International Journal of Food Microbiology*, 377, 109785.

Kale, Gaurav, Thitisilp Kijchavengkul, Rafael Auras, Maria Rubino, Susan E. Selke, and Sher Paul Singh. 2007. "Compostability of bioplastic packaging materials: An overview." *Macromolecular Bioscience*, 7 (3), 255–277.

Kamini, N. R., and H. Iefuji. 2001. "Lipase catalyzed methanolysis of vegetable oils in aqueous medium by *Cryptococcus* spp. S-2." *Process Biochemistry*, 37 (4), 405–410.

Karmanov, Anatoly Petrovich, Albert Vladimirovich Kanarsky, Lyudmila Sergeevna Kocheva, Vladimir Aleksandrovich Belyy, Eduard Ilyasovich Semenov, Natalia Gelievna Rachkova, Nikolai Ivanovich Bogdanovich, and Sergey Alexandrovich Pokryshkin. 2021. "Chemical structure and polymer properties of wheat and cabbage lignins-Valuable biopolymers for biomedical applications." *Polymer*, 220, 123571.

Keshavarz, Tajalli, and Ipsita Roy. 2010. "Polyhydroxyalkanoates: Bioplastics with a green agenda." *Current Opinion in Microbiology*, 13 (3), 321–326.

Khabbaz, Farideh, and Ann-Christine Albertsson. 2000. "Great advantages in using a natural rubber instead of a synthetic SBR in a pro-oxidant system for degradable LDPE." *Biomacromolecules*, 1 (4), 665–673.

Kijchavengkul, Thitisilp, and Rafael Auras. 2008. "Compostability of polymers." *Polymer International*, 57 (6), 793–804.

Kijchavengkul, Thitisilp, Rafael Auras, Maria Rubino, Mathieu Ngouajio, and R. Thomas Fernandez. 2008. "Assessment of aliphatic-aromatic copolyester biodegradable mulch films. Part I: Field study." *Chemosphere*, 71 (5), 942–953.

Kim, Hee-Soo, Hyun-Joong Kim, Jae-Won Lee, and In-Gyu Choi. 2006. "Biodegradability of bio-flour filled biodegradable poly (butylene succinate) bio-composites in natural and compost soil." *Polymer Degradation and Stability*, 91 (5), 1117–1127.

Kroschwitz, Jacqueline I., Mary Howe-Grant, Raymond Eller Kirk, and Donald Frederick Othmer. 1999. Encyclopedia of chemical technology. New York: Wiley.

Krzan, Andrej, Sarunya Hemjinda, Stanislav Miertus, Andrea Corti, and Emo Chiellini. 2006. "Standardization and certification in the area of environmentally degradable plastics." *Polymer Degradation and Stability*, 91 (12), 2819–2833.

Kubowicz, Stephan, and Andy M. Booth. 2017. Biodegradability of plastics: Challenges and misconceptions. *Environmental Science & Technology*, 51(21), 12058–12060. https://doi.org/10.1021/acs.est.7b04051

Kumar, Majeti N. V. Ravi. 2000. "A review of chitin and chitosan applications." *Reactive and Functional Polymers*, 46 (1), 1–27.

Kumar, Parveen, Diane M. Barrett, Michael J. Delwiche, and Pieter Stroeve. 2009. "Methods for pretreatment of lignocellulosic biomass for efficient hydrolysis and biofuel production." *Industrial & Engineering Chemistry Research*, 48 (8), 3713–3729.

Kyrikou, Ioanna, and Demetres Briassoulis. 2007. "Biodegradation of agricultural plastic films: A critical review." *Journal of Polymers and the Environment*, 15, 125–150.

Lauer, Moira K., Andrew G. Tennyson, and Rhett C. Smith. 2021. "Inverse vulcanization of octenyl succinate-modified corn starch as a route to biopolymer-sulfur composites." *Materials Advances*, 2 (7), 2391–2397.

Le Corre, Déborah, Julien Bras, and Alain Dufresne. 2010. "Starch nanoparticles: A review." *Biomacromolecules*, 11 (5), 1139–1153.

Lee, Hau L., Venkata Padmanabhan, and Seungjin Whang. 1997. "Information distortion in a supply chain: The bullwhip effect." *Management Science*, 43 (4), 546–558.

Lu, D. R., C. M. Xiao, and S. J. Xu. 2009. "Starch-based completely biodegradable polymer materials." *Express Polymer Letters*, 3 (6), 366–375.

Luckachan, Gisha E., and C. K. S. Pillai. 2011. "Biodegradable polymers-a review on recent trends and emerging perspectives." *Journal of Polymers and the Environment*, 19, 637–676.

Ma, X. Fa, J. G. Yu, and J. J. Wan. 2006. "Urea and ethanolamine as a mixed plasticizer for thermoplastic starch." *Carbohydrate Polymers*, 64 (2), 267–273.

Majid, R. Abdul, H. Ismail, and R. Mat Taib. 2009. "Effects of PE-g-MA on tensile properties, morphology and water absorption of LDPE/thermoplastic sago starch blends." *Polymer-Plastics Technology and Engineering*, 48 (9), 919–924.

Markit, I. H. S. 2018. *Technical report for psychologist workforce projections for 2015-2030: Addressing supply and demand.* Washington, DC: American Psychological Association.

Matzinos, P., D. Bikiaris, S. Kokkou, and C. Panayiotou. 2001. "Processing and characterization of LDPE/starch products." *Journal of Applied Polymer Science*, 79 (14), 2548–2557.

Maurus, Peter B., and Christopher C. Kaeding. 2004. "Bioabsorbable implant material review." *Operative Techniques in Sports Medicine*, 12 (3), 158–160.

Mendes, J. F., R. T. Paschoalin, V. B. Carmona, Alfredo R. Sena Neto, A. C. P. Marques, J. M. Marconcini, L. H. C. Mattoso, E. S. Medeiros, and J. E. Oliveira. 2016. "Biodegradable polymer blends based on corn starch and thermoplastic chitosan processed by extrusion." *Carbohydrate Polymers*, 137, 452–458.

Méndez-Hernández, M. L., C. S. Tena-Salcido, Z. Sandoval-Arellano, M. C. González-Cantú, M. Mondragón, and F. J. Rodríguez-González. 2011. "The effect of thermoplastic starch on the properties of HDPE/TPS blends during UV-accelerated aging." *Polymer Bulletin*, 67 (5), 903.

Mensitieri, Giuseppe, Ernesto Di Maio, Giovanna G. Buonocore, Irma Nedi, Maria Oliviero, Lucia Sansone, and Salvatore Iannace. 2011. "Processing and shelf life issues of selected food packaging materials and structures from renewable resources." *Trends in Food Science & Technology*, 22 (2–3), 72–80.

Mierzwa-Hersztek, M., K. Gondek, and M. Kopeć. 2019. "Degradation of polyethylene and biocomponent-derived polymer materials: An overview." *Journal of Polymers and the Environment*, 27 (3), 600–611.

Mohanty, S., and S. K. Nayak. 2012. "Biodegradable nanocomposites of poly (butylene adipate-co-terephthalate) (PBAT) and organically modified layered silicates." *Journal of Polymers and the Environment*, 20 (1), 195–207.

Molenveld, Karin, and Harriëtte Bos. 2017. "Bio-based and biodegradable plastics: Facts and figures: Focus on food packaging in the Netherlands." Report no. 1722. Wageningen University & Research.

Mooney, Brian P. 2009. "The second green revolution? Production of plant-based biodegradable plastics." *Biochemical Journal*, 418 (2), 219–232.

Mtibe, Asanda, Mpho Phillip Motloung, Jayita Bandyopadhyay, and Suprakas Sinha Ray. 2021. "Synthetic biopolymers and their composites: Advantages and limitations—An overview." *Macromolecular Rapid Communications*, 42 (15), 2100130.

Muniyasamy, Sudhakar. 2011. *Oxo-biodegradation of full carbon backbone polymers under different environmental conditions.*LAP LAMBERT Academic Publishing.

Muniyasamy, Sudhakar, Andrew Anstey, Murali M. Reddy, Manju Misra, and Amar Mohanty. 2013. "Biodegradability and compostability of lignocellulosic based composite materials." *Journal of Renewable Materials*, 1 (4), 253–272.

Muniyasamy, Sudhakar, and Maya Jacob John. 2017. "Biodegradability of biobased polymeric materials in natural environments." In *Handbook of composites from renewable materials*, V. K. Thakur, M. K. Thakur, and M. R. Kessler (eds.), 625–654. https://doi.org/10.1002/9781119441632.ch104

Muniyasamy, Sudhakar, Murali M. Reddy, Manjusri Misra, and Amar Mohanty. 2013. "Biodegradable green composites from bioethanol co-product and poly (butylene adipate-co-terephthalate)." *Industrial Crops and Products*, 43, 812–819.

Muzzarelli, R. A. A. 1997. "Human enzymatic activities related to the therapeutic administration of chitin derivatives." *Cellular and Molecular Life Sciences CMLS*, 53, 131–140.

Oksman, Kristiina, Mikael Skrifvars, and J. –F. Selin. 2003. "Natural fibres as reinforcement in polylactic acid (PLA) composites." *Composites Science and Technology*, 63 (9), 1317–1324.

Orhan, Yüksel, and Hanife Büyükgüngör. 2000. "Enhancement of biodegradability of disposable polyethylene in controlled biological soil." *International Biodeterioration & Biodegradation*, 45 (1–2), 49–55.

Perotti, Gustavo F., Rafael Auras, and Vera R. L. Constantino. 2013. "Bionanocomposites of cassava starch and synthetic clay." *Journal of Carbohydrate Chemistry*, 32 (8–9), 483–501.

Philp, James C., Rachael J. Ritchie, and Ken Guy. 2013. "Biobased plastics in a bioeconomy." *Trends in Biotechnology*, 31 (2), 65–67.

Pinchuk, Anatoliy, Gero Von Plessen, and Uwe Kreibig. 2004. "Influence of interband electronic transitions on the optical absorption in metallic nanoparticles." *Journal of Physics D: Applied Physics*, 37 (22), 3133.

Povolo, Silvana, Maria Giovanna Romanelli, Marina Basaglia, Vassilka Ivanova Ilieva, Andrea Corti, Andrea Morelli, Emo Chiellini, and Sergio Casella. 2013. "Polyhydroxyalkanoate biosynthesis by *Hydrogenophaga pseudoflava* DSM1034 from structurally unrelated carbon sources." *New Biotechnology*, 30 (6), 629–634.

Pradhan, Ranjan, Manjusri Misra, Larry Erickson, and Amar Mohanty. 2010. "Compostability and biodegradation study of PLA-wheat straw and PLA-soy straw based green composites in simulated composting bioreactor." *Bioresource Technology*, 101 (21), 8489–8491.

Pradhan, Ranjan, Murali Reddy, William Diebel, Larry Erickson, Manjusri Misra, and Amar Mohanty. 2010. "Comparative compostability and biodegradation studies of various components of green composites and their blends in simulated aerobic composting bioreactor." *International Journal of Plastics Technology*, 14, 45–50.

Prado, Karen S., Danielle Gonzales, and Marcia A. S. Spinace. 2019. "Recycling of viscose yarn waste through one-step extraction of nanocellulose." *International Journal of Biological Macromolecules*, 136, 729–737.

Prado, Karen S., and Márcia A. S. Spinacé. 2019. "Isolation and characterization of cellulose nanocrystals from pineapple crown waste and their potential uses." *International Journal of Biological Macromolecules*, 122, 410–416.

Raquez, Jean-Marie, Youssef Habibi, Marius Murariu, and Philippe Dubois. 2013. "Polylactide (PLA)-based nanocomposites." *Progress in Polymer Science*, 38 (10–11), 1504–1542.

Ratto, Jo Ann, Peter J. Stenhouse, Margaret Auerbach, John Mitchell, and Richard Farrell. 1999. "Processing, performance and biodegradability of a thermoplastic aliphatic polyester/starch system." *Polymer*, 40 (24), 6777–6788.

Rebouillat, Serge, and Fernand Pla. 2013. "State of the art manufacturing and engineering of nanocellulose: A review of available data and industrial applications." *Journal of Biomaterials and Nanobiotechnology*, 4 (2), 165–188. DOI: 10.4236/jbnb.2013.42022

Reddy, Murali M., Amar K. Mohanty, and Manjusri Misra. 2012. "Optimization of tensile properties thermoplastic blends from soy and biodegradable polyesters: Taguchi design of experiments approach." *Journal of Materials Science*, 47, 2591–2599.

Reddy, Murali M., Singaravelu Vivekanandhan, Manjusri Misra, Sujata K. Bhatia, and Amar K. Mohanty. 2013. "Biobased plastics and bionanocomposites: Current status and future opportunities." *Progress in Polymer Science*, 38 (10–11), 1653–1689.

Reddy, Murali, Amar K. Mohanty, and Manjusri Misra. 2010. "Thermoplastics from soy protein: A review on processing, blends and composites." *Journal of Biobased Materials and Bioenergy*, 4 (4), 298–316.

Redondo-Gómez, Carlos, Maricruz Rodríguez Quesada, Silvia Vallejo Astúa, José Pablo Murillo Zamora, Mary Lopretti, and José Roberto Vega-Baudrit. 2020. "Biorefinery of biomass of agro-industrial banana waste to obtain high-value biopolymers." *Molecules*, 25 (17), 3829.

Richards, Esther, Reza Rizvi, Andrew Chow, and Hani Naguib. 2008. "Biodegradable composite foams of PLA and PHBV using subcritical CO2." *Journal of Polymers and the Environment*, 16, 258–266.

Rodriguez-Gonzalez, F. J., B. A. Ramsay, and B. D. Favis. 2003. "High performance LDPE/thermoplastic starch blends: A sustainable alternative to pure polyethylene." *Polymer*, 44 (5), 1517–1526.

Rynk, Robert, Maarten Van de Kamp, George B. Willson, Mark E. Singley, Tom L. Richard, John J. Kolega, Francis R. Gouin, Lucien Laliberty, David Kay, and Dennis Murphy. 1992. *On-farm composting handbook (NRAES 54)*. Ithaca, NY: Northeast Regional Agricultural Engineering Service (NRAES).

Saadi, Zoubida, Aurore Rasmont, Guy Cesar, Hilaire Bewa, and Ludovic Benguigui. 2012. "Fungal degradation of poly (l-lactide) in soil and in compost." *Journal of Polymers and the Environment*, 20, 273–282.

Saha, Nabanita, Martin Zatloukal, and Petr Saha. 2003. "Modification of polymers by protein hydrolysate—A way to biodegradable materials." *Polymers for Advanced Technologies*, 14 (11–12), 854–860.

Sailaja, R. R. N., and Manas Chanda. 2000. "Use of maleic anhydride-grafted polyethylene as compatibilizer for polyethylene-starch blends: Effects on mechanical properties." *Journal of Polymer Materials*, 17 (2), 165–176.

Sailaja, R. R. N., A. Prasad Reddy, and Manas Chanda. 2001. "Effect of epoxy functionalized compatibilizer on the mechanical properties of low-density polyethylene/plasticized tapioca starch blends." *Polymer International*, 50 (12), 1352–1359.

Schnabel, Wolfram. 1982. *Polymer degradation*: Principles and practical applications. New York: MacMillan.

Sharmin, Nusrat, Ruhul A. Khan, Stephane Salmieri, Dominic Dussault, and Monique Lacroix. 2012. "Fabrication and characterization of biodegradable composite films made of using poly (caprolactone) reinforced with chitosan." *Journal of Polymers and the Environment*, 20 (3), 698–705.

Shi, Rui, Quanyong Liu, Tao Ding, Yanming Han, Liqun Zhang, Dafu Chen, and Wei Tian. 2007. "Ageing of soft thermoplastic starch with high glycerol content." *Journal of Applied Polymer Science*, 103 (1), 574–586.

Shimao, Masayuki. 2001. "Biodegradation of plastics." Current Opinion in Biotechnology, 12 (3), 242–247.

Shogren, Randal L. 1992. "Effect of moisture content on the melting and subsequent physical aging of cornstarch." *Carbohydrate Polymers*, 19 (2), 83–90.

Shogren, R. L., WM Doane, D Garlotta, JW Lawton, and JL Willett. 2003. "Biodegradation of starch/polylactic acid/poly (hydroxyester-ether) composite bars in soil." *Polymer Degradation and Stability*, 79 (3), 405–411.

Signori, Francesca, Alessia Boggioni, Francesco Ciardelli, and Simona Bronco. 2012. "Development of new PLA-based biodegradable compounds." *AIP Conference Proceedings*, 1459, 30–32. https://doi.org/10.1063/1.4738388

Signori, Francesca, Martina Pelagaggi, Simona Bronco, and Maria Cristina Righetti. 2012. "Amorphous/crystal and polymer/filler interphases in biocomposites from poly (butylene succinate)." *Thermochimica Acta*, 543, 74–81.

Siqueira, Gilberto, Julien Bras, and Alain Dufresne. 2008. "Cellulose whiskers versus microfibrils: Influence of the nature of the nanoparticle and its surface functionalization on the thermal and mechanical properties of nanocomposites." *Biomacromolecules*, 10 (2), 425–432.

Siracusa, Valentina. 2019. "Microbial degradation of synthetic biopolymers waste." *Polymers*, 11 (6), 1066.

Song, Fei, Dao-Lu Tang, Xiu-Li Wang, and Yu-Zhong Wang. 2011. "Biodegradable soy protein isolate-based materials: A review." *Biomacromolecules*, 12 (10), 3369–3380.

Song, J. H., R. J. Murphy, R. Narayan, and G. B. H. Davies. 2009. "Biodegradable and compostable alternatives to conventional plastics." *Philosophical Transactions of the Royal Society B: Biological Sciences*, 364 (1526), 2127–2139.

Sorrentino, Andrea, Giuliana Gorrasi, and Vittoria Vittoria. 2007. "Potential perspectives of bio-nanocomposites for food packaging applications." *Trends in Food Science & Technology*, 18 (2), 84–95.

Souza, Roberta C. R., and Cristina T. Andrade. 2001. "Processing and properties of thermoplastic starch and its blends with sodium alginate." *Journal of Applied Polymer Science*, 81 (2), 412–420.

SPI Bioplastics Council. 2013. "Development of biobased plastics independent of the future of biofuels." *Industrial Biotechnology*, 9 (5), 264–268.

Steller, Ryszard, and Wanda Meissner. 1998. "Structure and properties of degradable polyolefin-starch blends." *Polymer Degradation and Stability*, 60 (2–3), 471–480.

Stevens, E. S., and Nora Goldstein. 2002. "How green are green plastics?" *Biocycle*, 43 (12), 42–42.

Stoica, Maricica, Dimitrie Stoica, Angela Stela Ivan, and Carmelia Mariana Bălănică Dragomir. 2022. "Biopolymers: Regulatory and legislative issues." In *Biopolymers: Recent updates, challenges and opportunities*, 55–71. Springer. DOI: 10.1007/978-3-030-98392-5_4

St-Pierre, N., B. D. Favis, B. A. Ramsay, J. A. Ramsay, and H. Verhoogt. 1997. "Processing and characterization of thermoplastic starch/polyethylene blends." *Polymer*, 38 (3), 647–655.

Sudhakar, M., Mukesh Doble, P. Sriyutha Murthy, and Ramasamy Venkatesan. 2008. "Marine microbe-mediated biodegradation of low- and high-density polyethylenes." *International Biodeterioration & Biodegradation*, 61 (3), 203–213.

Sudhakar, M., A. Trishul, Mukesh Doble, K. Suresh Kumar, S. Syed Jahan, D. Inbakandan, R. R. Viduthalai, V. R. Umadevi, P. Sriyutha Murthy, and R. Venkatesan. 2007. "Biofouling and biodegradation of polyolefins in ocean waters." *Polymer Degradation and Stability*, 92 (9), 1743–1752.

Suh, J. –K. Francis, and Howard W. T. Matthew. 2000. "Application of chitosan-based polysaccharide biomaterials in cartilage tissue engineering: A review." *Biomaterials*, 21 (24), 2589–2598.

Tanamool, Varavut, Tsuyoshi Imai, Paiboon Danvirutai, and Pakawadee Kaewkannetra. 2013. "An alternative approach to the fermentation of sweet sorghum juice into biopolymer of poly-β-hydroxyalkanoates (PHAs) by newly isolated, *Bacillus aryabhattai* PKV01." *Biotechnology and Bioprocess Engineering*, 18, 65–74.

Tena-Salcido, C. S., F. J. Rodríguez-González, M. L. Méndez-Hernández, and J. C. Contreras-Esquivel. 2008. "Effect of morphology on the biodegradation of thermoplastic starch in LDPE/TPS blends." *Polymer Bulletin*, 60 (5), 677–688.

Teramoto, Naozumi, Kohei Urata, Koichi Ozawa, and Mitsuhiro Shibata. 2004. "Biodegradation of aliphatic polyester composites reinforced by abaca fiber." *Polymer Degradation and Stability*, 86 (3), 401–409.

Thakur, Manju Kumari, Vijay Kumar Thakur, Raju Kumar Gupta, and Asokan Pappu. 2016. "Synthesis and applications of biodegradable soy based graft copolymers: A review." *ACS Sustainable Chemistry & Engineering*, 4 (1), 1–17.

Thakur, Vijay Kumar, and Manju Kumari Thakur. 2014. "Processing and characterization of natural cellulose fibers/thermoset polymer composites." *Carbohydrate Polymers*, 109, 102–117.

Thakur, Vijay Kumar, Manju Kumari Thakur, Prasanth Raghavan, and Michael R. Kessler. 2014. "Progress in green polymer composites from lignin for multifunctional applications: A review." *ACS Sustainable Chemistry & Engineering*, 2 (5), 1072–1092.

Thakur, Vijay Kumar, Manju Kumari Thakur, and Raju Kumar Gupta. 2014. "Raw natural fiber-based polymer composites." *International Journal of Polymer Analysis and Characterization*, 19 (3), 256–271.

Thorén, Peter, Marie Åsberg, Leif Bertilsson, Britt Mellström, Folke Sjöqvist, and Lil Träskman. 1980. "Clomipramine treatment of obsessive-compulsive disorder: II. Biochemical aspects." *Archives of General Psychiatry*, 37 (11), 1289–1294.

Tokiwa, Yutaka, Buenaventurada P. Calabia, Charles U. Ugwu, and Seiichi Aiba. 2009. "Biodegradability of plastics." *International Journal of Molecular Sciences*, 10 (9), 3722–3742.

Touchaleaume, François, Lluís Martin-Closas, Helene Angellier-Coussy, Anne Chevillard, Guy Cesar, Nathalie Gontard, and Emmanuelle Gastaldi. 2016. "Performance and environmental impact of biodegradable polymers as agricultural mulching films." *Chemosphere*, 144, 433–439.

Van den Oever, Martien, Karin Molenveld, Maarten van der Zee, and Harriëtte Bos. 2017. *Bio-based and biodegradable plastics: Facts and figures: Focus on food packaging in the Netherlands*. Vol. 1722. Wageningen Food & Biobased Research. DOI: 10.18174/408350

Vargha, Viktória, Gabriella Rétháti, Tamás Heffner, Krisztina Pogácsás, László Korecz, and László Tolner. 2016. "Behavior of polyethylene films in soil." *Periodica Polytechnica-Chemical Engineering*, 60 (1), 60–68.

Verbeek, Casparus J. R., and Lisa E. van den Berg. 2010. "Extrusion processing and properties of protein-based thermoplastics." *Macromolecular Materials and Engineering*, 295 (1), 10–21.

Voeller, John G. 2010. *Wiley Handbook of Science and Technology for Homeland Security, 4 Volume Set*. John Wiley & Sons.

Volke-Sepúlveda, T., G. Saucedo-Castañeda, M. Gutiérrez-Rojas, A. Manzur, and E. Favela-Torres. 2002. "Thermally treated low density polyethylene biodegradation by *Penicillium pinophilum* and *Aspergillus niger*." *Journal of Applied Polymer Science*, 83 (2), 305–314.

Wang, Bei. 2010. *Dispersion of cellulose nanofibers in biopolymer based nanocomposites*. PhD Thesis, University of Toronto.

Wang, Ning, Jiugao Yu, Peter R. Chang, and Xiaofei Ma. 2007. "Influence of citric acid on the properties of glycerol-plasticized dry starch (DTPS) and DTPS/poly (lactic acid) blends." *Starch-Stärke*, 59 (9), 409–417.

Wang, Shujun, Jiugao Yu, and Jinglin Yu. 2004. "Influence of maleic anhydride on the compatibility of thermal plasticized starch and linear low-density polyethylene." *Journal of Applied Polymer Science*, 93 (2), 686–695.

Wani, Sonal J., Sohel S. Shaikh, and Riyaz Z. Sayyed. 2015. "Medium optimization for PHB depolymerase production by *Stenotrophomonas maltophilia* using Plackett Burman design & response surface methodology." *International Journal of Scientific and Engineering Research*, 6 (10), 818–829.

Way, Cameron, Katherine Dean, Dong Yang Wu, and Enzo Palombo. 2012. "Biodegradation of sequentially surface treated lignocellulose reinforced polylactic acid composites: Carbon dioxide evolution and morphology." *Polymer Degradation and Stability*, 97 (3), 430–438.

Weeg-Aerssens, Els. 1998. *Ecological assessment of polymers: Strategies for product stewardship and regulatory programs*. JSTOR.

Weiland, Michèle, Arlette Daro, and Christiane David. 1995. "Biodegradation of thermally oxidized polyethylene." *Polymer Degradation and Stability*, 48 (2), 275–289.

Wool, R. P., D. Raghavan, G. C. Wagner, and S. Billieux. 2000. "Biodegradation dynamics of polymer-starch composites." *Journal of Applied Polymer Science*, 77 (8), 1643–1657.

Yagi, Hisaaki, Fumi Ninomiya, Masahiro Funabashi, and Masao Kunioka. 2009a. "Anaerobic biodegradation tests of poly (lactic acid) and polycaprolactone using new evaluation system for methane fermentation in anaerobic sludge." *Polymer Degradation and Stability*, 94 (9), 1397–1404.

Yagi, Hisaaki, Fumi Ninomiya, Masahiro Funabashi, and Masao Kunioka. 2009b. "Anaerobic biodegradation tests of poly (lactic acid) under mesophilic and thermophilic conditions using a new evaluation system for methane fermentation in anaerobic sludge." *International Journal of Molecular Sciences*, 10 (9), 3824–3835.

Yamada-Onodera, Keiko, Hiroshi Mukumoto, Yuhji Katsuyaya, Atsushi Saiganji, and Yoshiki Tani. 2001. "Degradation of polyethylene by a fungus, *Penicillium simplicissimum* YK." *Polymer Degradation and Stability*, 72 (2), 323–327.

Yoshida, Shosuke, Kazumi Hiraga, Toshihiko Takehana, Ikuo Taniguchi, Hironao Yamaji, Yasuhito Maeda, Kiyotsuna Toyohara, Kenji Miyamoto, Yoshiharu Kimura, and Kohei Oda. 2016. "A bacterium that degrades and assimilates poly (ethylene terephthalate)." *Science*, 351 (6278), 1196–1199.

Youssef, Benyoussif, Aboulhrouz Soumia, Cherkaoui Omar, Lallam Abdelaziz, and Zahouily Mohamed. 2015. "Preparation and properties of bionanocomposite films reinforced with nanocellulose isolated from Moroccan alfa fibres." *Autex Research Journal*, 15 (3), 164–172.

Yu, Peiqiang. 2007. "Molecular chemical structure of barley proteins revealed by ultra-spatially resolved synchrotron light sourced FTIR microspectroscopy: Comparison of barley varieties." *Biopolymers: Original Research on Biomolecules*, 85 (4), 308–317.

Zhang, Jian-she, Peter R. Chang, Ying Wu, Jiu-gao Yu, and Xiao-fei Ma. 2008. "Aliphatic amidediol and glycerol as a mixed plasticizer for the preparation of thermoplastic starch." *Starch-Stärke*, 60 (11), 617–623.

Zhang, Xiaoqing, Yesim Gozukara, Parveen Sangwan, Dachao Gao, and Stuart Bateman. 2010. "Biodegradation of chemically modified wheat gluten-based natural polymer materials." *Polymer Degradation and Stability*, 95 (12), 2309–2317.

3 Global Plastic Waste Generation, Microfibres (MFs) Micro- and Nanoplastics (MNPs) Pollution and their Health-related Toxicities upon Exposure

A. Gada, Sudhakar Muniyasamy, S.P. Hlangothi, P. Melariri and C. Hoyo

3.1 INTRODUCTION

Plastics are categorized as either thermosets or thermoplastics based on their structure, chemical and physical properties, and durability (or non-durability). These items are products of synthetic materials formulated from organic polymers together with other essential chemical additives like flame retardants, phthalates, and bisphenols to further improve the plastic's unique material properties (Gigault et al., 2018). Thermosetting plastics are plastics that, upon heating, form new and irreversible cross-linking bonds (covalent bonds), thus creating a stable plastic material that is not easily decomposable (Rudyak et al., 2019). These plastics' common applications are adhesives, insulators, and plywood. The plastic species include alkyds, polyurethane, and epoxy. With thermoplastics, there are no newly formed chemical bonds upon heating; with this chemical formation and mechanical property, thermoplastics can be recycled and reshaped, thus making them easier materials for packaging applications for consumer goods than thermosets (Mattsson et al., 2015, Battulga et al., 2019, Raddadi and Fava, 2019, Xu et al., 2019).

Thermoplastic species include polypropylene (PP), polystyrene (PS), polyvinyl chloride (PVC) and polyethylene (PE). PS plastic is a regular application in the production of day-to-day items such as food containers, disposable cups, cutlery, plates, and plastic tableware. PE can applied in a diverse range of products such as plastic bags and bottles. Depending on the PE's chemical structure, it can be classifiable into two types: high-density polyethylene (HDPE) and low-density polyethylene (LDPE).

DOI: 10.1201/9781003304142-3

HDPE is popular in packaging containers such as moulded plastic cases, detergent bottles, and milk cans, whilst LDPE is useful mostly for various applications such as floor tiles, outdoor furniture, clamshell packaging, and shower curtains. PPs are primarily applicable for manufacturing bottle caps, fishing lines, plastic pressure pipe systems, car bumpers, appliances, and plastic drinking straws (Figure 3.1).

The continued production, trust, use, and consumption of these materials (since the 1950s) has led to a major cause of various environmental problems globally. In the year 1960 alone, about half a million metric ton has are disposed in the world yearly, and this annual tonnage increased to 359 million metric tonnes in 2018 (Wu et al., 2019, Europe, 2019). The estimated world's plastic tonnage was projected to be approximately 6.3 billion tons in 2015, just in a period of 65 years since 1950. If continued, the estimates may increase to twenty-six (26) billion tons by 2050 (Guglielmi, 2017, Geyer, 2020). It is estimated that 10% of the global annually produced plastics end up as waste in marine habitats, whilst 14% is incinerated, with the remaining 76% disposed to landfills where plastic waste proceeds further to other natural habitats, thus causing plastic pollution on both water and land areas (Chen, 2015, Liu et al., 2016, Rhodes, 2018, Geyer, 2020).

Plastic observed in these natural habitats include plastic species such as polyamide (PA), polypropylene (PP), polyethylene terephthalate (PET), polyethylene (PE), and polyvinyl chloride (PVC), polystyrene (PS) being the most predominant plastic type observed. PS, PE, and PVC plastics are the most mainly used in scientific research. PS and PE are the most famous consumer products with a shorter life cycle compared to other plastic species. PVC is applied in plastics for data cables (cable jacket) and wire insulators. Whilst the metal parts are recyclable after end-of-life service use, the plastic parts end disposed into the environment. This is due to the excessive cost of separation and insufficient recycling value. According to this study, only about 3% of PVC ends in recycle, whilst 82% is discarded in landfills, and the remaining

Elastomers	Thermoplastics	Thermosets
Styrene-butadiene-elastomer (SBR)	**Amorphous thermoplastics**	Phenol resin (PF)
Butadiene-elastomers (BR)	Polystyrene (PS)	Polyester resin (UP)
Polyurethane elastomers.(PUR)	Acrylonitrile-butadiene styrene (ABS)	Epoxy resin (EP)
Polyisoprene (IR)	Polycarbonate (PC)	Melamine resin (MF)
Fluoroelastomers (FKM)	Polyvinylchloride (PVC)	Polyimides (PI)
	Polyetheretherketone (PEEK)	Duroplast (PF)
	Semi-crystalline thermoplastics	Bakelite (PF)
	Polyphenylensulphone (PPS)	Urea-formaldehyde foam (UFF)
	Polyethylene terephthalate (PET, PETE)	
	Polyamide (PA)	
	Polyvinyldenfluoride (PVDF)	
	Polpropylene (PP)	
	Polyethylene (PE)	

FIGURE 3.1 Plastic types and their species.

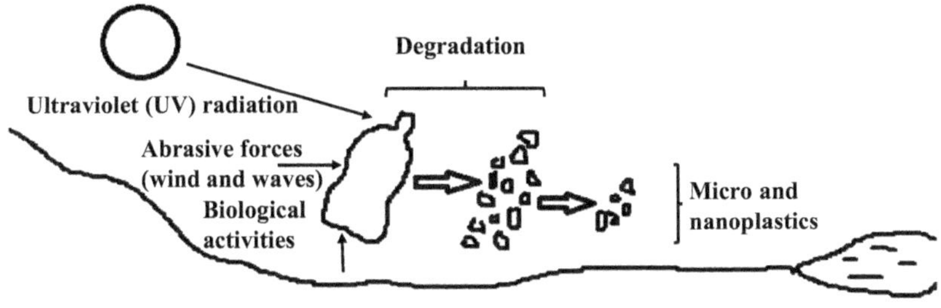

FIGURE 3.2 A schematic diagram of plastic degradation by UV light and environmental forces.

15% is incinerated (Suresh et al., 2017). Due to this varied relationship between their brief life process, huge output and infinite environmental display, these plastics have been a centre of scientific research (Van Cauwenberghe et al., 2015, Deng et al., 2017, Deng et al., 2018, Wu et al., 2019, Yang et al., 2019).

When these plastic materials enter the environment (air, land, and water), the environmental components such as UV (ultraviolet) radiation, wind, waves and other abrasive forces interact with these bulk plastics leading to degradation of larger fragments to smaller particles of plastic debris commonly found in varying sizes through a synergistic process of degradation composed of physical, chemical, and biological degrading activities leading to a production of secondary micro- and nanoplastics (MNPs) released from the plastic's surface through a delamination process (Figure 3.2) (Wagner et al., 2014, Avio et al., 2017, Kubowicz and Booth, 2017, Sharma and Chatterjee, 2017, Adawi et al., 2018, Waring et al., 2018, Lehner et al., 2019).

Microplastics and nanoplastics both inherit their names from the sizes of their diameters, microplastics with 0.1 µm to 5 mm diameter, and nanoplastics being plastics with the diameter range below 100 nm (Andrady, 2017, Karbalaei et al., 2018, Alimba and Faggio, 2019, Hernandez et al., 2019), whilst some claim they could also be anywhere between 1 and 100 nm or 1000 nm (Lambert and Wagner, 2016, Auta et al., 2017, Chae and An, 2017, Alimi et al., 2018, Peeken et al., 2018, Hernandez et al., 2019). According to Thompson et al. (2004), Andrady (2017), and De Sá et al. (2018), these MNPs can be found in all types of aquatic habitats, and their pollution claimed to be 'ubiquitous' in terrestrial environments such as home gardens, industrial, coastal and floodplain soils, greenhouses, and agricultural or farmlands (Xu et al., 2020) (Figure 3.3).

These microplastics observed in terrestrial environments are consequences of unsustainable plastic use and improper management of plastic waste procured from single-use applications such as packaging. Additionally, microplastics in the land may be due to farming or agricultural activities or processes such as 'plastic mulches and the use of sewerage sludge as means of soil nutrient supplementation methods, that is, fertilisers (Wang et al., 2019). Whilst preventative measures, such as the removal of microplastics from wastewaters to avoid microplastic contaminations in aquatic

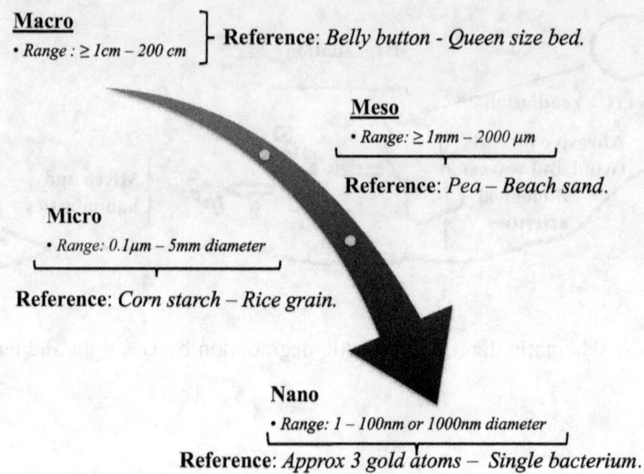

FIGURE 3.3 Size classification of plastic particles.

environments, wastewater treatment plants concentrate all the extracted microplastics into sewerage sludges subsequently used in agricultural soils as fertilisers inevitably entering marine environments (Corradini, Meza et al., 2019). The approximated yearly measure of plastic waste entering aquatic habitats is between 4.8 and 12.7 million metric tons (Jambeck et al., 2015), 80% of it coming from land (Schwarz, Ligthart et al., 2019). Other sources include fisheries, textiles, health care products, shipping, tourism, oil, and gas platforms (Bouwmeester et al., 2015).

Primary micro and nanoplastics are micro- and nano-sized plastics intentionally produced for various commercial applications such as paints, detergents, fabrics, exfoliants in cosmetic products, and products of personal care such as facial scrubs, cleansers, and drug carriers. In the view that primary MNPs are intentionally blended to a diverse range of consumer care products with the intent to provide physical stimulus and cleaning carriers, these, after a life of service, can further be disposed leading to an additional source of plastic pollution (Cheung and Fok, 2016, Sharma and Chatterjee, 2017, Adawi et al., 2018, Lehner, Weder et al., 2019). Another sub-classification group of intentionally produced plastics is microfibres (MFs). These plastic types (MFs) are derived from polyester (PES) or polypropylene (PP) and are by far extremely abundant in the environment (Cole, 2016, Hu et al., 2020). These MFs find use mostly in items ranging from textile products to clothing, household textiles, agricultural, industrial, and semi-finished or ancillary items used in different fields (Liu, Yang et al., 2019).

Generally, these MNPs and microfibres can also act as carriers or sources of plastic chemicals such as additives, consequently and subsequently adsorbing added environmental chemical contaminants, namely, pathogenic microbes, antibiotics, metals, and persistent organic pollutants (POPs) (Koelmans et al., 2013, Koelmans et al., 2016, Li et al., 2018, Waring et al., 2018, Gerdes et al., 2019). Majority of these adsorbents are known to be endocrine disrupters. The extent to which MNPs will affect or enhance the toxicity of these chemical additives, and environmental chemical adsorbents remains unknown (Figure 3.4).

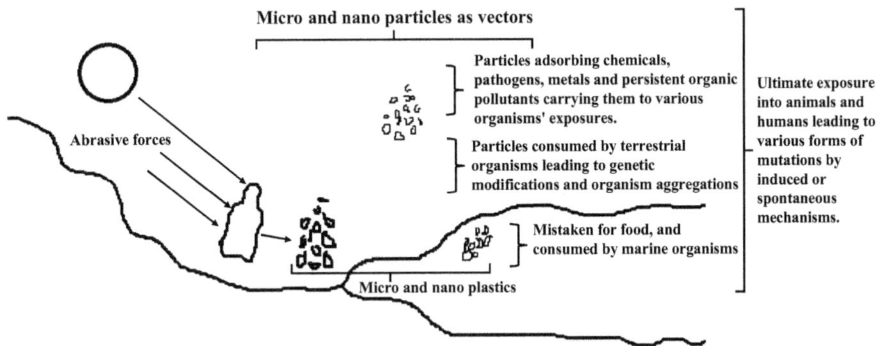

FIGURE 3.4 An illustrative plastic degradation diagram showing secondary particles as various vectors (Shen et al., 2019).

Genes contribute significantly to life development and function (Health); this is equally the same regarding our behaviour (diet) and surrounding environments (physical activeness) (CDC, 2020). Epigenetics is the study of how life behaviour and surrounding can influence modifications that alter genetic functions. Epigenetics, contrary to genetic change, are reversible and do not change DNA sequence, even though they may change how the body reads the sequence of the DNA (Udayangani, 2019, CDC, 2020). Whilst changes in DNA can influence instruction of specific proteins to be produced, epigenetics focuses on the expressions of the genes, that is, 'On' and 'Off', that is, which expression to be loud or which expression to be low or silent. Thus, from this correlation between the genes and the behaviour and surrounding environment, it is extremely easy to observe the connection between the two (CDC, 2020).

When changes take place in the DNA nucleotides, this phenomenon is called mutation. When a mutation takes place, the precise order of the nucleotide of the DNA sequence is altered, consequently changing genetic information in that sequence. When a nucleotide changes, dramatic effects in an organism can be created, ranging from physical disorders to diseases, even though sometimes, it may also not pose any health threats (Udayangani, 2019). Due to exposure to radiation, errors in the DNA replication process, carcinogens, strong chemicals and UV light, these DNA sequence mutations can occur (Clancy, 2008, Udayangani, 2019). The procured mutations may be chromosomal or point mutations (single nucleotide). Point mutations can be caused by deletions, substitutions, and insertions. Mutations responsible for structural alterations in chromosomes are a result of genetic duplications, elimination of chromosomal parts, chromosomal rearrangements, etc. Epigenetic modifications have a strong contribution factor in the expression and regulation of genes and subsequently play a pivotal role in cellular role processes like differentiation, development, and tumorigenesis (Udayangani, 2019).

Examples of epigenetic modifications include (i) DNA methylation (when a methyl or hydroxymethyl group is added to the structural sequence of the DNA), (ii) histone modification (changes in the chromatin protein, i.e., histone) – leading

TABLE 3.1
A List of Examples of Varying Genetic Mutations (Roberti et al., 2019).

Genetic Mutation Type	Case
Deletion (point mutation)	Cystic fibrosis
Trisomy 21	Down syndrome
Insertion	Beta thalassemia
Frameshift mutation	Crohn's disease
Chromosomal translocation	Philadelphia chromosome
Chromosomal duplication	A type of cancer
Chromosomal deletion	Cri-do-chat
Base substitution	Sickle cell anaemia

to changes of chromatin structures, and gene expressions with no impairment procured in the genetic DNA sequence, and (iii) microRNA-mediated genetic silencing (Udayangani, 2019, CDC, 2020). With various epigenetic modifications/mutations observed in literature such as ocular (eye disease) (Oliver et al., 2016), diabetes (Nano et al., 2018), cardiovascular diseases (Qayyum et al., 2019), infectious diseases (Niller and Minarovits, 2018), and biomarkers of asthma and allergic disorders (Akhtari and Mahmoudi, 2019), a study of microfibres, MNP-related epigenetic mutations or modifications is necessary. Genetic mutation types are listed in Table 3.1 with their appropriate occurrences.

3.2 MICRO- AND NANOPLASTIC EXPOSURE AND PREVALENCE

3.2.1 Occupational and Indoor

3.2.1.1 Microfibres

Occupational indoor *exposure* levels are reported to potentially extend to 0.5 and 0.8 particles/ml for plastic species such as nylon (PA) and polyvinyl chloride (PVC). Crucial particle concentrations are reported to be 0.06 and 0.02 mL (Bahners et al., 1994). At the flocking area, a place where deposition of small fibres onto the surface of the flock manufacturing plant takes place was reported to have concentrations of airborne particles reached 7 mg/m^3 (Burkhart et al., 1999), and during the processing operations period, concentration of polyester reached 700,000 (up to 1,000,000) total fibres/m^3 and 10,000 critical fibres/m^3 (Bahners et al., 1994). An estimated 2 million tons of microfibres enters the ocean ecosystems yearly, with a total tonnage of 1.3 million microfibres as deposits in the oceans (Mishra et al., 2019). It is claimed that the existing literature claims suggest that in the total of all primary source microplastics in the ocean, a total of 20%–35% are from synthetic textiles (Laitala et al., 2018). In another article, it is claimed that microfibres constitutes a compelling fraction greater than 85% of microplastic fragments observed in worldwide shorelines (Carr, 2017). Fibres, together with resins, are alleged to have increased globally from a production of 2–380 Mt in the period from 1950 to 2015, with a constant compound annual growth rate of 8.4% (PlasticsEurope, 2006, 2016). If the plastic production projections, consistent use and waste management patterns remain the same, it is predicted that humankind can potentially reach 6000 Mt of fibre production by the year 2050 (Hoornweg, 2012). From these observations, it was deduced that fibre size and fibre subjection concentrations may contribute towards the risks that microplastics may expose to human health. Additionally, he suggests that fibres are airborne, and due to this atmospheric prevalence, they are likely to be inhaled. When these fibres enter the respiratory tract, the majority are prone to be trapped in the lung fluid (Figure 3.5). This is primarily for individuals with impaired clearance mechanisms (Hirai et al., 2011; Mizukawa et al., 2013).

When lung biopsies through histopathological analysis were done on employees of a textile firm working with polyester, acrylic, nylon, and polyolefin, it was discovered that their lungs exhibited interstitial fibrosis and foreign body-containing granulomatous injuries presumed to be sustained from acrylic, polyester and or nylon dust. These exhibited symptoms were also observed to be like those of allergic alveolitis (lung inflammation) (Fazen et al., 2020). Health assessment results proved that microplastics can elicit restrained biological responses due to their absorption and persistence, even though concentration levels varied between occupational and environmental exposures. When patients of several types of lung cancers were assessed, the cellulosic and plastic fibres were spotted in the non-tumorous and virulent lung tissue (Pauly et al., 1998). The fibres were found to exhibit small-scale disintegrations, consequently proving the impression that they are bio-persistent.

The obtained results hinted that the human airways have sufficient sizes to allow plastic fibres access in the deep lungs through penetrative pulmonary mechanisms. One strand distinguished was of 135 μm fibre length, a fourth of a generation 17 respiratory bronchiole, that is, 540 μm diameter and 1410 μm length (Pauly et al., 1998). Based on the obtained observations, it can be asserted that a certain group of fibres can evade clearance mechanisms and consequently, may later exert acute or chronic inflammations if they persist. Furthermore, fibre dimensions were also

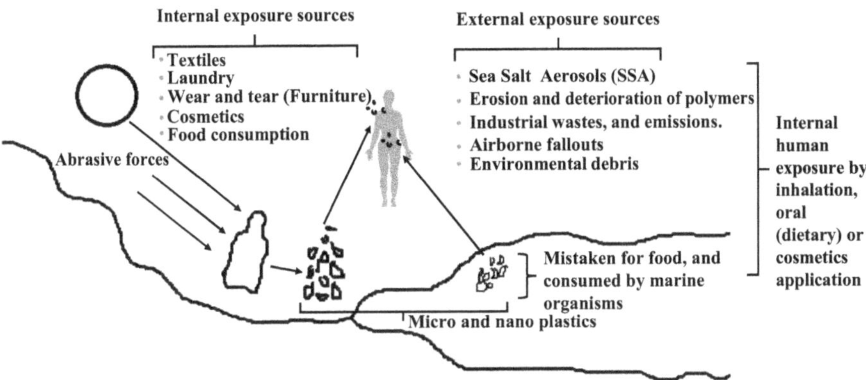

FIGURE 3.5 Indoor and outdoor varying plastic particle exposure sources and methods (Habibi et al., 2022).

observed to be an added attribute of toxicity, in addition to their bio-persistence. Fibres of attenuated surfaces was found to be respirable, whilst longer fibres were found to be endurant and harmful to pulmonary cells. It is argued by Warheitet al. (2001) that it is impossible to efficiently clear fibres ranging between 15 and 20 μm from the lungs by use of alveolar macrophages and mucociliary escalators. Omenn et al. (1986) added that fibres of less than 0.3 μm thickness and those greater than 10 μm long were carcinogenic. Industries such as sports clothing industry have increased use of plastic fibres of fine diameter sized between 1 and 5 μm (Warheit et al., 2001).

It was discovered that nylon fibres of average diametric size and length, that is, 2 and 14 μm on average, were respirable (Porter et al., 1999). Consequently, these fibres altered the alveolar macrophages of exposed test mice and were kept hold of them for 29 days after exposure, causing severe inflammatory responses. In another study, fibres derived from nylon with length and diameters of 9.8 and 1.6 μm, respectively, did not reveal any impacts on pulmonary weight changes, inflammation, or male rat macrophage functions even at an elevated fibre concentration (57 fibres/cm^3) compared to their control test animals (Warheit et al., 2003). It was stated that the burden of fibres, the deposition site, and the chemical potential to desorb from fibre surfaces also play roles in toxicity (Pauly et al., 1998). For example, PAH affinity for plastic hydrophobic surfaces may potentially represent a path of carcinogenicity (Hirai et al., 2011, Mizukawa et al., 2013)

3.2.2 Outdoor

3.2.2.1 Urban and Suburban

Megacities (Liu et al., 2019, Wright et al., 2020) and sparsely populated environments (Allen et al., 2019) have been reported to measure out a fallout of microplastics in their atmospheric environment. A Paris-based study revealed that prevalent microparticles determined in the atmosphere were fibres, with 30% of those being plastic fibres (Dris et al., 2016). The study further alleged that areas with high populations (urban) tended to have greater microparticle concentrations than areas with sparse populations (suburban), and some of the highest concentrations observed corresponded to seasons or times of heavy rainfall. The organic fibres found from both natural and man-made, included polyester, polyvinyl alcohol, polypropylene, polyethylene, polytetrafluoroethylene, polyamide, natural cellulose, and carbon. There were no distinctions made with regard to the compositions of the sampled fibres obtained.

3.2.2.2 Wastewater Treatment Plants and Soil Fertilisers

Microplastics that are released by municipal effluent can be retained in the sludges that may later be used as a fertiliser, representing a persistent contaminant in terrene up to between 5 and 15 years post-application (Zubris and Richards, 2005). Samples of synthetic clothing fibres have been found persistent in columns of soil media and sludge for up to five years post-application. This follows the use of WWTP (Wastewater Treatment Plant) sludge by-products in agricultural lands. Even after 15 years of application, soil samples of some field sites tested positive for particulate synthetic fibres. The transportation of microplastics from fertilisers derived from both (i) dried sludge and (ii) fibres released from dried laundry/clothes, which further contribute to the cause of microplastics in the atmosphere, commonly known as airborne microplastics (Kasirajan and Ngouajio, 2012, Liebezeit and Liebezeit, 2014). This also includes microplastics derived from degradation and fragmentation of PE sheets used in agriculture (Kasirajan and Ngouajio, 2012).

3.2.2.3 Sea Salt Aerosols (SSAs)

These range between less than 0.2 and greater than 2000 μm in diameter size and ambient particle mass in the range of 1–10 μm. SSAs are transported to urban terrestrial environments near the coasts by action of onshore winds during their season of onshore wind. It is alleged that SSA particles of smaller than 50 μm tend to have more atmospheric lifetime, as claimed by Athanasopoulou et al. (2008). Due to their properties and understanding that many microplastics possess a specific gravity lower than that of general seawater, the action of winds and sea spraying may potentially cause sea surfaces to be aerosoled by microplastics of appropriate size.

3.2.2.4 Additional Sources

Other additional sources in which microplastics can enter the wider environment include the atmosphere (Prata, 2018, Chen et al., 2020). Additional studies (Vianello, Jensen et al., 2019, Zhang et al., 2020) have reported a detection of suspended microplastics in indoor air measured from the atmosphere. These observations have led to increased concerns regarding possible adverse effects of humans potentially inhaling nanoplastics particles (Stapleton, 2019). The United States releases about 8 trillion microbeads into aquatic habitats daily (Rochman et al., 2015). So far, microplastics and nanoplastics have been identified globally in aquatic (drinking water, rivers, oceans, and marine), terrestrial (mountain tops, tropics, and the poles) (Lamb et al., 2018, Bergmann et al., 2019, Borunda, 2019, Laganà et al., 2019), atmospheric (air) (Prata, 2018, Allen et al., 2019), sediments, and foodstuffs (Van Cauwenberghe et al., 2013, Cozar et al., 2014, Bouwmeester et al., 2015, Hoellein et al., 2017, Sharma and Chatterjee, 2017, Chiba et al., 2018, Kelly et al., 2020).

3.2.3 Dietary Exposure Pathway

3.2.3.1 Fish

Fish are a dietary source of microplastics for humans, namely shellfish (bivalves' mollusc). According to FAO (2014), shellfish are also quite an important dietary source. Bivalve fish are known to be filter feeders, pumping up enormous quantities of water into their pallial cavity located within the shells. Particles within the pumped water are then retained from gill suspension, where they are simultaneously ingested (Ward and Shumway, 2004), and it is by the water column that shellfish are exposed to microplastics. Potential consumption of microplastics by humans has also been detected in wild and aquaculture fishes.

The consumption of bivalves as seafood is popular in China (Li et al., 2015), and this is also where the production of global aquaculture volume is greater than 60%. Coincidentally, China is also the largest producer of plastic waste volumes that eventually infiltrate the oceans from inner land sources (Jambeck et al., 2015). It is reported that because of this phenomenon, the concentration of sediments of microplastics in China is reaching 8720 microplastic/kg (Qiu et al., 2015). These concentration sediments include plastics such as PET, polystyrene (PS), and polyethylene (PE) particles found in oceans. With these observed increasing concentrations, it is estimated that the exposure of Chinese shellfish to microplastics could be 100,000s yearly.

Microplastic contamination in shellfish is not only limited to China but also in some other areas such as Belgium and Canada. Both Belgian and Canadian mussels (wild and commercial) tested positive for microplastic fibres. It is estimated that an average shellfish consumer (in Europe) would potentially consume about 11,000 microplastics every year, judging from the established average estimates (Van Cauwenberghe et al., 2015). Microplastics contaminate farmed mussels from used polypropylene (PP) lines normally used during culture whilst Belgium recovered microplastics from Pacific oysters bought from the shops and farmed mussels. These oysters and mussels would then be subjected to a three-day purifying rinsing (depuration) period. Microplastics were said to remain in the bivalves, suggesting that the acceptable (three-day) depuration period is not sufficient to ensure microplastics have been totally cleared. This phenomenon has led to a rise of shellfish food concerns, respectively, with microplastics.

3.2.3.2 Sugar

Foods such as honey and sugar have also tested positive for the potential presence of microplastics. Particles observed and reported were synthetic fibres, with 40 μm minimal length and fragments, which mostly ranged between 10 and 20 μm in size (Liebezeit and Liebezeit, 2013). According to Liebezeit and Liebezeit (2013), about 174 (on average) fibres (660 maximum units quantified) and eight fragments per kilogram (38 maximum units quantified) in honey were found whilst sugar samples quantified an average of 217 fibres (388 maximum units quantified) and 32 fragments per kilogram (270 units quantified). It was, however, unfortunate that no methods were thus used to determine if particles were plastic (Liebezeit and Liebezeit, 2013).

3.2.3.3 German Beer

Another study (Liebezeit and Liebezeit, 2014) showed a contamination of German beer by microplastics. A conducted test series of 24 samples showed potential contamination by microplastics. Microplastic fragments were found to be the most prevalent contaminants discovered, about up to 109 fragments/L. The alleged origin of contamination was suggested to be the atmosphere through depositions, and the second suggested source was the production materials used in the brewing process of the German beer (Liebezeit and Liebezeit, 2014).

3.2.3.4 Sea Salt

According to Yang's study, about 15 varying brands of purchased sea salt reported positive for microplastics. About 681 microplastics per kilogram of sea salt are down to 45 μm size. The most prevalent plastic types found were PET and PE. It is reported that from this deduction, coastal waters were the possible sources of contamination, even though atmospheric depositions of microplastics at the sites were not excluded, concluded the study (Yang et al., 2015).

From these reports, consumable food products, both commercially and farmed or cultured, prove to harbour microplastics, which may potentially be consumed by human beings and cause unknown health effects due to absence of reports on such toxicities.

3.2.4 Adsorbent Chemicals and Additives

3.2.4.1 Adsorbed Chemicals – Heavy Metals

Studies have revealed that MNPs can be vehicles of very toxic chemicals which could be excessively higher in magnitudes of concentrations compared to their surrounding environment, and new studies allege that these adsorbed chemicals may possess more hazardous traits than the neat ones (Deng et al., 2018). Due to limited evidence on nanoplastic-related studies, adsorption-toxicity studies were focused on heavy metals and hydrophobic organic chemicals (HOCs) to investigate environmental chemical adsorption on microplastics. HOCs can be found in numerous environmental chemical forms, with a substantial number of them being endocrine-disrupting chemicals (EDCs) (Karwacka et al., 2019). Examples include phthalates, polychlorinated biphenyls (PCBs), bisphenol A (BPA), and perfluorinated chemicals (PFCs).

Microplastics are also known to sorb metals from muddy and aquatic habitats, consequently enabling accumulation of these metals (Munier and Bendell, 2018). It has been argued that the longer the microplastics remain at sea, the more susceptible they are to accumulate more metals (Rochman et al., 2014). This suggestion followed an observed pattern of increased accumulation of heavy metals, namely lead, nickel, cadmium, and zinc, on microplastic materials over time. A range of metals can be observed in plastic fragments simultaneously with the chemical pollutants regarded as 'priority pollutants' under the Environmental Protection Agency of the United States of America (USA) (Rochman et al., 2014).

Previous studies have shown that heavy metals, such as mercury, are a health hazard to both fauna and humans (Barboza et al., 2018). This follows an analysis of mercury where bioaccumulation, together with microplastics in *Dicentrarchus labrax* (sea bass fish), was observed in fish brain and muscle tissues, detecting prominent correlation between the heavy metal (mercury) and microplastic. These studies have been a trigger for concerns regarding both microplastic and adsorbed hazardous chemical's potential uptake by the body.

It is also alleged that the dual combination of microplastic and adsorbed hazardous chemicals is more likely to be more lethal than its correlative pure elements alone. An additional study examining a dual combination exposure of microplastics with OPFRs (organophosphorus flame retardants) revealed that co-exposure conditions exerted severe oxidative stress, metabolic disruptions, and neurotoxicity in mice organisms than the individual test elements (Deng et al., 2018). This observed phenomenon can be caused by either (i) an increase in toxic chemical uptake with microplastics or (ii) HOC and microplastic sorption and desorption equilibrium slowing down the mice metabolism, thus consequently leading to higher toxicity. With these studies, though, understanding microplastics from the lab can be more complex because studying neat plastics and their consequences does not explain how chemicals are adsorbed to plastics.

A *Scenedesmus obliquus* and *Danio rerio* (zebrafish) study of jointly exposed PS nanoplastics together with NAOPs (natural acidic organic polymers), namely, humic acid and fulvic acid, revealed that the dual combo can result in impaired function of the mitochondria, conditioned cell oxidative stress, including severe consequences of oxidative stress and enzymatic antioxidant defence of zebrafish whilst it causes stunted growth toxicity in algae (Wang et al., 2019). Factors that determine chemical adsorption, degree of adsorption, and how they will access humans and aquatic animals vary from plastic type to plastic size, environmental temperature, and mode of transportation.

Other studies and obtained results also revealed that microplastic can potentially cross intestinal barriers and decrease the diversity of microorganisms in the intestines (microbiome), thus suggesting that the biological hazards posed by combined exposures require urgent attention. Furthermore, it is suggested that the biofilm developed on the MNP surfaces can possibly facilitate distribution, invasion, and mutation, and continued adhesion of varying microbial populations to MNP surfaces may improve energy flow rates, matter, and information in the habitat, carrying long-lasting and extensive toxic impacts (Shen et al., 2019).

The possibility of how particles will pose harm (cytotoxicity) to cells and tissues in vivo, all depends on the surface charge, size, solubility, and shape (Nel et al., 2006). Bio-persistence and chemical effects of microplastics are the physical effects that may lead to human health effects. Bio-persistence could lead to biological responses like inflammation, necrosis, oxidative stress, genotoxicity, and apoptosis. If these circumstances remain unaddressed, a series of potential effects could develop, such as damage of tissues, carcinogenesis, and fibrosis. Chemical transformations and alterations like the polymer compositions, unreacted residual monomers, desorption of HOCs or leaching of unbound chemicals could be established, and these make the list of priority pollutants known to have health effects in humans.

The uptake of microplastics by cellular means allows endogenous or adhered contaminants entry into the cells (Khan et al., 2015). Earlier investigations of inhalation exposure revealed that the best particle toxicity paradigms are presented by oxidative stress and subsequent inflammation. Airway inflammation and intestinal fibrosis are caused by oxidative stress due to nanoparticles such as TiO_2, PM and quartz (Nel et al., 2006). Ascribing to the miniscule sizes of microplastics and nanoplastics, a similar toxicity mechanism may be observed, and thus a larger surface area for functional sites.

The weathering of microplastics and nanoplastics leads to formation of free radicals through C-H bond dissociation reactions (Gewert et al., 2015, White and Turnbull, 1994). These free radicals may pose danger due to their continued reactions. To terminate the continued reactions of these free radical reactions, reaction of ROS (reactive oxygen species) pairs or oxidation of the target substrate (e.g. tissues) aids in achieving the achievement of this objection (White and Turnbull, 1994). This is because all plastics possess ROS in their polymer structure, which is borne from their polymerisation and processing history. After light or transition metal exposure, the concentration of free radicals is most likely to increase.

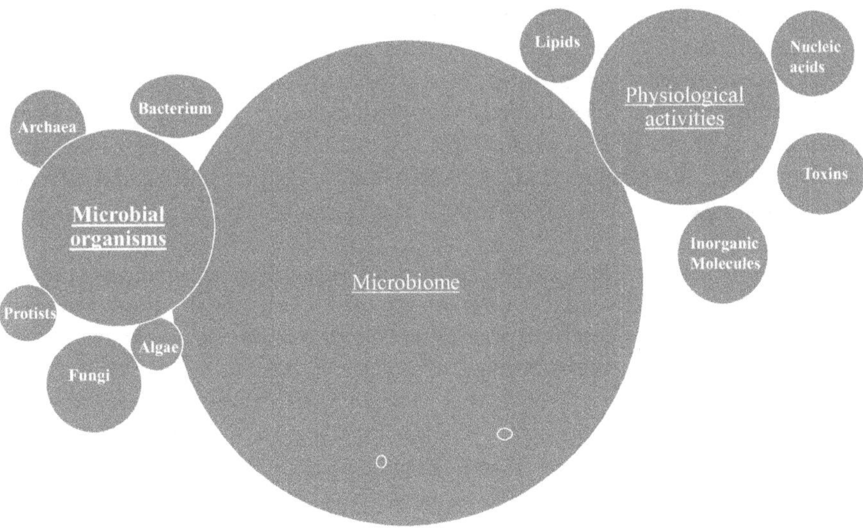

FIGURE 3.6 A diagrammatic summary of a microbiome.

3.2.5 Microbiome

This refers to the collected community of microbial organisms that live in or on a body with a physiological activity that influences the well-being of the host organism (Kirstein et al., 2016, Zettler et al., 2013). In the open environment, microplastic surfaces become colonised rapidly by microbes. During this microbiome process, well-developed biofilms are established on the polymer surface following seven days of water or sediment covering or exposure (Harrison et al., 2014, Lobelle and Cunliffe, 2011). This biofilm possesses a significant difference from atmospheric habitat (Zettler et al., 2013), and can potentially contain or include harmful human microbes like *Vibrio* sp. strains (Zettler et al., 2013, Kirstein et al., 2016).

The GIT microbiome composition differs significantly between solid and liquid phases. With this established statement, it can be determined that the composition of the microbiome will vary to the neighbouring habitat when microplastics are being colonised during the GIT transit. This phenomenon is attributable to the unusual cluster of microbes that are attracted by plastics (Zettler et al., 2013, Kirstein et al., 2016). The body's microplastic response may be influenced by the unique coating, thereby improving bioavailability or triggering an immune response (Figure 3.6).

3.3 MICRO- AND NANOPLASTIC HEALTH IMPACTS

3.3.1 Marine Animals

Several studies have resolved to purse studies of MNP in sea animals since oceans are ultimate depository sites of waste plastics (Guzzetti et al., 2018, Franzellitti et al., 2019). Some of the organisms studied mostly have been bivalves because they function as very important food sources for humans, being another potential MNP route exposure to human beings. Bivalves are filter feeders, lacking radula and odontophores (chewing organs for molluscs), thus ingesting all their food (including harmful plastic particles) directly into their digestive system, a good trait for research exploration (Von Moos et al., 2012, Van Cauwenberghe and Janssen, 2014, Capolupo et al., 2018, Magara et al., 2018, Rist et al., 2019, Fernandez and Albentosa, 2019). These bivalve marine animals include mussels, oysters, shellfish, and clams (see Figure 3.3).

Some studies have already discovered that plastic particles as small as 4 μm can easily be retained in blue mussel (*Mytilus edulis*) bodies, whilst 10 μm-sized particles can collect in the gut system, thus eventually drawn into the circulatory system (Riisgård, 1988, Bouwmeester et al., 2015). An investigation by Rist et al. (2019) studying blue mussel larvae revealed that when 2 μm and 100 nm plastics (diameter) were exposed to these organisms (0.42, 28.3 and 282 g/L), the absorption of 2 μm particles was more than that of 100 nm when compared with the same measure of plastic test particles. The absorption variations observed are believed to be due to variations in particle sizes. For example, 2 μm particles were mistaken for food (usually sized 1–9 μm), whilst 100 nm particles occurred passively afloat in water, thus eventually entering the digestive tract and system. Results showed that even though blue mussel larvae were not affected, malformation appeared, and abnormalities in development increased across all treatment groups and of both plastic sizes (Rist et al., 2019).

A second study devoted to the investigation of oyster (*Crassostrea gigas*) larvae revealed that these organism's larvae can consume plastic particles of 160–7.3 μm. Once exposed to 1 and 10 μm PS particles at 0.11×10^{-3} μg/mL and 0.18 g/mL concentrations for eight days, no severe observable effects were observed in growth, feeding capacity or growth (Cole and Galloway, 2015). In another additional study investigating adult oysters that ingested PS microspheres, a concentration of 0.023 mg/L revealed that these adult oysters preferred particles sized between 6 and 2 μm (Sussarellu et al., 2016). The postulation is that 6 μm diameter size of these particles is closer to their dietary particle sizes and shape. From this study, microplastics were observed to greatly reduce follicle number and sperm motility together with production and offspring larvae development after maternal exposure in a two-month experiment. A 50 nm sized nanoparticle study, similar to the above study, revealed that this particle size can lead to drastic reduction in fertilisation rates in oysters, development in embryo and larvae, subsequently including deformities which can ultimately lead to permanent stunted or stagnated development (Tallec et al., 2018).

Some studies have revealed that *Mytilus edulis* and *Crassostrea gigas* are the two most common bivalves subject to human consumption, and from the soft tissues of these organisms, microplastics have been found present (Van Cauwenberghe et al., 2015, Van Cauwenberghe and Janssen, 2014, Li et al., 2016). According to a European study (Van Cauwenberghe and Janssen, 2014), following these observations and the abundance of microplastics in these consumable bivalves, it is alleged that the European consumption of shellfish leads to an estimated 11,000 types of microplastic to humans from their dietary intake. This indicates that microplastics that have accumulated in bivalves' bodies and tissues can be a vital exposure route to humans consuming them, thus requiring extensive research investigation. Studies investigating microplastic exposure in clams, that is, *Venerupis philippinarum*, of various sources, revealed that there were no elaborate differences in MNPs absorption and collection of plastic particles amongst wild and farmed clams, detected plastic concentrations ranging between 0.07 and 5.47 particles/g (Davidson and Dudas, 2016).

Other studies investigating biological changes due to plastic particle exposure in marine animals such as lugworms, that is, *Arenicola marina* – deposit feeders (large marine worms) – revealed that there were 1.2 ± 2.8 particles/g (Van Cauwenberghe et al., 2015). The obtained findings also revealed that these observed particles in lugworms posed no severe health effects on the animals, nor did they improve or weaken the animal's foreign chemical bioaccumulation (Besseling et al., 2017). Another study investigating microplastics of 200 μm size (larger than a cell) in *Dunaliella salina* at concentrations of 200, 250, 300, and 350 mg/L revealed that these microplastics promoted growth and photosynthesis in *D. salina*. It was also observed that severe effects grew with a decrease in the diameter of microplastics, that is, the smaller the microplastic diameter, the more severe the adverse effects. This observation revealed that there is relation between plastic particle sizes and biological effects, i.e., nano-sized plastic particles are more likely to induce severe biological toxicity than microplastics (Chae et al., 2019).

With sea invertebrates (*Crepidula onyx*), the larval and young ones tended to grow at a slower rate when exposed to excessive concentrations of PS microplastics, that is, 1.4×10^5 particles/ml. This suggests that exposure to microplastics may cause irregular energy consumption (Lo and Chan, 2018). These observed results indicate that MNP particles of varying sizes can potentially be absorbed and accumulated by sea animals, thus causing various health impacts on varying sea species due to ingested plastics. Additionally, many sea filter feeders are more susceptible to absorb and ingest 10 μm-sized particles, nanoplastics being smaller than microplastics with further likelihood of ingestion, pose a greater risk of particle absorption and accumulation in the body, thus creating increased particle concentrations that consequently lead to elevated toxicity levels in the circulation system.

3.3.2 Fresh Water Animals

In a study investigating the impacts of plastic particles (100 mg/L concentration, 48 hours period of exposure) on aquatic animals using *Daphnia magna*, a planktonic crustacean classified under subclass Phyllopoda, about four plastic species of 'microplastic-size environmental relevance' were observed in the organism's gut but no severe health and behavioural impacts (Kokalj et al., 2018). In another investigation studying the impacts of PE microplastic particles of 1 and 100 μm diameter at 12.5–400 mg/L concentrations for 96 hours exposure periods in *D. magna* organism, it was discovered that the immobilisation of test organism by microplastic (1 μm diameter) varied and was dependent on exposure time and particle dose. The observations regarding 100 μm-sized particles showed that these particles were hard to ingest, and there were no severe health effects observed (Rehse et al., 2016).

In an additional study investigating *D. magna* feeding and reproduction rates using fluorescent PS plastic particles sized 100 nm and 2 μm, the study was partitioned into two experimental parts: (i) examining microplastic intake in animals and (ii) establishing microplastic toxic effects on *D. magna* reproduction. In the first study, organisms were subjected to fluorescent PS particles (100 nm and 2 μm) at 1 mg/L concentrations for 24 hours treatment period. The second group was subjected to three varying particle concentrations, respectively, that is, 1, 0.5, and 0.1 mg/L. From the obtained outcomes, the observations showed that the ingestion of both PS sizes was easier to consume, and the absorption of 2 μm particles was easier to absorb five times than that of 100 nm particles. No severe effects were observed in reproduction after 21 days of 1, 0.5, and 0.1 mg/L microparticle exposure. With 100 nm particles, however, excretion and feeding rates were observed to have reduced, thus revealing that particles of greater health hazard to *D. magna* are particles at nanoscale than micro (Rist et al., 2017).

Theoretically, this could be because nano-sized particles are easily retainable in the digestive tract due to their small diameters. The more the nanoplastic particles accumulate in the gut, the more they give a false impression that the animal is full, subsequently leading to a prompted lesser forage or eating from the organisms. Other studies investigating toxicological effects related to *D. magna* plastic exposure revealed that there were no morphological alterations (body length and width and tailbone size), no hazardous changes in reproductive parameters, and no mortalities observed in adult organisms. These results together reveal that MNPs may pass through the *D. magna* gut, but the extent to which MNPs may induce severe health hazards still needs further investigations (Rehse et al., 2016, Imhof et al., 2017, Rist et al., 2017, Canniff and Hoang, 2018, Kokalj et al., 2018).

Additional toxicological and effects of MNP studies were conducted in *Danio rerio* (zebrafish), a freshwater fish commonly used in research. To explore the uptake and cumulation model of zebrafish, 5 and 20 μm-sized PS particles were subjected at 20 mg/L for an exposure period of seven days. Simultaneously, liver toxicity was also examined using 70 nm and 5 μm PS particles at 20, 200, and 2000 mg/L concentrations for an exposure period of three days. Obtained results revealed that the 5 μm PS particles could easily accumulate through the zebrafish organs such as liver, gut, and gills, whilst 20 μm PS particles could not (gill tissues). Organ toxicity results suggested that 70 nm and 5 μm PS particles possess the potential to cause lipid accumulation and inflammation in the liver.

To examine variations in lipid energy metabolism, and oxidative stress, increase or decrease of some enzymatic activities was analysed (Lu et al., 2016). Another ten-day study investigating the effects of microplastics at concentrations between 0.001 and 10.0 mg/L, revealed that microplastics do not, or hardly, cause any mortalities in zebrafish. However, after exposing zebrafish to all four common microplastic types, that is, PVC, PA (polyamide), PP, and PE, intestinal damages were observed. This included villi cracks and splits in the enterocytes (Lei et al., 2018). When exposing zebrafish embryos to nanoparticle-sized PS plastics of 51 nm average diameter at varying concentrations, that is, 0.1, 1, or 10 ppm 24 hours post-fertilisation (hpf), the results showed developmental toxicities. The nanoparticle-sized PS particles were observed to cumulate in the embryos (yolk sacs), progressing to the GIT (gastrointestinal tract), pancreas, liver, heart, gallbladder, and to the brain within 48–120 hpf. During the purification period of all involved organs, PS nanoparticles were observed to decrease in accumulation within 120–168 hpf, even though the pancreatic and GIT clearance rates were sluggish compared to all other organs. According to the observations of this study, no mortalities, nor severe malformations, or bioenergy changes of the mitochondria were observed when PS nanoparticles were exposed on zebrafish, even though the heart rates of the embryos were reduced. Lastly, the obtained data infer that nano-sized plastic particles can potentially pass through the developing zebrafish choroid membranes, piling up in the embryonic tissues, and subsequently affecting behaviour and physiology thus leading to inter- or transgenerational toxic harm (Pitt et al., 2018, Lee et al., 2019).

Additionally, it has been reported that the use of isolated MNPs from the environmental samples is far more convincing and more reflective than using commercial regular particles in conducting exposure experiments (Hu et al., 2020). This follows a 21-day exposure test conducted on *Oryzias latipes* (Japanese medaka) using microfibres at 10,000 particles/L. In these findings, it was discovered that secondary patellar aneurysms and oviposition were increased due to microfibre exposure. Additionally, the study alleges to provide better insight that both MNP samples harvested naturally can be better used in taxicological tests. It is believed that, compared to commercially synthesised microspheres, natural MNP fibres separated from the environment possess value in research and significance in analysis and research.

3.3.3 Mammalian Animals

The use of mammalian animal models in research studies, and prediction of MNP potential impacts on humans have already started. Some findings have suggested that pre-exposed mice model to PS microplastics (diameter 5 and 20 μm) for 28 days revealed that microplastic particles were present in the gut, kidney, and liver (Deng et al., 2017). Additionally, the results obtained from lipid metabolisms, energy, etc. showed that there are possible toxic effects sustained or suffered after microplastic exposure. The microplastic-treated group revealed that energy levels (TG and T-CHO) dropped significantly. It was further observed that droplets of lipids were noted in the liver; this suggested that microplastic-exposed animals suffered lipid metabolism disruptions and liver inflammations (Figure 3.7.).

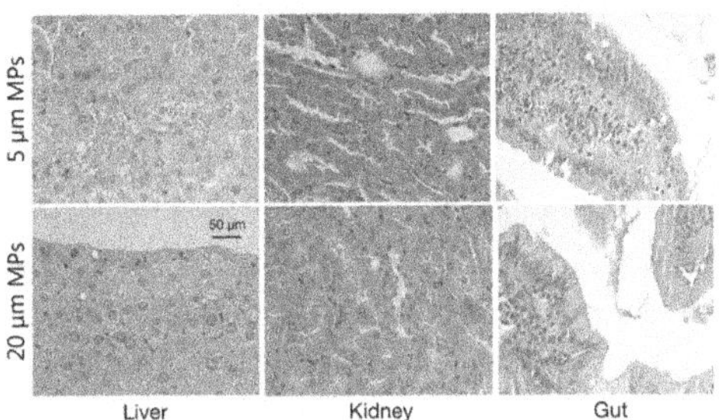

FIGURE 3.7 Microplastics observed in gut, kidney and liver of mice (Deng et al., 2017).

Two additional studies exploring exposure of mice to PS microplastics revealed that there was a decrease in intestinal mucus, and there was a vast variation in the diversity and richness of the intestinal biota (Lu et al., 2018, Jin et al., 2019). In a study investigating long-term exposure of PS nanoplastics (1, 3, 6, and 10 mg PS nanoparticles/kg daily body weight, diameter 38.9 nm) on Wistar adult mice, the results revealed that no significant effects of neural behaviour were observed in all behavioural tests. Nevertheless, a very subtle toxic effect observed was decreased locomotor activity (Rafiee et al., 2018).

From the obtained mammalian results, it is logical to assume that plastic particle accumulation and its associated health effects are a threat to humans. With the findings above, it was also determined that human drinking water constitutes an average of 193 particles per litre of micro and/or nanoplastic abundance, and people are said to drink between 1200 and 1600 mL of water daily. Considering these and their long-term effects, it is also sensible to study micro and nanoplastic exposure. As things stand, there are no conclusive reports recorded to assist in predicting daily MNP intake and exposure. Secondly, the point of entrance to absorb micro- and nanoparticles used to access internal circulation is unclear. The assertion that it is through the GIT and to the target organs is still under speculation. More studies that will evaluate exposure and toxicity by applying experimental models appropriate to humans are necessary.

3.3.4 Human (Health and Toxicological Pathways)

3.3.4.1 Plastic Debris and Prosthetic Implants Wear on Human Health

There have been a series of literature reports regarding inflammation borne from wear particles of degraded plastic prosthetic implants thus indicating biological reactions that could be expected if microplastics would potentially cross the epithelium or the gastrointestinal tract. Joint capsules, surrounding tissues and cavities of patients who have undergone plastic endoprosthesis implantation, have tested positive for PET and PE wear and tear plastic particles (Willert et al., 1996). According to Urban et al. (2000), cellular responses anticipated could be a few scattered cells or extensively aggravated macrophages, and these particles may be in granule or spear form in shape (Willert et al., 1996). PE particles of 0.5–50 µm were found to provoke foreign body responses that are non-immunological (Doorn et al., 1996).

Once particles were exposed and phagocytosed, cellular aggravations that looked like foreign body granulations were observed, and by lymphatic perivascular spaces, PE particles can travel to surrounding vessels (Willert et al., 1996). In a rabbit study by Kubo et al. (1999), it was revealed that smaller PE particles sized 11 µm were more potent than particles of larger size sized 99 µm. This was marked by a high number of histiocytes surrounding the particles of lesser size. The PET plastic particles are in the sizes between 0.5 and 20 µm in the joint capsule histiocytes as cytoplasm deposits, whilst larger PET particles sized up to 100 µm located in the tissues extracellularly. The surrounding tissues were found to undergo substantial changes due to the reaction to PET particles. Cavities at the joints (joint cavities) that carry substantial amounts of fibrin showed or displayed necrosis and strong necrotic behaviour and formation of scars in the joint capsules.

PET particles occurring in abundance can also be found saturated in the granulation tissues of the joint capsule through phagocytosis (Willert et al., 1996). This has proved the endowed incapacity to eliminate phagocytosed particles in the lymphatic system (Willert et al., 1996). The same reaction to microplastics would be observed if they could potentially cross the epithelia after exposure and uptake. In a test study by Mendes (1974), using dogs as test subjects, some abrasion PE particles were identified in the para-aortic lymph nodes of dogs that received a full surface hip replacement 18 months after the procedure. The observations were also similar for human patients who received joint replacements; abrasion PE particles were observed to have accumulated in the lymphatic nodes (Morawski et al., 1995, Leugering and Püschner, 1978). It was also found that wear particles can accumulate so much that PE particles containing macrophages completely replace the lymph nodes (Urban et al., 2000). These PE particles–filled lymph nodes present histiocytic infiltration, that is, granulomatous inflammation, the PE wear particle containing histiocytes inducing severe responses of the macrophage in the surrounding tissues (Morawski et al., 1995).

Lung alveolar walls of dogs that received a total hip replacement are tested positive for PE particles, even though the deposits were found in smaller quantities, suggesting a redistribution to the secondary tissues (Walker, 1973, Mendes, 1974). Human abdominal lymph nodes tested positive for smaller than 50 μm sized PE abrasion particles, with 14% of the tested human subjects testing positive for additional abrasion PE particles in the liver and spleen. Most particles in the less than 1 μm range accumulated in the mobile macrophages of the liver portal tracts delivered through lymphatic transport (Urban et al., 2000).

Hick's study suggests that inflammation response to plastic abrasion particles observed in the lymph nodes involved immune activation of the macrophage and was also associated with the production of cytokines. Subsequently, these studies revealed that plastic microparticles would disseminate through the body if released, and also highlighted that plastic chemical composition will determine immunological responses, with PET observed to be more eliciting of immunological responses than PE polymer (Hicks et al., 1996).

3.3.4.2 Drinking Tap Water and Human Stool

According to a Lancet planetary health article, the issue of microplastics has been brought to sharp attention after Orb media's investigative study conducted a few years ago. According to findings of a study on tap water collected worldwide, the drinking water showed high proportions of contamination by microscopic fragments. The samples collected showed up to 83% of microscopic fragment contaminations from the worldwide samples and 94% for those collected in the US. From these findings, it was also discovered that the widespread extent of these contaminations was greater than what people may think posing a daily possibility of ingestion by humans.

The little knowledge possessed regarding these microplastic contaminants, and their effects posed to human health when consumed is concerning. The obtained water results reflecting microplastic contaminations are not positive findings to reflect contaminations. The most concerning attribute of tap water contamination by microplastics is that this whole phenomenon suggests the possible substantial contamination of the whole ecosystem through freshwater, terrestrial water, and marine water. From these reports, the omnipresence of microplastics is just undeniable.

In another investigation, it was discovered that microplastics were found deep inside human bodies (Schwabl, 2018). This was after a researcher from the Medical University of Vienna had requested eight healthy human subjects from four continents to take part in an experimental test. The subjects were requested to record their weekly dietary routines, including whether they drank water from plastic bottles, chewed any gums, and what types or brands of cosmetics and toothpaste they used. These test subjects were all non-vegetarian and consumed food wrapped in plastics, and most of them had fish in one of their meals. The test subjects were also requested to send pieces of their stools to the Austrian government laboratory, where they were analysed for visible particles of microplastics.

It was in this study that the presence of microplastic particles was confirmed to be present in the human body, with all volunteers testing positive. About all the nine common types of plastics were found including PET and polypropylene (PP). In this study, it was confirmed that 20 particles for every 10 g of stool were found (Schwabl, 2018). Schwable alleges that as data have showed microplastic contamination of fish foods such as oysters, fishes, mussels, and shrimps' contaminations transferrable to humans was procured during the varying food processing steps or through packaging. The smaller particles are even most likely to enter through the blood circulation system and the lymphatic system through to the liver, this is also where (Schwabl, 2018) suggests that further research aiming at understanding what this may potentially mean with regards to humans in the events of exposure.

Some other growing concerns are the added synthetic chemicals for purposes of material stiffness or transparency. These synthetic chemicals are alleged to be hormone altering, and most are rarely experimented. The toxic chemical bisphenol A has been revealed to be found in detectable levels, with 95% of American adults testing positive in their urine samples (Vidal, 2018). About 95% of the adult population of Americans tested positive in their urine samples. It is further suggested that due to the minute sizes of MNPs and their unperceivable nature in the naked eye, they pose greater potential for human health disruptions, and above all, they remain under study. A total of 4000 varying chemicals are identified in plastic packaging, and 148 of those chemicals are alleged to be potent to human health due to their possibility to disrupt hormones (Warhurst, 2018) (Figure 3.8).

It is added that most additional chemicals used in plastic packaging have not been tested for potential to disrupt endocrine systems. This is also because the currently existing methods are not good enough for the identification requests. The most common amongst these chemicals are bisphenols and phthalates (Warhurst, 2018). These have been associated with obesity in children, cancers, cardiovascular diseases, and asthma. The United States of America National Toxicology Program has taken the initiative to raise concern against bisphenol A's effects on the brain, behaviour, and prostate gland in both children and infants. Regardless of this action from the National Toxicology Program, bisphenol A remains in use in Europe.

Several other phthalates and bisphenols are not banned and remain in normal use to produce goods purchasable by consumers. It is suggested that many of these may be like those banned, or even more toxic than them. Warhurst further suggests

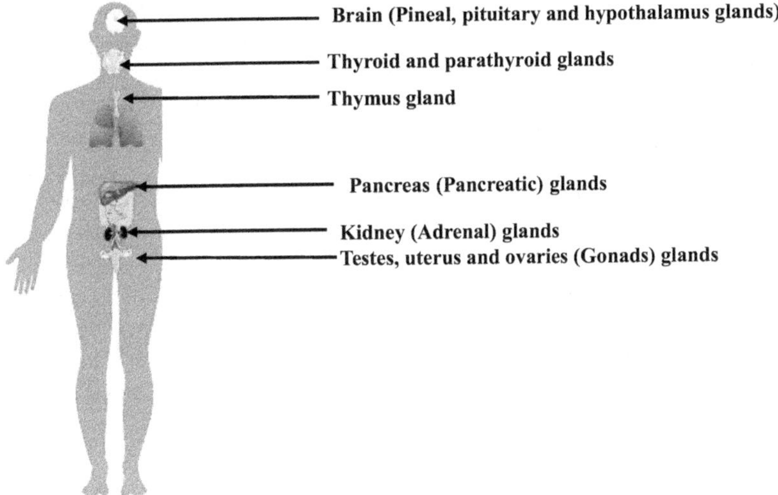

FIGURE 3.8 Various location spots as primary targets of endocrine disruption in the human body (Pironti et al., 2021).

that there are many chemicals like known chemicals that disrupt the endocrine. Even if one may be banned, another is left behind. There are requirements placed on companies for the provision of data, but sadly that is not always comprehensive (Warhurst, 2018).

3.3.5 Additives/Adsorbents

Equally to adsorbed toxic chemicals, leached plastic additives such as OPFRs, phthalates, and BPA are critical issues to be reckoned with. These have been observed to show endocrine-disrupting consequences and other hazards (Diamanti-Kandarakis et al., 2009, Gore et al., 2015). Leaching, or release of plastic additives, is expected during the plastic products' service life span, and/or after disposal. For additives to migrate, or leach, several factors must be considered, that is, plastic pore diameter, additive size, and additive partition coefficient at the plastic's surface. All these factors influence the migration potential of the additive. It is believed that additives of low molecular weight would diffuse effortlessly through larger-sized polymer pores (Teuten et al., 2009).

With significance, the surrounding plastic environment can potentially influence plastic chemical properties together with their additives. For instance, with increased temperature surrounding the plastic material, more additive diffusion or movement would be promoted, and with more UV radiation exposure, rate of plastic degradation would be increased (Bjorn et al., 2007, Coffin et al., 2018). Additionally, it was noted that polymers exposed to sea salts tend to release oestrogenic plasticisers (Coffin et al., 2018).

Bacteria, present within living organisms, also play a role in transmitting ingested microplastic additives, posing ultimate consequences to the exposed organism. Even though previous studies have demonstrated how environmental chemistry, or polymer degradation susceptibility to bacterial presence, the understanding of the combination of these factors and how they affect the organisms who have ingested the microplastics remains limited. This also includes understanding the degree to which the ingested polymer particles can subsequently degrade and release chemical additives to the exposed host organisms. With advanced knowledge about endocrine-disrupting impacts and other organ toxicities suffered from plastic additives, further studies are required to understand micro and nanoplastic additive leaching abilities from plastic polymer to host organisms.

Using experimental animal models, it was discovered that chemical additives can have adverse health effects (Talsness et al., 2009). Chemical concentration levels found in plastics from manufacturing have a correlative role with adverse health effects in humans; this also includes abnormalities in reproduction (Swan et al., 2005, Lang et al., 2008, Swan, 2008).

Extensive research evidence shows that traditional approaches for toxicological determinations are unfit or ineffective in revealing some outcomes such as molecular system's reprogramming in the cells caused by very low exposure doses during development (Myers et al., 2009). Through animal research studies, epidemiologists have been able to determine that there are adverse potential health risks posed by chemical additives in humans, further playing a crucial role in the risk assessment of these toxicological chemicals. It has been reported that amending the approach to test chemicals for risk assessment is a necessity (Talsness et al., 2009). These observations suggest that there is a room open to integrate concepts of endocrinology that embody chemical risk assessments.

Phthalates, PBA, and other plastic additives, together with their metabolites, have been demonstrated to be present in the human population by use of the biomonitoring approach. Observations from biomonitoring approaches have also revealed that most common human exposures are chemical additives, or adsorbents concurrently. The data indicated variations according to geographic locations and age, with considerable chemical additive concentrations of some of these chemicals found in children. Exposures from house dust are broad (Rudel et al., 2008), suggesting that phthalates like diethyl hexyl phthalate (DEHP), edibles and use of oral drugs (even though to a lesser extent) show considerable pathways of uptake (Wormuth et al., 2006). A similar extensive trend is observed with BPA exposure data, even though they are a little less extensive.

Ingestion, dermal contact and inhalation remain significant exposure pathways for the general population of chemical toxins (Adibi et al., 2003, Rudel et al., 2003). It has been revealed that even if general population exposure mean levels were below the stipulated acceptable daily exposure doses (recommended tolerable daily intake), the dibutyl phthalate upper percentiles and the concentrations of DEHP urinary metabolite showed that daily intake levels would vary, with some people ranking higher than previously set daily limits (Koch and Calafat, 2009). It is also projected that they could also exceed the estimated safe daily exposure limits too. The currently set safety exposure measures are derived from established standard toxicological presuppositions concerning acute toxicants to determine chemical exposures within a scope of commonly used plastic objects. It is still unclear what toxicological impacts the above-mentioned exposures can have, especially for more vulnerable subpopulations such as pregnant women and children. This therefore warrants a need for further investigation.

Evidence regarding the relationship between some phthalate metabolite's urinary concentrations and biological effects has been reported (Swan et al., 2005, Swan, 2008). In women, there was an inverse connection between DEHP metabolite concentrations found in women's urine and their perineum distance, whilst in men, the observed biological effects were in the penis diameter and testicular descent of the masculine offspring (Swan et al., 2005, Swan, 2008). Adults exhibited negative relations between phthalate metabolites and quality of semen (Meeker et al., 2009). Free testosterone levels were observed in PVC workers in association with high phthalate exposure. It has been additionally proven that there is a relationship between BPA concentration in urine and sugar II diabetes, heart and liver-related abnormalities (Lang et al., 2008).

It was revealed that American adults were exposed to BPA from various sources. It was also added that the half-life projected for BPA was longer than originally estimated (Stahlhut et al., 2009). Therefore, concerns were raised regarding the extreme BPA and phthalates exposure to premature infants under neonatal care (Calafat, Weuve et al., 2009). From these findings, it is indisputable that the general population's unpleasant effects are due to long-term low-dose exposures and acute high-dose exposures. The in-depth extent to which chemicals are transported by plastics to humans remains unclear and warrants further investigations.

Mice male reproductive areas are noted to be hypersensitive to phthalates exposures, even though most effects observed are not imposed or induced by phthalates diesters but by monoester by-products formed in the liver. Most of these examinations have been evaluated using laboratory mice model organisms. The minimum doses used in these studies and model organisms are a degree higher than the real-time exposures human populations are subjected to. These inflated mice exposure have resulted in immediate and adverse transformations in mice testis. These reproductive transformations have also been noted in guinea pigs. Other concerning issues have included the 'before' and 'after' birth early development effects. Certain investigative studies have revealed that certain phthalate exposures have the capability to induce adverse abnormalities in male reproductive systems that are still under development.

It is important to note that many of these investigative studies have used inflated phthalates exposure concentrations compared to the real time approximated exposures general humans would be subjected to, and only recently have there been investigations focused on the biological transformations within the feasible average phthalate exposures subjected to general human populations (Talsness et al., 2009).

Investigating biological health transformations within the feasible human phthalate exposure concentrations is very vital because there have been observed relations between phthalate concentrations and outbreaks of some severe human health issues following certain epidemiological studies (Swan et al., 2005). It has been further added that due to these unrealistic phthalates exposure investigations, it may be discoverable that general human populations may possibly be highly sensitive to phthalates comparable to laboratory mice model organisms or, the used analytical model in the investigation of conventional toxins examining a single phthalate a time does not accurately account for the accurate prediction of adverse effects a mixture of phthalates would have on humans upon exposure (Andrade et al., 2006).

Surprisingly, mice model organisms also suffered noticeable insulin secretion stimulations, resulting in insulin resistance at low BPA dose exposures. Rats were observed to exhibit a great reduction in sperm production whilst mice were observed to exhibit decreased maternal behaviour and interferences in the hippocampal joints, leading to a brain appearance like that observed in rats and monkey senilities. Another concern brought forward regarding BPA exposure was brain development seemed to be affected during development, resulting in a loss in sex differentiation in behaviour and brain structure (Talsness et al., 2009). Further observation of significance regarding extreme responses to exposures of animal development to BPA at low doses is that many of those responses were relatable to human disease trends. No substantial literature has been published regarding flame retardant TBBPA effects, but evidence shows that effects can be on thyroid hormones, the success function of both the pituitary and reproduction in animals (Talsness et al., 2009).

With regard to plastic chemicals used in plastic processing and their effects, despite environmental concerns, evidence relating to human effects due to plastic chemicals is still extremely limited. Further research studies are required for extensive studies to understand temporal relations between plastics and their chemicals leached out (Adibi et al., 2008). Current conventional approaches used to determine chemical toxicities focus on individual chemicals and their caused diseases or health abnormalities. Due to the complex integrated endocrine system's natural set-up, future studies that involve plastic chemicals that leach to disrupt the endocrine should be focused on mixtures like plastic chemical mixture leaches found in plastic products such as those commonly used in household plastic products. This would be a much more accurate approach resembling the real-life plastic chemical exposures which humans are exposed to.

A fitting case study was when a substantial number of babies were exposed to nine varying phthalates metabolite measurable levels, and the health outcomes of those cumulative exposures had to be established (Meeker et al., 2009). According to the initial attempt to study multiple phthalates as abnormal genital development contributors to babies, findings have shown why this approach is so important (Swan, 2008). A variation or modification to chemical mixture studies must be developed and extended beyond individual class chemicals but rather varied classes of chemicals, for example, different PCBs, different phthalates, and suchlike. An exemplary species would be a PVC-based product, applicable in house-hold water pipes, and it would contain flame retardants in the likes of lead, organotin, cadmium, PBDEs or TBBPA, BPA and phthalates. In myriad studies, these chemicals have been listed as causing obesity (Heindel and vom Saal, 2009). Additionally, the vinyl chloride monomer, applied in PVC plastic manufactory, is known to be a carcinogen, and vinyl chloride exposures could cause liver angiosarcoma in manufactory employees (Bolt, 2005, Gennaro et al., 2008). In another study, PVC medical tubes found in neonatal intensive care units or nurseries were observed to be a high-concentration source of DEHPs amongst newborn infants (Green et al., 2005). In view that BPA is also an additive in PVC manufacturing, PVC is suspected to be a probable contributor to high BPA volumes detected in babies (Calafat et al., 2009).

The attempts to determine associations between adverse human health issues and polymer additives can give rise to several difficulties. Considering the frequent production pattern, the classified industrial specifications, and the polymer application and its additives thereof makes exposure studies quite distressing. The evolution of technology, statistical approaches and methodologies may potentially assist in unravelling the mystery between these chemicals and their adverse effects on humans. However, even with the most statistically significant variations of hormones credited to occupational and environmental exposures, the definite degree to which hormonal variation occurs is regarded as mild observable abnormality symptoms. This then substantiates why further information is needed on biological techniques that may have impact on polymer-based additives, particularly those at low-dose chronic exposures.

3.3.6 Microbiome

Considering that microbes possess ability to metabolise several environmental toxicants, it is suggested that these environmental pollutants may affect the microbiome (Vanhaecke et al., 2006). These modifications in the microbiome can exert cumulative effects on the host organism, subsequently weakening immunity and triggering inflammation (Cho and Blaser, 2012). When mice were exposed to lavage administered PM10 for extended periods of time, it was noticeable that their colonic microbiome exhibited changes in composition and function (Kish et al., 2013). This phenomenon potentially contributed to the host's induced proinflammatory cytokines. It is, however, still undetermined as to whether this occurrence is an absolute PM10 reason or the observed immune changes were PM10 induced or even both (Kish et al., 2013).

It is claimed that the lungs provide a microbial population sustained via alveolar macrophages and other environmental conditions added to antibacterial surfactants (Adar et al., 2016). Colonisation is alleged to be low compared to the gut, even though development and microbial collective shift correspond to abnormality and illness (Marri et al., 2013). Pathogenesis of inhaled contaminants and modification of nearby surroundings result from inflammation and oxidative stress which consequently concludes to severed microbiome. Consequently, shift may be caused in microbial composition colonising the gut or lung due to response to ingested or inhaled PM10 including microplastic. Inflammation or leach HOCs may be caused by microplastics and the microbial metabolism, causing oxidative stress. Microplastics can potentially transport pathogen strains or an additional lung or gut layer to promote 'specific groups' development, thus modifying the community.

By this, microbial population composition, lung performances or gut microbiome can be modified, consequently creating a series of sequential effects that will affect the host and its well-being. This is also true for human health.

REFERENCES

Agriculture Organization (FAO) (Ed.). (2014). *FAO Yearbook: Fishery and Aquaculture Statistics: 2012*. Food & Agriculture Organization of the UN (FAO).

Adar, S. D., G. B. Huffnagle and J. L. Curtis (2016). "The respiratory microbiome: an underappreciated player in the human response to inhaled pollutants?" *Annals of Epidemiology*, 26(5), 355–359.

Adawi, H. I., M. A. Newbold, J. M. Reed, M. E. Vance, I. L. Feitshans, L. R. Bickford and N. A. Lewinski (2018). "Nano-enabled personal care products: current developments in consumer safety." *NanoImpact*, 11, 170–179.

Adibi, J. J., F. P. Perera, W. Jedrychowski, D. E. Camann, D. Barr, R. Jacek and R. M. Whyatt (2003). "Prenatal exposures to phthalates among women in New York City and Krakow, Poland." *Environmental Health Perspectives*, 111(14), 1719–1722.

Adibi, J. J., R. M. Whyatt, P. L. Williams, A. M. Calafat, D. Camann, R. Herrick, H. Nelson, H. K. Bhat, F. P. Perera and M. J. Silva (2008). "Characterization of phthalate exposure among pregnant women assessed by repeat air and urine samples." *Environmental Health Perspectives*, 116(4), 467–473.

Akhtari, M., & Mahmoudi, M. (2019). Epigenetic biomarkers of asthma and allergic disorders. In *Prognostic Epigenetics*, Volume 15 in Translational Epigenetics (pp. 139–169). Academic Press.

Alimba, C. G. and C. Faggio (2019). "Microplastics in the marine environment: current trends in environmental pollution and mechanisms of toxicological profile." *Environmental Toxicology and Pharmacology*, 68, 61–74.

Alimi, O. S., J. Farner Budarz, L. M. Hernandez and N. Tufenkji (2018). "Microplastics and nanoplastics in aquatic environments: aggregation, deposition, and enhanced contaminant transport." *Environmental Science & Technology*, 52(4), 1704–1724.

Allen, S., D. Allen, V. R. Phoenix, G. Le Roux, P. D. Jiménez, A. Simonneau, S. Binet and D. Galop (2019). "Atmospheric transport and deposition of microplastics in a remote mountain catchment." *Nature Geoscience*, 12(5), 339–344.

Andrade, A. J. M., S. W. Grande, C. E. Talsness, K. Grote and I. Chahoud (2006). "A dose-response study following in utero and lactational exposure to di-(2-ethylhexyl)-phthalate (DEHP): non-monotonic dose-response and low dose effects on rat brain aromatase activity." *Toxicology*, 227, 185–192.

Andrady, A. L. (2017). "The plastic in microplastics: a review." *Marine Pollution Bulletin*, 119(1), 12–22.

Athanasopoulou, E., M. Tombrou, S. Pandis and A. Russell (2008). "The role of sea-salt emissions and heterogeneous chemistry in the air quality of polluted coastal areas." *Atmospheric Chemistry and Physics*, 8(19), 5755–5769.

Auta, H. S., C. Emenike and S. Fauziah (2017). "Distribution and importance of microplastics in the marine environment: a review of the sources, fate, effects, and potential solutions." *Environment International*, 102, 165–176.

Avio, C. G., S. Gorbi and F. Regoli (2017). "Plastics and microplastics in the oceans: from emerging pollutants to emerged threat." *Marine Environmental Research*, 128, 2–11.

Bahners, T., P. Ehrler and M. Hengstberger (1994). "Erste Untersuchungen zur Erfassung und Charakterisierung textiler Feinstäube." *Melliand Textilberichte*, 1(1994), 24–30.

Barboza, L. G. A., L. R. Vieira, V. Branco, N. Figueiredo, F. Carvalho, C. Carvalho and L. Guilhermino (2018). "Microplastics cause neurotoxicity, oxidative damage and energy-related changes and interact with the bioaccumulation of mercury in the European seabass, *Dicentrarchus labrax* (Linnaeus, 1758)." *Aquatic Toxicology*, 195, 49–57.

Battulga, B., M. Kawahigashi and B. Oyuntsetseg (2019). "Distribution and composition of plastic debris along the river shore in the Selenga River basin in Mongolia." *Environmental Science and Pollution Research*, 26(14), 14059–14072.

Bergmann, M., S. Mützel, S. Primpke, M. B. Tekman, J. Trachsel and G. Gerdts (2019). "White and wonderful? Microplastics prevail in snow from the Alps to the Arctic." *Science Advances*, 5(8), eaax1157.

Besseling, E., E. M. Foekema, M. J. van den Heuvel-Greve and A. A. Koelmans (2017). "The effect of microplastic on the uptake of chemicals by the lugworm *Arenicola marina* (L.) under environmentally relevant exposure conditions." *Environmental Science & Technology*, 51(15), 8795–8804.

Bjorn, A., H. M., A. Karlsson, I. Mersiowsky and J. Ejlertsson (2007). "Impacts of temperature on the leaching of organotin compounds from poly(vinylchloride) plastics – A study conducted under simulated landfill conditions." *Journal of Vinyl and Additive Technology*, (13), 176–188.

Bolt, H. M. (2005). "Vinyl chloride – a classical industrial toxicant of new interest." *Critical Reviews in Toxicology*, 35(4), 307–323.

Borunda, A. (2019). "This young whale died with 88 pounds of plastic in its stomach." *National Geographic*, 18 March 2019.

Bouwmeester, H., P. C. Hollman and R. J. Peters (2015). "Potential health impact of environmentally released micro-and nanoplastics in the human food production chain: experiences from nanotoxicology." *Environmental Science & Technology*, 49(15), 8932–8947.

Burkhart, J., C. Piacitelli, D. Schwegler-Berry and W. Jones (1999). "Environmental study of nylon flocking process." *Journal of Toxicology and Environmental Health Part A*, 57(1), 1–23.

Calafat, A. M., J. Weuve, X. Ye, L. T. Jia, H. Hu, S. Ringer, K. Huttner and R. Hauser (2009). "Exposure to bisphenol A and other phenols in neonatal intensive care unit premature infants." *Environmental Health Perspectives*, 117(4), 639–644.

Canniff, P. M. and T. C. Hoang (2018). "Microplastic ingestion by Daphnia magna and its enhancement on algal growth." *Science of the Total Environment*, 633, 500–507.

Capolupo, M., S. Franzellitti, P. Valbonesi, C. S. Lanzas and E. Fabbri (2018). "Uptake and transcriptional effects of polystyrene microplastics in larval stages of the Mediterranean mussel Mytilus galloprovincialis." *Environ Pollution*, 241, 1038–1047.

Carr, S. A. (2017). "Sources and dispersive modes of micro-fibers in the environment." *Integrated Environmental Assessment and Management*, 13(3), 466–469.

CDC. (2020). "Genomics and precision health." from https://www.cdc.gov/genomics/disease/epigenetics.htm.

Chae, Y. and Y.-J. An (2017). "Effects of micro-and nanoplastics on aquatic ecosystems: current research trends and perspectives." *Marine Pollution Bulletin*, 124(2), 624–632.

Chae, Y., D. Kim and Y.-J. An (2019). "Effects of micro-sized polyethylene spheres on the marine microalga *Dunaliella salina*: focusing on the algal cell to plastic particle size ratio." *Aquatic Toxicology*, 216, 105296.

Chen, C.-L. (2015). *Regulation and management of marine litter. Marine anthropogenic litter.* Springer, Cham, 395–428.

Chen, G., Q. Feng and J. Wang (2020). "Mini-review of microplastics in the atmosphere and their risks to humans." *Science of the Total Environment*, 703, 135504.

Cheung, P. K. and L. Fok (2016). "Evidence of microbeads from personal care product contaminating the sea." *Marine Pollution Bulletin*, 109(1), 582–585.

Chiba, S., H. Saito, R. Fletcher, T. Yogi, M. Kayo, S. Miyagi, M. Ogido and K. Fujikura (2018). "Human footprint in the abyss: 30 year records of deep-sea plastic debris." *Marine Policy*, 96, 204–212.

Cho, I. and M. J. Blaser (2012). "The human microbiome: at the interface of health and disease." *Nature Reviews Genetics*, 13(4), 260–270.

Clancy, S. (2008). "Genetic Mutation." *Nature Education*, 2020, from https://www.nature.com/scitable/topicpage/genetic-mutation-441/.

Coffin, S., S. Dudley, A. Taylor, D. Wolf, J. Wang, I. Lee and D. Schlenk (2018). "Comparisons of analytical chemistry and biological activities of extracts from North Pacific gyre plastics with UV-treated and untreated plastics using in vitro and in vivo models." *Environment International*, 121, 942–954.

Cole, M. (2016). "A novel method for preparing microplastic fibers." *Scientific Reports*, 6(1), 1–7.

Cole, M. and T. S. Galloway (2015). "Ingestion of nanoplastics and microplastics by Pacific oyster larvae." *Environmental Science & Technology*, 49(24), 14625–14632.

Corradini, F., P. Meza, R. Eguiluz, F. Casado, E. Huerta-Lwanga and V. Geissen (2019). "Evidence of microplastic accumulation in agricultural soils from sewage sludge disposal." *Science of the Total Environment*, 671, 411–420.

Cozar, A., F. Echevarria, J. I. Gonzalez-Gordillo, X. Irigoien, B. Ubeda, S. Hernandez-Leon, A. T. Palma, S. Navarro, J. Lomas, A. Ruiz, M. L. Fernǻndez-de-Puelles and C. M. Duarte (2014). "Plastic debris in the open ocean." *Proceedings of the National Academy of Sciences of the United States of America*, 111, 10239–10244.

Davidson, K. and S. E. Dudas (2016). "Microplastic ingestion by wild and cultured Manila clams (*Venerupis philippinarum*) from Baynes Sound, British Columbia." *Archives of Environmental Contamination and Toxicology*, 71(2), 147–156.

De Sá, L. C., M. Oliveira, F. Ribeiro, T. L. Rocha and M. N. Futter (2018). "Studies of the effects of microplastics on aquatic organisms: what do we know and where should we focus our efforts in the future?" *Science of the Total Environment*, 645, 1029–1039.

Deng, Y., Y. Zhang, B. Lemos and H. Ren (2017). "Tissue accumulation of microplastics in mice and biomarker responses suggest widespread health risks of exposure." *Scientific Reports*, 7(1), 1–10.

Deng, Y., Y. Zhang, R. Qiao, M. M. Bonilla, X. Yang, H. Ren and B. Lemos (2018). "Evidence that microplastics aggravate the toxicity of organophosphorus flame retardants in mice (Mus musculus)." *Journal of Hazardous Materials*, 357, 348–354.

Diamanti-Kandarakis, E., J.-P. Bourguignon, L. C. Giudice, R. Hauser, G. S. Prins, A. M. Soto, R. T. Zoeller and A. C. Gore (2009). "Endocrine-disrupting chemicals: an Endocrine Society scientific statement." *Endocrine Reviews*, 30(4), 293–342.

Doorn, P. F., P. A. Campbell and H. C. Amstutz (1996). "Metal versus polyethylene wear particles in total hip replacements: a review." *Clinical Orthopaedics and Related Research* (1976–2007), 329, S206–S216.

Dris, R., J. Gasperi, M. Saad, C. Mirande and B. Tassin (2016). "Synthetic fibers in atmospheric fallout: A source of microplastics in the environment?" *Marine Pollution Bulletin*, 104(1–2), 290–293.

Europe, P. (2019). "Plastics - the facts 2019." Retrieved 12 May 2021, from https://www.plasticseurope.org/fr/resources/publications/1804-plasticsfacts.

Fazen, L. E., B. Linde and C. A. Redlich (2020). "Occupational lung diseases in the 21st century: the changing landscape and future challenges." *Current Opinion in Pulmonary Medicine*, 26(2), 142–148.

Fernandez, B. and M. Albentosa (2019). "Insights into the uptake, elimination and accumulation of microplastics in mussel." *Environmental Pollution*, 249, 321–329.

Franzellitti, S., L. Canesi, M. Auguste, R. H. Wathsala and E. Fabbri (2019). "Microplastic exposure and effects in aquatic organisms: a physiological perspective." *Environmental Toxicology and Pharmacology*, 68, 37–51.

Gennaro, V., M. Ceppi, P. Crosignani and F. Montanaro (2008). "Reanalysis of updated mortality among vinyl and polyvinyl chloride workers: confirmation of historical evidence and new findings." *BMC Public Health*, 8(1), 1–8.

Gerdes, Z., M. Ogonowski, I. Nybom, C. Ek, M. Adolfsson-Erici, A. Barth and E. Gorokhova (2019). "Microplastic-mediated transport of PCBs? A depuration study with Daphnia magna." *PloS One*, 14(2), e0205378.

Gewert, B., M. M. Plassmann and M. MacLeod (2015). "Pathways for degradation of plastic polymers floating in the marine environment." *Environmental Science: Processes & Impacts*, 17(9), 1513–1521.

Geyer, R. (2020). Production, use, and fate of synthetic polymers. In *Plastic waste and recycling*. Letcher TM (ed.). Oxford: Academic Press, 13–32.

Gigault, J., Ter Halle, A., Baudrimont, M., Pascal, P. Y., Gauffre, F., Phi, T. L., ... & Reynaud, S. (2018). Current opinion: what is a nanoplastic?. *Environmental Pollution*, 235, 1030–1034.

Gore, A. C., V. A. Chappell, S. E. Fenton, J. A. Flaws, A. Nadal, G. S. Prins, J. Toppari and R. T. Zoeller (2015). "EDC-2: the Endocrine Society's second scientific statement on endocrine-disrupting chemicals." *Endocrine Reviews*, 36(6), E1–E150.

Green, R., R. Hauser, A. M. Calafat, J. Weuve, T. Schettler, S. Ringer, K. Huttner and H. Hu (2005). "Use of di (2-ethylhexyl) phthalate-containing medical products and urinary levels of mono (2-ethylhexyl) phthalate in neonatal intensive care unit infants." *Environmental Health Perspectives*, 113(9), 1222–1225.

Guglielmi, G. (2017). "In the next 30 years, we'll make four times more plastic waste than we ever have." *Science*, 19 July 2017.

Guzzetti, E., A. Sureda, S. Tejada and C. Faggio (2018). "Microplastic in marine organism: environmental and toxicological effects." *Environmental Toxicology and Pharmacology*, 64, 164–171.

Habibi, N., S. Uddin, S. W. Fowler and M. Behbehani (2022). "Microplastics in the atmosphere: a review." *Journal of Environmental Exposure Assessment*, 1(1), 6.

Harrison, J. P., M. Schratzberger, M. Sapp and A. M. Osborn (2014). "Rapid bacterial colonization of low-density polyethylene microplastics in coastal sediment microcosms." *BMC Microbiology*, 14(1), 1–15.

Heindel, J. J. and F. S. vom Saal (2009). "Role of nutrition and environmental endocrine disrupting chemicals during the perinatal period on the aetiology of obesity." *Molecular and Cellular Endocrinology*, 304(1–2), 90–96.

Hernandez, L. M., E. G. Xu, H. C. Larsson, R. Tahara, V. B. Maisuria and N. Tufenkji (2019). "Plastic teabags release billions of microparticles and nanoparticles into tea." *Environmental Science & Technology*, 53(21), 12300–12310.

Hicks, D. G., A. R. Judkins, J. Z. Sickel, R. N. Rosier, J. E. Puzas and R. J. O'Keefe (1996). "Granular histiocytosis of pelvic lymph nodes following total hip arthroplasty. The presence of wear debris, cytokine production, and immunologically activated macrophages." *JBJS*, 78(4), 482–496.

Hirai, H., H. Takada, Y. Ogata, R. Yamashita, K. Mizukawa, M. Saha, C. Kwan, C. Moore, H. Gray and D. Laursen (2011). "Organic micropollutants in marine plastics debris from the open ocean and remote and urban beaches." *Marine Pollution Bulletin*, 62(8), 1683–1692.

Hoellein, T. J., A. R. McCormick, J. Hittie, M. G. London, J. W. Scott and J. J. Kelly (2017). "Longitudinal patterns of microplastic concentration and bacterial assemblages in surface and benthic habitats of an urban river." *Freshwater Science*, 36(3), 491–507.

Hoornweg, Daniel A. (2012). "What a waste: a global review of solid waste management." Urban Development Series Knowledge Papers (World Bank). Washington, DC: World Bank Group.

Hu, L., M. Chernick, A. M. Lewis, P. L. Ferguson and D. E. Hinton (2020). "Chronic microfiber exposure in adult Japanese medaka (Oryzias latipes)." *PloS One*, 15(3), e0229962.

Imhof, H. K., J. Rusek, M. Thiel, J. Wolinska and C. Laforsch (2017). "Do microplastic particles affect Daphnia magna at the morphological, life history and molecular level?" *PLoS One*, 12(11), e0187590.

Jambeck, J. R., R. Geyer, C. Wilcox, T. R. Siegler, M. Perryman, A. Andrady, R. Narayan and K. L. Law (2015). "Plastic waste inputs from land into the ocean." *Science*, 347(6223), 768–771.

Jin, Y., L. Lu, W. Tu, T. Luo and Z. Fu (2019). "Impacts of polystyrene microplastic on the gut barrier, microbiota and metabolism of mice." *Science of the Total Environment*, 649, 308–317.

Karbalaei, S., P. Hanachi, T. R. Walker and M. Cole (2018). "Occurrence, sources, human health impacts and mitigation of microplastic pollution." *Environmental Science and Pollution Research*, 25(36), 36046–36063.

Karwacka, A., D. Zamkowska, M. Radwan and J. Jurewicz (2019). "Exposure to modern, widespread environmental endocrine disrupting chemicals and their effect on the reproductive potential of women: an overview of current epidemiological evidence." *Human Fertility*, 22(1), 2–25.

Kasirajan, S. and M. Ngouajio (2012). "Polyethylene and biodegradable mulches for agricultural applications: a review." *Agronomy for Sustainable Development*, 32(2), 501–529.

Kelly, J. J., M. G. London, N. Oforji, A. Ogunsola and T. J. Hoellein (2020). "Microplastic selects for convergent microbiomes from distinct riverine sources." *Freshwater Science*, 39(2), 281–291.

Khan, F. R., K. Syberg, Y. Shashoua and N. R. Bury (2015). "Influence of polyethylene microplastic beads on the uptake and localization of silver in zebrafish (*Danio rerio*)." *Environmental Pollution*, 206, 73–79.

Kirstein, I. V., S. Kirmizi, A. Wichels, A. Garin-Fernandez, R. Erler, M. Löder and G. Gerdts (2016). "Dangerous hitchhikers? Evidence for potentially pathogenic *Vibrio* spp. on microplastic particles." *Marine Environmental Research*, 120, 1–8.

Kish, L., N. Hotte, G. G. Kaplan, R. Vincent, R. Tso, M. Gänzle, K. P. Rioux, A. Thiesen, H. W. Barkema and E. Wine (2013). "Environmental particulate matter induces murine intestinal inflammatory responses and alters the gut microbiome." *PloS One*, 8(4), e62220.

Koch, H. M. and A. M. Calafat (2009). "Human body burdens of chemicals used in plastic manufacture." *Philosophical Transactions of the Royal Society B: Biological Sciences*, 364(1526), 2063–2078.

Koelmans, A. A., A. Bakir, G. A. Burton and C. R. Janssen (2016). "Microplastic as a vector for chemicals in the aquatic environment: critical review and model-supported reinterpretation of empirical studies." *Environmental Science & Technology*, 50(7), 3315–3326.

Koelmans, A. A., E. Besseling, A. Wegner and E. M. Foekema (2013). "Plastic as a carrier of POPs to aquatic organisms: a model analysis." *Environmental Science & Technology*, 47(14), 7812–7820.

Kokalj, A. J., U. Kunej and T. Skalar (2018). "Screening study of four environmentally relevant microplastic pollutants: uptake and effects on Daphnia magna and Artemia franciscana." *Chemosphere*, 208, 522–529.

Kubo, T., K. Sawada, K. Hirakawa, C. Shimizu, T. Takamatsu and Y. Hirasawa (1999). "Histiocyte reaction in rabbit femurs to UHMWPE, metal, and ceramic particles in different sizes." *Journal of Biomedical Materials Research*, 45(4), 363–369.

Kubowicz, S., & Booth, A. M. (2017). *Biodegradability of plastics: challenges and misconceptions*. https://doi.org/10.1021/acs.est.7b04051

Laganà, P., G. Caruso, I. Corsi, E. Bergami, V. Venuti, D. Majolino, R. La Ferla, M. Azzaro and S. Cappello (2019). "Do plastics serve as a possible vector for the spread of antibiotic resistance? First insights from bacteria associated to a polystyrene piece from King George Island (Antarctica)." *International Journal of Hygiene and Environmental Health*, 222(1), 89–100.

Laitala, K., I. G. Klepp and B. Henry (2018). "Does use matter? Comparison of environmental impacts of clothing based on fiber type." *Sustainability*, 10(7), 2524.

Lamb, J. B., B. L. Willis, E. A. Fiorenza, C. S. Couch, R. Howard, D. N. Rader, J. D. True, L. A. Kelly, A. Ahmad and J. Jompa (2018). "Plastic waste associated with disease on coral reefs." *Science*, 359(6374), 460–462.

Lambert, S. and M. Wagner (2016). "Characterisation of nanoplastics during the degradation of polystyrene." *Chemosphere*, 145, 265–268.

Lang, I. A., T. S. Galloway, A. Scarlett, W. E. Henley, M. Depledge, R. B. Wallace and D. Melzer (2008). "Association of urinary bisphenol A concentration with medical disorders and laboratory abnormalities in adults." *Journal of the American Medical Association*, 300(11), 1303–1310.

Lee, W. S., H.-J. Cho, E. Kim, Y. H. Huh, H.-J. Kim, B. Kim, T. Kang, J.-S. Lee and J. Jeong (2019). "Bioaccumulation of polystyrene nanoplastics and their effect on the toxicity of Au ions in zebrafish embryos." *Nanoscale*, 11(7), 3173–3185.

Lehner, R., C. Weder, A. Petri-Fink and B. Rothen-Rutishauser (2019). "Emergence of nanoplastic in the environment and possible impact on human health." *Environmental Science & Technology*, 53(4), 1748–1765.

Lei, L., S. Wu, S. Lu, M. Liu, Y. Song, Z. Fu, H. Shi, K. M. Raley-Susman and D. He (2018). "Microplastic particles cause intestinal damage and other adverse effects in zebrafish Danio rerio and nematode Caenorhabditis elegans." *Science of the Total Environment*, 619, 1–8.

Leugering, H. and H. Püschner (1978). "Identification of wear particles in tissue after implantation of different plastic materials." *Journal of Biomedical Materials Research*, 12(4), 571–578.

Li, J., D. Yang, L. Li, K. Jabeen and H. Shi (2015). "Microplastics in commercial bivalves from China." *Environmental Pollution*, 207, 190–195.

Li, J., K. Zhang and H. Zhang (2018). "Adsorption of antibiotics on microplastics." *Environmental Pollution*, 237, 460–467.

Li, J., X. Qu, L. Su, W. Zhang, D. Yang, P. Kolandhasamy, D. Li and H. Shi (2016). "Microplastics in mussels along the coastal waters of China." *Environmental Pollution*, 214, 177–184.

Liebezeit, G. and E. Liebezeit (2013). "Non-pollen particulates in honey and sugar." *Food Additives & Contaminants: Part A*, 30(12), 2136–2140.

Liebezeit, G. and E. Liebezeit (2014). "Synthetic particles as contaminants in German beers." *Food Additives & Contaminants: Part A*, 31(9), 1574–1578.

Liu, J., N. Lezama, J. Gasper, J. Kawata, S. Morley, D. Helmer and P. Ciminera (2016). "Burn pit emissions exposure and respiratory and cardiovascular conditions among airborne hazards and open burn pit registry participants." *Journal of Occupational and Environmental Medicine*, 58(7), e249–e255.

Liu, J., Y. Yang, J. Ding, B. Zhu and W. Gao (2019). "Microfibers: a preliminary discussion on their definition and sources." *Environmental Science and Pollution Research*, 26(28), 29497–29501.

Liu, K., X. Wang, T. Fang, P. Xu, L. Zhu and D. Li (2019). "Source and potential risk assessment of suspended atmospheric microplastics in Shanghai." *Science of the Total Environment*, 675, 462–471.

Lo, H. K. A. and K. Y. K. Chan (2018). "Negative effects of microplastic exposure on growth and development of Crepidula onyx." *Environmental Pollution*, 233, 588–595.

Lobelle, D. and M. Cunliffe (2011). "Early microbial biofilm formation on marine plastic debris." *Marine Pollution Bulletin*, 62(1), 197–200.

Lu, L., Z. Wan, T. Luo, Z. Fu and Y. Jin (2018). "Polystyrene microplastics induce gut microbiota dysbiosis and hepatic lipid metabolism disorder in mice." *Science of the Total Environment*, 631, 449–458.

Lu, Y., Y. Zhang, Y. Deng, W. Jiang, Y. Zhao, J. Geng, L. Ding and H. Ren (2016). "Uptake and accumulation of polystyrene microplastics in zebrafish (*Danio rerio*) and toxic effects in liver." *Environmental Science & Technology*, 50(7), 4054–4060.

Magara G, E. A., K. Syberg, F. R. Khan (2018). "Single contaminant and combined exposures of polyethylene microplastics and fluoranthene: accumulation and oxidative stress response in the blue mussel, Mytilus edulis." *Journal of Toxicology and Environmental Health Part A*, 81, 761–773.

Marri, P. R., D. A. Stern, A. L. Wright, D. Billheimer and F. D. Martinez (2013). "Asthma-associated differences in microbial composition of induced sputum." *Journal of Allergy and Clinical Immunology*, 131(2), 346–352.e343.

Mattsson, K., L. A. Hansson and T. Cedervall (2015). "Nano-plastics in the aquatic environment international." *Environ Sci Process Impacts*, 17, 1712–1721.

Meeker, J. D., S. Sathyanarayana and S. H. Swan (2009). "Phthalates and other additives in plastics: human exposure and associated health outcomes." *Philosophical Transactions of the Royal Society B: Biological Sciences*, 364(1526), 2097–2113.

Mendes, D. (1974). "Total surface hip replacement in the dog: a preliminary study of local tissue reaction." *Clin Orthop*, 100, 256–264.

Mishra, S., C. Charan Rath and A. P. Das (2019). "Marine microfiber pollution: a review on present status and future challenges." *Marine Pollution Bulletin*, 140, 188–197.

Mizukawa, K., H. Takada, M. Ito, Y. B. Geok, J. Hosoda, R. Yamashita, M. Saha, S. Suzuki, C. Miguez and J. Frias (2013). "Monitoring of a wide range of organic micropollutants on the Portuguese coast using plastic resin pellets." *Marine Pollution Bulletin*, 70(1–2), 296–302.

Morawski, D. R., R. D. Coutts, E. G. Handal, F. J. Luibel, R. F. Santore and J. L. Ricci (1995). "Polyethylene debris in lymph nodes after a total hip arthroplasty. A report of two cases." *JBJS*, 77(5), 772–776.

Munier, B. and L. Bendell (2018). "Macro and micro plastics sorb and desorb metals and act as a point source of trace metals to coastal ecosystems." *PLoS One*, 13(2), e0191759.

Myers, J. P., F. S. vom Saal, B. T. Akingbemi, K. Arizono, S. Belcher, T. Colborn, I. Chahoud, D. A. Crain, F. Farabollini and L. J. Guillette Jr (2009). "Why public health agencies cannot depend on good laboratory practices as a criterion for selecting data: the case of bisphenol A." *Environmental Health Perspectives*, 117(3), 309–315.

Nano, J., Fernandez, E. P., Troup, J., Ghanbari, M., Franco, O. H., & Muka, T. (2018). Epigenetics of diabetes in humans. In *Epigenetics in human disease* (pp. 457–488). Academic Press.

Nel, A., T. Xia, L. Mädler and N. Li (2006). "Toxic potential of materials at the nanolevel." *Science*, 311(5761), 622–627.

Niller, H. H., & Minarovits, J. (2024). Epigenetics and human infectious diseases. In *Epigenetics in human disease* (pp. 779–852). Academic Press.

Oliver, V. F., van Bysterveldt, K. A., and Merbs, S. L. (2016). Epigenetics in ocular medicine. In *Medical epigenetics* (pp. 391–412). Academic Press.

Omenn, G. S., J. Merchant, E. Boatman, J. M. Dement, M. Kuschner, W. Nicholson, J. Peto and L. Rosenstock (1986). "Contribution of environmental fibers to respiratory cancer." *Environmental Health Perspectives*, 70, 51–56.

Pauly, J. L., S. J. Stegmeier, H. A. Allaart, R. T. Cheney, P. J. Zhang, A. G. Mayer and R. J. Streck (1998). "Inhaled cellulosic and plastic fibers found in human lung tissue." *Cancer Epidemiology and Prevention Biomarkers*, 7(5), 419–428.

Peeken, I., S. Primpke, B. Beyer, J. Gütermann, C. Katlein, T. Krumpen, M. Bergmann, L. Hehemann and G. Gerdts (2018). "Arctic sea ice is an important temporal sink and means of transport for microplastic." *Nature Communications*, 9(1), 1–12.

Pironti, C., M. Ricciardi, A. Proto, P. M. Bianco, L. Montano and O. Motta (2021). "Endocrine-disrupting compounds: an overview on their occurrence in the aquatic environment and human exposure." *Water*, 13(10), 1347.

Pitt, J. A., J. S. Kozal, N. Jayasundara, A. Massarsky, R. Trevisan, N. Geitner, M. Wiesner, E. D. Levin and R. T. Di Giulio (2018). "Uptake, tissue distribution, and toxicity of polystyrene nanoparticles in developing zebrafish (Danio rerio)." *Aquatic Toxicology*, 194, 185–194.

PlasticsEurope (2006). "The compelling facts about plastics: an analysis of plastic production, demand and recovery for 2006 in Europe."

PlasticsEurope (2016). "Plastics-The Facts 2016: an Analysis of European Plastics Production, Demand and Waste Data.".

Porter, D. W., V. Castranova, V. Robinson, A. F. Hubbs, R. R. Mercer, J. Scabilloni, T. Goldsmith, D. Schwegler-Berry, L. Battelli and R. Washko (1999). "Acute inflammatory reaction in rats after intratracheal instillation of material collected from a nylon flocking plant." *Journal of Toxicology and Environmental Health Part A*, 57(1), 25–45.

Prata, J. C. (2018). "Airborne microplastics: consequences to human health?" *Environmental Pollution*, 234, 115–126.

Qayyum, H., Munir, A., Fayyaz, S. M., Aftab, A., Butt, H. A., Soomro, N. I., ... & Bakhtiar, S. M. (2019). Single-Cell Omics in CVDs. In *Single-cell omics* (pp. 129–152). Academic Press.

Qiu, Q., J. Peng, X. Yu, F. Chen, J. Wang and F. Dong (2015). "Occurrence of microplastics in the coastal marine environment: first observation on sediment of China." *Marine Pollution Bulletin*, 98(1–2), 274–280.

Raddadi, N. and F. Fava (2019). "Biodegradation of oil-based plastics in the environment: existing knowledge and needs of research and innovation." *Science of the Total Environment*, 679, 148–158.

Rafiee M, D. L., Eslami A, Beirami E, Jahangiri-Rad M, Sabour S, et al. (2018). "Neurobehavioral assessment of rats exposed to pristine polystyrene nanoplastics upon oral exposure." *Chemosphere*, (193), 745–753.

Rehse, S., W. Kloas and C. Zarfl (2016). "Short-term exposure with high concentrations of pristine microplastic particles leads to immobilisation of Daphnia magna." *Chemosphere*, 153, 91–99.

Rhodes, C. J. (2018). "Plastic pollution and potential solutions." *Science Progress*, 101, 207–260.

Riisgård, H. U. (1988). "Efficiency of particle retention and filtration rate in 6 species of Northeast American bivalves." *Marine Ecology Progress Series Oldendorf*, 45(3), 217–223.

Rist, S., A. Baun, R. Almeda and N. B. Hartmann (2019). "Ingestion and effects of micro-and nanoplastics in blue mussel (*Mytilus edulis*) larvae." *Marine Pollution Bulletin*, 140, 423–430.

Rist, S., A. Baun and N. B. Hartmann (2017). "Ingestion of micro-and nanoplastics in Daphnia magna – quantification of body burdens and assessment of feeding rates and reproduction." *Environmental Pollution*, 228, 398–407.

Roberti, A., A. F. Valdes, R. Torrecillas, M. F. Fraga and A. F. Fernandez (2019). "Epigenetics in cancer therapy and nanomedicine." *Clinical Epigenetics*, 11(1), 1–18.

Rochman, C. M., B. T. Hentschel and S. J. Teh (2014). "Long-term sorption of metals is similar among plastic types: implications for plastic debris in aquatic environments." *PloS One*, 9(1), e85433.

Rochman, C. M., Kross, S. M., Armstrong, J. B., Bogan, M. T., Darling, E. S., Green, S. J., ... & Veríssimo, D. (2015). *Scientific evidence supports a ban on microbeads.*

Rudel, R., R. Dodson, E. Newton, A. Zota and J. Brody (2008). "Correlations between urinary phthalate metabolites and phthalates, estrogenic compounds 4-butyl phenol and o-phenyl phenol, and some pesticides in home indoor air and house dust." *Epidemiology*, 19(6), S332.

Rudel, R. A., D. E. Camann, J. D. Spengler, L. R. Korn and J. G. Brody (2003). "Phthalates, alkylphenols, pesticides, polybrominated diphenyl ethers, and other endocrine-disrupting compounds in indoor air and dust." *Environmental Science & Technology*, 37(20), 4543–4553.

Rudyak, V. Y., E. A. Efimova, D. V. Guseva and A. V. Chertovich (2019). "Thermoset polymer matrix structure and properties: coarse-grained simulations." *Polymers*, 11(1), 36.

Schwabl, P. (2018). "Microplastics discovered in human stools across the globe in 'first study of its kind'." Retrieved 13 March, 2020, from https://www.eurekalert.org/pub_releases/2018-10/sh-mdi101518.php.

Schwarz, A., T. Ligthart, E. Boukris and T. Van Harmelen (2019). "Sources, transport, and accumulation of different types of plastic litter in aquatic environments: a review study." *Marine Pollution Bulletin*, 143, 92–100.

Sharma, S. and S. Chatterjee (2017). "Microplastic pollution, a threat to marine ecosystem and human health: a short review." *Environmental Science and Pollution Research*, 24(27), 21530–21547.

Shen, M., Y. Zhu, Y. Zhang, G. Zeng, X. Wen, H. Yi, S. Ye, X. Ren and B. Song (2019). "Micro (nano) plastics: unignorable vectors for organisms." *Marine Pollution Bulletin*, 139, 328–331.

Stahlhut, R. W., W. V. Welshons and S. H. Swan (2009). "Bisphenol A data in NHANES suggest longer than expected half-life, substantial nonfood exposure, or both." Environmental *Health Perspectives*, 117(5), 784–789.

Stapleton, P. (2019). "Toxicological considerations of nano-sized plastics." *AIMS Environmental Science*, 6(5), 367.

Suresh, S. S., S. Mohanty and S. K. Nayak (2017). "Composition analysis and characterization of waste polyvinyl chloride (PVC) recovered from data cables." *Waste Management*, 60, 100–111.

Sussarellu, R., M. Suquet, Y. Thomas, C. Lambert, C. Fabioux, M. E. J. Pernet, N. Le Goïc, V. Quillien, C. Mingant and Y. Epelboin (2016). "Oyster reproduction is affected by exposure to polystyrene microplastics." *Proceedings of the National Academy of Sciences*, 113(9), 2430–2435.

Swan, S. H. (2008). "Environmental phthalate exposure in relation to reproductive outcomes and other health endpoints in humans." *Environmental Research*, 108(2), 177–184.

Swan, S. H., K. M. Main, F. Liu, S. L. Stewart, R. L. Kruse, A. M. Calafat, C. S. Mao, J. B. Redmon, C. L. Ternand and S. Sullivan (2005). "Decrease in anogenital distance among male infants with prenatal phthalate exposure." *Environmental Health Perspectives*, 113(8), 1056–1061.

Tallec, K., A. Huvet, C. Di Poi, C. González-Fernández, C. Lambert, B. Petton, N. Le Goïc, M. Berchel, P. Soudant and I. Paul-Pont (2018). "Nanoplastics impaired oyster free living stages, gametes and embryos." *Environmental Pollution*, 242, 1226–1235.

Talsness, C. E., A. J. Andrade, S. N. Kuriyama, J. A. Taylor and F. S. Vom Saal (2009). "Components of plastic: experimental studies in animals and relevance for human health." *Philosophical Transactions of the Royal Society B: Biological Sciences*, 364(1526), 2079–2096.

Teuten, E. L., J. M. Saquing, D. R. Knappe, M. A. Barlaz, S. Jonsson, A. Björn, S. J. Rowland, R. C. Thompson, T. S. Galloway and R. Yamashita (2009). "Transport and release of chemicals from plastics to the environment and to wildlife." *Philosophical Transactions of the Royal Society B: Biological Sciences*, 364(1526), 2027–2045.

Thompson, R. C., Y. Olsen, R. P. Mitchell, A. Davis, S. J. Rowland, A. W. John, D. McGonigle and A. E. Russell (2004). "Lost at sea: where is all the plastic?" *Science (Washington)*, 304(5672), 838.

Udayangani, S. (2019). "Difference between DNA sequence mutations and epigenetic modifications." Differencebetween.com Retrieved 22 June, 2020, from https://www.differencebetween.com/difference-between-dna-sequence-mutations-and-epigenetic-modifications/.

Urban, R. M., J. J. Jacobs, M. J. Tomlinson, J. Gavrilovic, J. Black and M. Peoc'h (2000). "Dissemination of wear particles to the liver, spleen, and abdominal lymph nodes of patients with hip or knee replacement." *The Journal of Bone and Joint Surgery*, 82(4), 457.

Van Cauwenberghe, L., M. Claessens, M. B. Vandegehuchte and C. R. Janssen (2015). "Microplastics are taken up by mussels (*Mytilus edulis*) and lugworms (*Arenicola marina*) living in natural habitats." *Environmental Pollution*, 199, 10–17.

Van Cauwenberghe, L. and C. R. Janssen (2014). "Microplastics in bivalves cultured for human consumption." *Environmental Pollution*, 193, 65–70.

Van Cauwenberghe, L., A. Vanreusel, J. Mees and C. R. Janssen (2013). "Microplastic pollution in deep-sea sediments." *Environmental Pollution*, 182, 495–499.

Vanhaecke, L., N. Van Hoof, W. Van Brabandt, B. Soenen, A. Heyerick, N. De Kimpe, D. De Keukeleire, W. Verstraete and T. Van de Wiele (2006). "Metabolism of the food-associated carcinogen 2-amino-1-methyl-6-phenylimidazo [4, 5-b] pyridine by human intestinal microbiota." *Journal of Agricultural and Food Chemistry*, 54(9), 3454–3461.

Vianello, A., R. L. Jensen, L. Liu and J. Vollertsen (2019). "Simulating human exposure to indoor airborne microplastics using a Breathing Thermal Manikin." *Scientific Reports*, 9(1), 1–11.

Vidal, J. (2018). "Humans are pooping plastic, and no one's certain how bad that is." Retrieved 18 March, 2020, from https://www.huffpost.com/entry/plastics-and-human-health_n_5c097d7de4b0b6cdaf5d3263.

Von Moos, N., P. Burkhardt-Holm and A. Köhler (2012). "Uptake and effects of microplastics on cells and tissue of the blue mussel *Mytilus edulis* L. after an experimental exposure." *Environmental Science & Technology*, 46(20), 11327–11335.

Wagner, M., M. Engwall and H. Hollert (2014). (Micro) Plastics and the environment, SpringerOpen.

Walker, P. (1973). "The effects of friction and wear in artificial joints." *Oprthop Clin Nirth Am*, 4, 275–293.

Wang, J., X. Liu, Y. Li, T. Powell, X. Wang, G. Wang and P. Zhang (2019). "Microplastics as contaminants in the soil environment: a mini-review." *Science of the Total Environment*, 691, 848–857.

Ward, J. E. and S. E. Shumway (2004). "Separating the grain from the chaff: particle selection in suspension-and deposit-feeding bivalves." *Journal of Experimental Marine Biology and Ecology*, 300(1–2), 83–130.

Warheit, D., G. Hart, T. Hesterberg, J. Collins, W. Dyer, G. Swaen, V. Castranova, A. Soiefer and G. Kennedy (2001). "Potential pulmonary effects of man-made organic fiber (MMOF) dusts." *Critical Reviews in Toxicology*, 31(6), 697–736.

Warheit, D. B., T. R. Webb, K. L. Reed, J. F. Hansen and G. L. Kennedy Jr (2003). "Four-week inhalation toxicity study in rats with nylon respirable fibers: rapid lung clearance." *Toxicology*, 192(2–3), 189–210.

Warhurst, M. (2018). "Hazardous chemicals in plastic packaging: how can we prioritise substances for action?" ChemTrust Retrieved 16 March 2020, from https://chemtrust.org/hazardous-chemicals-plastic-packaging-how-prioritise/.

Waring, R. H., R. M. Harris and S. C. Mitchell (2018). "Plastic contamination of the food chain: a threat to human health?" *Maturitas*, 115, 64–68.

White, J. and A. Turnbull (1994). "Weathering of polymers: mechanisms of degradation and stabilization, testing strategies and modelling." *Journal of Materials Science*, 29(3), 584–613.

Willert, H. G., M. Semlitsch and L. F. Peltier (1996). "Tissue reactions to plastic and metallic wear products of joint endoprostheses." *Clinical Orthopaedics and Related Research®*, 333, 4–14.

Wormuth, M., M. Scheringer, M. Vollenweider and K. Hungerbühler (2006). "What are the sources of exposure to eight frequently used phthalic acid esters in Europeans?" *Risk Analysis*, 26(3), 803–824.

Wright, S. L., J. Ulke, A. Font, K. L. A. Chan and F. J. Kelly (2020). "Atmospheric microplastic deposition in an urban environment and an evaluation of transport." *Environment International*, 136, 105411.

Wu, P., J. Huang, Y. Zheng, Y. Yang, Y. Zhang, F. He, H. Chen, G. Quan, J. Yan and T. Li (2019). "Environmental occurrences, fate, and impacts of microplastics." *Ecotoxicology and Environmental Safety*, 184, 109612.

Xu, B., F. Liu, Z. Cryder, D. Huang, Z. Lu, Y. He, H. Wang, Z. Lu, P. C. Brookes and C. Tang (2020). "Microplastics in the soil environment: occurrence, risks, interactions and fate-a review." *Critical Reviews in Environmental Science and Technology*, 50(21), 2175–2222.

Xu, Y., He, Q., Liu, C., & Huangfu, X. (2019). *Are micro-or nanoplastics leached from drinking water distribution systems?*

Yang, D., H. Shi, L. Li, J. Li, K. Jabeen and P. Kolandhasamy (2015). "Microplastic pollution in table salts from China." *Environmental Science & Technology*, 49(22), 13622–13627.

Yang, Y.-F., C.-Y. Chen, T.-H. Lu and C.-M. Liao (2019). "Toxicity-based toxicokinetic/toxicodynamic assessment for bioaccumulation of polystyrene microplastics in mice." *Journal of Hazardous Materials*, 366, 703–713.

Zettler, E. R., T. J. Mincer and L. A. Amaral-Zettler (2013). "Life in the 'plastisphere': microbial communities on plastic marine debris." *Environmental Science & Technology*, 47(13), 7137–7146.

Zhang, Q., Y. Zhao, F. Du, H. Cai, G. Wang and H. Shi (2020). "Microplastic fallout in different indoor environments." *Environmental Science & Technology*, 54(11), 6530–6539.

Zubris, K. A. V. and B. K. Richards (2005). "Synthetic fibers as an indicator of land application of sludge." *Environmental Pollution*, 138(2), 201–211.

4 Sustainability of Bioplastics

Principle, Recent Advances and Applications

T. Angelin Swetha, Abhispa Bora, Kunyu Zhang, Dong PoSong, Sudhakar Muniyasamy and A. Arun

4.1 INTRODUCTION

Plastic has been a part of human life for a long time. In the 1950s, plastic became an industrial-scale product because of its good durability, less bulk density, great mechanical quality and easy manufacturing (Geyer et al., 2017). These petrochemical plastics typically have a high molecular weight with a recurring monomeric unit of 1000 to 10,000 (Brydson, 1999; Crawford & Martin, 2020; Kuhn et al., 2007). Petrochemical plastics are primarily made by crude oil refiners via a series of steps in which crude oil is separated and fractionated into lighter elements known as segments. These segments consist of various polymeric hydrocarbon chains that differ in their structure and size. Naphtha is one of the fractions that produce monomers like styrene, propylene, and ethylene, which are used in plastics. Catalysts are used to convert these monomers into plastics by polycondensation or polyaddition (Saleh & Gupta, 2016; Armand, 1994). This conversion process generates pollutants and greenhouse gases (carbon dioxide (CO_2)), which results in global warming and environmental pollution (Buis, 2019). Likewise, certain petrochemical plastics are non-biodegradable, causing them to remain at the place of discharge and pollute the environment (Tokiwa et al., 2009). Several types of research in the last two decades have been done to find alternatives to petrochemical polymers.

Bioplastics are one of the polymeric materials that perform a similar function to synthetic plastics. Bioplastics account for approximately 1% of global plastic production, which totals 370 million tonnes (Di Bartolo et al., 2021). However, the yearly growth rate of bioplastic is predicted to be about 30% until 2025. According to the International Union of Pure and Applied Chemistry, bioplastic is defined as a monomer that is produced using plant biomass and can be altered by processing. When bioplastic is exposed to the environment, microbes naturally degrade it (Di Bartolo et al., 2021).

Most of the biobased polymers are biodegradable. Because of their biodegradability, they may be a viable option for achieving long-term growth in the plastics industry. Additionally, biobased plastic provides a viable alternative to petrochemical plastics

DOI: 10.1201/9781003304142-4

by diverting a huge volume of plastic waste management systems and plastics that are hard to recycle (Steven et al., 2020). Biodegradable plastics might serve as a solution for overloaded landfills and also help in recycling non-renewable materials and protecting the environment (Sintim et al., 2019; Boisvert et al., 2016). Biodegradation of bioplastic occurs in the presence of sufficient oxygen, humidity and microorganisms, which can be obtained from manure or land soils. The CO_2 emissions from the decomposition of bioplastic are relatively low, enhancing the need to use biodegradable plastic regularly. Bioplastics are decomposed within 20–45 days (Ecosystem, 2011). The end product of bioplastic decomposition is carbon dioxide and water, but the average life span of petrochemical plastic lasts for about 100–1000 years (Wang et al., 2016; Ghimire et al., 2020; Scaffaro et al., 2016).

Producing bioplastics is the first crucial step in ensuring that consumers are well-informed and that the proper disposal pathway for biopolymers is determined. Biobased plastics are designed to perform the necessary functions while in use. If discarded, it is expected to biodegrade within a specified time frame while leaving no harmful residues (Song et al., 2009; Jain & Tiwari, 2015). Without a doubt, biodegradable polymers are environmentally safe, but they also have significant drawbacks, such as a high production cost and a poor mechanical tendency (Davis & Song, 2006; Yates & Barlow, 2013; Thakur et al., 2017a, 2017b). In recent years, the government has established numerous criteria to minimise the use of non-biodegradable plastic, and several scientists have attempted to produce biodegradable polymers such as polyhydroxyalkanoates (PHA), polylactic acid (PLA) and starch-based bioplastic (Tabasi & Ajji, 2015). Producing bioplastic from biorenewable resources is preferable because their biodegradability makes them the most useful materials (Thakur & Arotiba, 2018; Thakur, 2016). This chapter discusses the role of bioplastic in comparison to conventional plastic, as well as its biodegradability and sustainability. It also explains the principles of bioplastic identification and its applications in various fields.

4.2 ROLE OF BIOPLASTIC AS AN ALTERNATIVE SOURCE FOR CONVENTIONAL PLASTIC POLLUTION

Plastics have increasingly permeated every part of human life due to the fast expansion of the petrochemical sector, and this has resulted in the problem of "white pollution," which cannot be neglected (Shen et al., 2019, 2020). Currently, the most common methods for treating plastic waste are recycling, incineration and landfilling. Landfill and incineration are the most commonly used methods for disposing of plastic waste, which leads to pollution in the environment and releases harmful gases (Ellen MacArthur Foundation, 2017). Multi-layer plastic such as packaging films is hard to dispose of, so they are recycled two to three times. Recycling plastic reduces the length and mass of polymer chains that are needed for further treatment (Sedaghat, 2018).

A bioplastic is a type of plastic that can be easily degraded in the environment by the microorganisms present and releases carbon dioxide (CO_2), water (H_2O) and methane (CH_4). These compounds (CO_2, H_2O and CH_4) are easily absorbed by the ecosystem with a less toxic effect, which takes part in the carbon cycle (Pico & Barcelo, 2019). Because of their biodegradability, bioplastics are commonly used as hard and soft food packaging materials. Garbage bags, disposable ceramics and hard and soft packaging are the most popular goods. Bioplastics' primary markets in the future will be disposable plastic tableware, disposable plastic bags, and disposable agricultural film (Rujnic-sokele & Pilipovic, 2017).

Based on the rate of degradation bioplastic can be categorised into two types: bioplastic with complete degradability and bioplastic with harmful degradability. Completely degradable bioplastics are produced from sources like chitin, cellulose, starch, the biomass of agriculture and microbial fermentation (Iannotti et al., 1990). Starch-based biopolymers such as thermoplastic starch that is synthesized from plants (stored in the form of energy) are easily biodegradable. Biopolymers produced by microbial fermentation using sugars and organic acid as a source of raw material have the main chain of the aliphatic ester group (polyhydroxybutyrate) (Belal & Farid, 2016). It is the coupling of a natural polymer, such as starch, with a synthetic polymer with the intention of degrading the copolymer's structure by the biodegradation of natural elements, including oxo-biodegradable polymers (Pico & Barcelo, 2019; Gere & Czigany, 2020; Thomas et al., 2012). The polymer's molecular chain will be destroyed by the addition of additives, resulting in its biodegradability. So, these polymers are produced by blending natural and synthetic materials in a liquid or molten state by copolymerisation of monomers in a heterogeneous dispersion system. The difference between the two is that the natural component in the former is given as a supplement to the generated conventional plastics and does not make a chemical link with the polymer structure of the plastics; the latter is attached to the chain by a chemical bond as a plastics monomer (Napper & Thompson, 2019; European Bioplastics, 2019). Biodegradation can be categorised into three stages, as shown in Figure 4.1.

Stage 1: Bio-deterioration

It occurs whenever live microorganisms aggregate on the surface of bioplastic to form biofilms (colonisation of microbes), which leads to some changes in the properties of bioplastic.

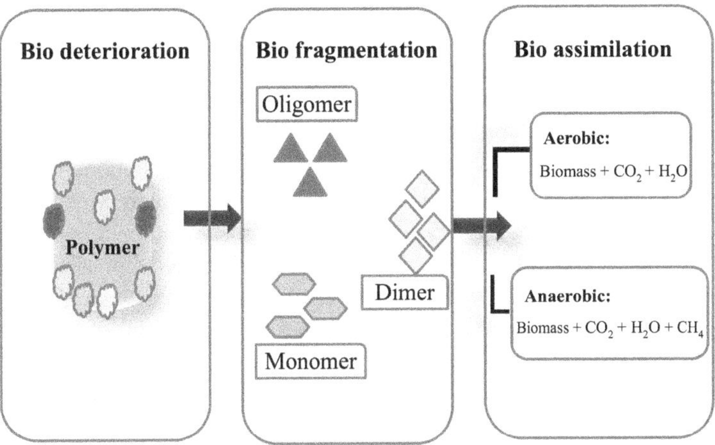

FIGURE 4.1 Schematic representation of biodegradation of polymer.

Stage 2: Bio-fragmentation

In the second stage, after continuous colonisation of microorganisms, these microbes tend to produce an enzyme called depolymerase that converts bioplastic into a monomer and oligomer.

Stage 3: Bio-assimilation

In the last stage, oligomers and monomers of bioplastic are consumed by microbes and release CO_2, H_2O and other substances. The release of these metabolic compounds into the environment indicates that biodegradation of bioplastic is completed (Nazareth et al., 2019). The rate of degradation of bioplastics is determined by their qualities, which govern the use of bioplastics in management until their end-of-life in the environment. The biodegradability of polymers is determined by their chemical structure, complexity and crystallinity. Bioplastic containing functional groups (alcohols, carboxylic acid and carboxylate) and suitable binding sites degrade quicker because these binding groups get attached to enzyme more rapidly than rigid bioplastic. Biopolymers possessing simple side chains degrade more rapidly than biopolymers with complex side chains because the complex side chain requires more coenzymes for their degradation (Narancic & O'Connor, 2019). The rate of biodegradation is influenced by pH and atmospheric temperature because the surface cracking of bioplastic is induced by temperature and pH fluctuations that enhance the degradation process. Incineration, biological treatment – anaerobic digestion and composting, landfill and recycling are the most commonly used methods for disposing of large amounts of plastic waste. Nowadays, industrial composting is the primary function of bioplastic management and end-of-life. During composting, bioplastics are collected in different schemes, segregated and finally sent to industrial composting plants (Narancic & O'Connor, 2019).

4.3 BIODEGRADABLE PLASTIC AND ITS SUSTAINABILITY

The term sustainable was originally used to refer to natural resources, but it is now used to describe a set of processes that provide resources to be consumed and conserved at a set rate. Then, the same term was applied to agriculture, suggesting a shift in their pattern. One of the sustainability economic model principles is that changes in consumer behaviour should be somewhat futuristic (Behm, 2011). At this point, it is clear that assessing people's current plastic consumption without making us needy in the future is necessary. The growth of sustainability is widely accepted as a fundamental objective of global policy. The well-known illustration defines sustainable development as "the development that fulfils current needs without compromising future generations" (WCED: World Commission on Environment and Development) (WCED, 1987). This makes the concept connect environmental preservation, economic progress and social health. However, the implementation of these relationships is a significant issue for a broad range of associates, such as NGOs, the business community, the government and people. In Stockholm, the agenda of the United Nations Conference on the Human Environment defined the term sustainability as a method that fulfils the needs of people and achieves this by protecting natural resources for future generations (Wu et al., 2020).

According to Ayar and Gürbü (2021), economic growth and development were discovered and sustained through interaction in the most remote ecological limits. Roundtable (2019) defined "sustainable consumption" as utilisation that fulfils basic needs and reduces the usage of toxic materials, natural resources, waste emissions and environmental pollutants across the life cycle of the product.

4.4 PRINCIPLES FOR IDENTIFYING THE SUSTAINABILITY OF BIOPLASTICS

The idea of sustainability is based on the belief that society is expected to accept an actual and feasible level that future generations may also use. As a result, achieving sustainable growth necessitates the development of precise efforts that involve all sectors of society and human activities to attain the goal (Wang et al., 2008). The sustainable development of biodegradable plastics is crucial in regulating the relationship between the environment and human needs so that non-renewable resources with serious ecological constraints are not overburdened unnecessarily. Simultaneously, the present ethics of social integrity and essential civil rights are unaffected. It might also imply avoiding environmental and sociological failures to secure the existence of the modern way of life and future generations (Du Plessis, 2007). A set of comprehensive, process-oriented principles as well as economic, social, technical and biophysical aspects are necessary for the adoption of sustainable, biodegradable plastics (Hill and Bowen, 1997). According to their research, social sustainability was intended to improve human life quality, implement training and offer reasonable costs for biodegradable plastics in order to pursue intergenerational rights and ethnic identity in plastics production. Sustainable bioplastic projects are economically advantageous to the targeted market. It is also necessary to increase the number of environmentally conscious employees, as well as to reinforce and ensure that biodegradable plastic products can meet future needs (Tate & Bals, 2018). In terms of biophysical characteristics, the sustainability of biodegradable plastics includes the collection of renewable resources over their slow rate of regeneration, the reduction of intake of four basic materials (resources, energy, land and water) and the development of resource reuse or recycling. It gives more importance to renewable resources and minimising pollution (land, water and air). Apart from this, importance is also given to conserving ecological biodiversity and preventing the destruction of sensitive land (Álvarez-Chávez et al., 2012). The bioplastic industry and biodegradable plastic goods both require contributions from the earth's resources. These resources are known as development materials and include energy in encapsulated form. The resources should be managed properly, which is the key to recycling, reusing and reducing waste by bioplastics companies, and it helps to build a better environment. Biodegradable polymers can be used for food packaging, packaging ware, shopping bags and agriculture. They are available in several forms and have different properties, so they can be utilised for a wide range of applications such as packaging and agro-agriculture (Kumar, 2011). The use of the above characteristics influences the sustainability of biodegradable polymers. Nonetheless, when it is compared to conventional fossil-based plastic, the form of biobased plastic products is always equivalent (van den Oever et al., 2017). In 2015, the manufacturing capabilities of biobased and biodegradable polymers accounted for about one per cent of total global plastics output. Biodegradable plastics (biobased polyethyelene terephthalate (PET), polybutylene succinate (PBS), and polylactic acid (PLA)) are known to be relatively profitable in the future, whereas other polymers will be incorporated. It was expected both biobased and biodegradable plastics would account for 2.5 per cent of worldwide plastics production in 2020 (Moshood et al., 2022).

Several manufacturers made biodegradable plastics and biobased plastics widely available, although they were more expensive in terms of weight than fossil-based plastics. However, once the product has been used, specific physical qualities can result in cost savings. Many biodegradable plastic items are now affordable. Furthermore, oil price variations affect the cost of fossil-based polymers. The price of biodegradable polymers is primarily determined by the durability of biomass. Biodegradable plastic prices are predicted to drop as the economic scale of production improves. The discovery of sustainability principles for biodegradable plastics is critical for modern management as it focuses on long-term growth (Boukherroub et al., 2015). Economic performance, environmental protection and social responsibility are the three factors that the triple bottom line concept commonly differentiates for biodegradable plastics sustainability evaluations. As a result of previous research and professional perspectives, the proposed standard includes 26 principles covering the three dimensions of sustainability. There are nine social, economic and environmental dimensions, each with eight elements (Moshood et al., 2022).

4.4.1 Sustainability Based on Environmental Aspect

Biodegradable polymers' environmental awareness is tested through environmental assessment (Chardine-Baumann & Botta-Genoulaz, 2014). However, the social and environmental consequences are not universally accepted because they vary depending on the industry and location of the activities (Boukherroub et al., 2015). International standards, such as the International Organization for Standardization (ISO26000, ISO14001), Global reporting initiative, guidelines of the Organization for Economic Co-operation and Development, Social accounting (SA) 8000 and scientific studies, have suggested a variety of more or less general standards. These criteria do not include all aspects of sustainability.

Energy consumption, carbon emissions, waste management systems and environmental management systems are the five environmental aspects of biodegradable plastics management. One of the most widely cited environmental variables is energy use. The use of natural gas, electricity and energy sources in the production and processing operations is referred to as energy consumption in the plastics biodegradable industry. Although climate change is recognised as a severe concern in modern communities, the minimisation of carbon emissions has become a major environmental concern in the last decade. Biodegradable plastic waste disposal often includes the removal of items from garbage dumps, the recovery of products to raw materials and replacing the product for reuse in the future (Xu et al., 2016). The environmental management system facilitates firms to make decisions that increase resource efficiency, assure long-term success, and minimise negative effects on the environment (Chardine-Baumann & Botta-Genoulaz, 2014).

The environmental protection framework for the biodegradable plastics business includes issues such as environmental approval, assessments, budgeting and compliance (Xu et al., 2016). Establishing waste management systems for biobased goods is important for both the environment and the use of waste as a resource in recycling and reuse. Environmental assessment is essential for developing existing goods and services in diplomatic ways (White & Noble, 2013). The focus of ecological management is to minimise the negative environmental impact of corporate activities. According to recent sustainable environmental research, uncontrolled industrialisation is one of the most serious risks to the natural world and ecological processes (Kopnina, 2017). The preservation and renewal of the current biosphere and the biosphere for future generations is defined as environmental sustainability (Kasayanond et al., 2019). Climate change, global warming, deforestation and biodiversity loss are recent environmental threats. To improve their effectiveness, companies are continuously pressuring businesses to embrace environmentally friendly tactics (Kabir et al., 2020).

4.4.2 Sustainability Based on Economical Aspect

Economic outcomes will have a substantial impact on biodegradable plastic goods, showing the quality of market operations. Quality, reliability, flexibility, responsiveness and financial efficiency have always been incorporated into measurement analysis (Boukherroub et al., 2015). Reliability in bioplastic film refers to the supply of necessary products at the right place at the right time, in good condition, using suitable packaging and with the proper user certification (Spierling et al., 2018). From the perspective of biodegradable plastic products, reactivity refers to the characteristics of evidence flows, monetary flows and material flows from the source to the user. Flexibility refers to the capacity of the biodegradable plastics distribution network to respond to market developments toward sustainability and gain or gain market share (Xu et al., 2016). The film of biodegradable polymers may adjust and evolve constantly with changes in time and the environment due to its adaptability. Design costs, supply chain costs, manufacturing costs, procurement costs, investment return costs and new features are included in the financial recital (Chardine-Baumann & Botta-Genoulaz, 2014). One of the most important aspects of quality is the customer-seller association. Poor output has an impact on company's financial results and reputation. In the biodegradable plastics sector, quality is defined by customer satisfaction and service quality (Xu et al., 2016). During the process, sustainable materials are also employed to save waste, energy and manufacturing costs (Dilkes-Hoffman, 2020).

To reduce hazardous trash at disposal sites, inks, metallic objects, waste oil, glass, chemicals, paper, wood and plastics must be recycled, which directly reduces waste disposal costs. The use of biodegradable polymers will reduce plastic manufacturing costs by utilising minimum energy pumps, vehicles and lighting fixtures, which can directly minimise energy expenses in the process of plastic production (Akadiri et al., 2019). According to Montini (2020), the theory of sustainability does not imply economic decline. However, the advanced and efficient service markets, qualitative development, cooperation, attractiveness and ecological constraints are required for sustainability. A smaller economy is healthier than a larger one. Economic viability in the business sector refers to a rise in the value of short- or long-term owners and the management of a stable financial basis for long-term organisational survival (Moshood et al., 2022).

4.4.3 Sustainability Based on Social Aspect

Many researchers have found that good social capital management is essential for societal sustainability (Döhler et al., 2022; Akadiri et al., 2019). Social capital may be viewed as a long-term commodity of an organisation that is not destroyed but improved and retained for a long time (Baland et al., 2018). Social capital development within a company entails management creating a desired working environment in which employees can improve their social and other abilities. This may be accomplished by altering things such as investing in human resources, improving worker capability, building a collaborative working community, networking opportunities, access to relevant information, and obtaining new knowledge more productively and efficiently (Magni et al., 2020). Social capital also assists an organisation in raising the quality of education on a bigger scale, addressing poverty and reducing growth and other significant public issues on a sustainable basis (Akadiri et al., 2019). Organisations strive to improve living circumstances by providing full-time and active work. The amount of biodegradable plastic items and employee transfer rate are included as social measures (Moshood et al., 2022).

In health and safety, the impact of a technique or substance on the safety and health of employees and consumers is measured (Chardine-Baumann & Botta-Genoulaz, 2014). The goal is to develop and maintain a high level of physical, intellectual, and social advantage for providers and customers in order to minimise work grievances and ensure that customers do not gain from limited goods. Relaxation periods, working hours, human resource growth, seasonal festivities, maternity leave, wages, benefits, leave of absence and management procedures are all factors that affect employee retention (Xu et al., 2016). The corporation's social responsibility to the community includes its stakeholders, as well as improvements in the fields of culture, education, health, technology and social investment. Any issues that affect each consumer separately are referred to as customer complaints. Their main focus is on consumer protection and safety, data security, consumer behaviour and marketing information, and other critical resources. Environmentally friendly goods should be actively developed (Döhler et al., 2022). Biodegradable plastic materials have been created to reduce the amount of litter made by recycled plastics, either as biodegradable materials or as manure replacements. Biodegradable polymers are widely utilised in disposable products and in agriculture where biodegradability is necessary. In fact, people are becoming more aware of the social impact of "Biodegradable Plastics" packaging. Several consumers want to switch to a less polluting or renewable energy-based packaging material. This is the major reason for improving the management of biodegradable plastics in order to decrease plastic waste in the environment (Wang et al., 2020).

4.5 APPLICATION OF BIOPLASTICS ON VARIOUS FIELD

Bioplastics are commonly used in food packaging, surgical equipment and pharmaceuticals. Bioplastics could be made from polyhydroxyalkanoates (PHA), polylactic acid (PLA) and their nanocomposites. In the production of bioplastics, starch is the most commonly used component (Bakar & Othman, 2019). It is made with corn or potato starch because it has biodegradable properties and can be produced in high volume at a low cost. It is one of the most promising bioplastic manufacturing options. PLA is used as packaging materials-films, containers, cups and bottles (Bakar & Othman, 2019). It is also used in the textile industry to produce furniture fabrics, diapers and shirts. Using kenaf fibre, PLA is also used to enhance the casing of smartphones (Ramesh Kumar et al., 2020). In the industrial sector, PHA has a wide range of uses. The materials used in surgical devices, drug carriers and even granules of surface proteins were all studied. To accomplish this, a variety of polyhydoxyalkonate (PHA) structures have been synthesised, such as polyhydoxybutyrate (PHB), poly(3-hydroxyoctanoate) (P3HO), poly (4-hydroxybutyrate) (P4HB) and poly(3-hydroxybutyrate-co-3-hydroxyvalerate) (PHBV). Synthetic oesophagus, repair devices, wound dressings, repair patches, sutures and tend on repair devices were some of the items currently studied. The nutritional and therapeutic effects of PHA oligomers have also been identified (Bakar & Othman, 2019). Polyhydoxyalkonates (PHA) are also used as a medicine carrier because of their biocompatibility and biodegradability, and they can be easily broken down via surface erosion (Bakar & Othman, 2019).

4.5.1 MEDICAL

Medical devices, gloves and blood containers are all made of bioplastics. They are also used in implants due to their biodegradability. Other applications include wound dressings, cardiovascular, dental implants, burn and medicine delivery systems. Biodegradable plastic material innovations in biomedical applications led to the development of improved tissue engineering, drug delivery systems, and medical devices including scaffolds and implants (Narancic et al., 2020). Polymers are employed in a wide range of biological and pharmaceutical applications. As a key green bioplastic, cellulose can help in these areas. Because of its nontoxicity, absence of mutagenicity and biocompatibility in pharmaceuticals, cellulose is being intensively explored in the fields of implants, neurological and tissue engineering (Vieira et al., 2015). Fibrils are the primary structural units present with cell widths of 10 nm and are arranged macroscopically to form cellulose fibres. Cellulosic membranes made from bacterial cellulose are being used in tissue repair. The diameter of the holes in these membranes varies from 60 to 300 m. Bacterial nano-networks and altered cellulose matrixes were also evaluated (Coppola et al., 2021). Nano cellulose and its composites are extensively used in green plastic study based on the fabrication of implant materials, whether in biomedicine, dental or orthopaedic surgery. 3D printing and magnetically active nanocellulose-based materials are being produced more in recent surveys. Wound dressing with nano-cellulosic membranes is one of the best methods because it has benefits such as accelerated re-epithelialisation, reduced wound pain, reduced infection and extrusion retention. PHAs are also biocompatible, which makes them useful in a wide range of medical applications such as post-surgical ulcer care, anticancer therapy, bone tissue engineering and wound therapeutic dressing (Narancic et al., 2020).

4.5.2 Food Industry

In recent years, the food industry has focused heavily on plastic packaging issues, which seems to be a completely separate industry in itself (Coppola et al., 2021). To satisfy the demands and needs of the food production business, these industries are continually expanding. It mainly focuses on implementing creative polymer-based packaging, which is essential for the food industry and serves as a stable, sustainable practice and quality control, resulting in environmentally friendly and more resource-efficient production facilities, storage space systems, and transportation facilities, including other factors (Ramadhan & Handayani, 2020). The desire for food packaging can be fulfilled by degradable and compostable biomaterials because they are inexpensive, have a minimal environmental impact, are fully customisable, and have a minimal carbon impact, in relation to complying with the requirements for increased storage qualities (Jabeen et al., 2015). Although efficient packaging in the food industry is restricted in comparison with other industries and has to be developed, today's most essential food distribution companies are alert and are ready to switch to biopolymers as much as possible (Pei et al., 2011). When developing packaging material, it was essential to construct many technologies such as multi-layer films, customised ambient packaging, smart and dynamic packaging because of the importance of understanding that various food products demand distinct packaging aspects (Muthusamy & Pramasivam, 2019).

4.5.3 Agriculture

In the field of agriculture, PHA-based bioplastics are used in the form of mulch films, nets, and plant grow bags. Bioplastic-based nets are becoming more popular as an alternative to high-density polyethene (HDPE) and have been mainly used to improve the quality and yield of crops while safeguarding them from insects, birds and the environment (Coppola et al., 2021). Plant grow bags are made primarily from low-density polyethene, which is widely available. Polyhydroxyalkanoate (PHA) plant growth sacks are eco-friendly, biodegradable and non-toxic to water bodies. In order to replace mulch film with fossil-based polymers, bioplastics are utilised to enhance the superior quality of soil and control weeds, water content and pollution reduction (Urbina et al., 2020). Agriculture and horticulture both utilise bioplastics frequently (do Val Siqueira et al., 2021). Using a biodegradable planter for flowers or plants saves time and money because they do not need to be disposed of. It may be sown next to the flower and will decay eventually. Other applications include fragrance traps, flower bulb packing, connecting technologies, manure strips and mulching film.

4.6 CONCLUSION

Bioplastics, which are defined as biodegradable plastics made from renewable resources, can be used as a substitute for conventional plastics. Biodegradable bioplastics offer a variety of waste disposal options, reducing the quantity of plastic garbage that ultimately ends up in the ecosystem, while biobased bioplastics have less greenhouse gas emissions at the resource refinement process. Food packaging with biodegradable plastics has been found to be very efficient. Due to the advancement of biodegradable plastics, various issues were being solved, and the environment was protected for a while. The development of sustainable, biodegradable plastics necessitates analysis of the economic, social and environmental framework with its techniques. The goal of social sustainability was to improve the quality of human life through the implementation of training programmes and the provision of efficient and affordable biodegradable plastics in order to achieve intergenerational equity and cultural diversity in the production of plastics. The financial aspect of sustainability is making bioplastics efforts accessible to the target market. It also requires increasing employment generation, improving productivity, retaining ecologically aware workers and providing biodegradable polymer products to fulfil rising needs. Collecting renewable resources, conserving the energy of four generic resource management (power, land, water and materials), enhancing energy utilisation or recycling, optimising renewable resources over non-renewable resources, reducing air, land, and water pollution, and preserving and renovating are all biological qualities of sustainable, biodegradable plastics. The effectiveness, durability and reliability of biodegradable polymers are considered to be part of environmental sustainability. Given that bioplastics production is increasing and that these materials will coexist with conventional plastics for many years, it is critical to identify the best end-of-life options for each of the most popular bioplastics, regardless of whether they degrade through biodegradation, in order to encourage genuine shift in the entire plastics significance sequence towards sustainability.

ACKNOWLEDGEMENTS

Authors acknowledge the financial support provided by the Technology Mission Division (Energy, Water and All Others), Department of Science and Technology, Ministry of Science and Technology, Government of India (Reference Number: DST/TMD/IC-MAP/2K20/02 Project Title: DST-IIT Hyderabad Integrated Clean Energy Material Acceleration Platform on Bioenergy and Hydrogen); Department of Science and Technology-Science and Engineering Research Board (DST-SERB-No. SB/YS/LS-47/2013), India; Department of Science and Technology-Science and Engineering Research Board (DST-SERB-No. SB/YS/LS-47/2013), India; Department of Science and Technology-Promotion of University Research and Scientific Excellence (DST-PURSE) [DST letter No. SR/PURSE phase 2/38(G); dated: February 21, 2017], India; RUSA – Phase 2.0 grant [Letter No. F. 24-51/ 2014-U, Policy (TN Multi-Gen), Department of Education, Government of India; dated: October 09, 2018]; Scheme for Promotion of Academic and Research Collaboration (SPARC) (No. SPARC/2018-2019/P485/SL; dated: March 15, 2019) and University Science Instrumentation Centre (USIC), Alagappa University, Karaikudi, Tamil Nadu, India.

REFERENCES

Akadiri, S., Alola, A. A., Akadiri, A. C., & Alola, U. V. (2019). Renewable energy consumption in EU-28 countries: policy toward pollution mitigation and economic sustainability. *Energy Policy*, *132*, 803–810.

Álvarez-Chávez, C. R., Edwards, S., Moure-Eraso, R., & Geiser, K. (2012). Sustainability of bio-based plastics: general comparative analysis and recommendations for improvement. *Journal of Cleaner Production*, *23*(1), 47–56.

Armand, M. (1994). The history of polymer electrolytes. *Solid State Ionics*, *69*(3–4), 309–319.

Ayar, I., & Gürbüz, A. (2021). Sustainable consumption intentions of consumers in Turkey: a research within the theory of planned behavior. *SAGE Open*, *11*(3), 21582440211047563.

Bakar, N. F. A., & Othman, S. A. (2019). Corn bio-plastics for packaging application. *Journal of Design for Sustainable and Environment*, *1*(1), 1–3.

Baland, J. M., Bardhan, P., & Bowles, S. (Eds.). (2018). *Inequality, Cooperation, and Environmental Sustainability*. Princeton University Press.

Behm, C. L. (2011). *Student Perceptions and Definitions of Sustainability*. Master's Thesis. University of Illinois at Urbana, IL.

Belal, E. B., & Farid, M. A. (2016). Production of poly-β-hydroxybutyric acid (PHB) by *Bacillus cereus*. *International Journal of Current Microbiology and Applied Sciences*, *5*, 442–460.

Boisvert, A., Jones, S., Issop, L., Erythropel, H. C., Papadopoulos, V., & Culty, M. (2016). In vitro functional screening as a means to identify new plasticizers devoid of reproductive toxicity. *Environmental Research*, *150*, 496–512.

Boukherroub, T., Ruiz, A., Guinet, A., & Fondrevelle, J. (2015). An integrated approach for sustainable supply chain planning. *Computers & Operations Research*, *54*, 180–194.

Brydson, J. A. (1999). *Plastics Materials*. Elsevier. Oxford; Boston : Butter worth-Mannheim.

Buis, A. (2019). The atmosphere: getting a handle on carbon dioxide. *NASA.Global Climate Change*. Retrieved May, 1, 2020.

Chardine-Baumann, E., & Botta-Genoulaz, V. (2014). A framework for sustainable performance assessment of supply chain management practices. *Computers & Industrial Engineering*, *76*, 138–147.

Coppola, G., Gaudio, M. T., Lopresto, C. G., Calabro, V., Curcio, S., & Chakraborty, S. (2021). Bioplastic from renewable biomass: a facile solution for a greener environment. *Earth Systems and Environment*, *5*(2), 231–251.

Crawford, R. J., & Martin, P. (2020). *Plastics Engineering*. Butterworth-Heinemann.

Davis, G., & Song, J. H. (2006). Biodegradable packaging based on raw materials from crops and their impact on waste management. *Industrial Crops and Products*, *23*(2), 147–161.

Di Bartolo, A., Infurna, G., & Dintcheva, N. T. (2021). A review of bioplastics and their adoption in the circular economy. *Polymers*, *13*(8), 1229.

Dilkes-Hoffman, L. S. (2020). *Exploring the Role of Biodegradable Plastics*. Doctoral Thesis, The University of Queensland.

do Val Siqueira, L., Arias, C. I. L. F., Maniglia, B. C., & Tadini, C. C. (2021). Starch-based biodegradable plastics: methods of production, challenges and future perspectives. *Current Opinion in Food Science*, *38*, 122–130.

Döhler, N., Wellenreuther, C., & Wolf, A. (2022). Market dynamics of biodegradable bio-based plastics: projections and linkages to European policies. *EFB Bioeconomy Journal*, *2*, 100028.

Du Plessis, C. (2007). A strategic framework for sustainable construction in developing countries. *Construction Management and Economics*, *25*(1), 67–76.

Ecosystem, G. (2011). A study on plastic management in peninsular Malaysia. National Solid Waste Management Department Ministry of Housing and Local Government Malaysia, 1–282.

Gere, D., & Czigany, T. (2020). Future trends of plastic bottle recycling: compatibilization of PET and PLA. *Polymer Testing*, *81*, 106160.

Geyer, R., Jambeck, J. R., & Law, K. L. (2017). Production, use, and fate of all plastics ever made. *Science Advances*, *3*(7), e1700782.

Ghimire, S., Flury, M., Scheenstra, E. J., & Miles, C. A. (2020). Sampling and degradation of biodegradable plastic and paper mulches in field after tillage incorporation. *Science of the Total Environment*, *703*, 135577.

Hill, R. C., & Bowen, P. A. (1997). Sustainable construction: principles and a framework for attainment. *Construction Management & Economics*, *15*(3), 223–239.

Iannotti, G., Fair, N., Tempesta, M., Neibling, H., Hsieh, F. H., & Mueller, R. (1990). *Studies on the Environmental Degradation of Starch-Based Plastics* (pp. 425–439). CRC Press.

Jabeen, N., Majid, I., & Nayik, G. A. (2015). Bioplastics and food packaging: a review. *Cogent Food & Agriculture*, *1*(1), 1117749.

Jain, R., & Tiwari, A. (2015). Biosynthesis of planet friendly bioplastics using renewable carbon source. *Journal of Environmental Health Science and Engineering*, *13*(1), 1–5.

Kabir, E., Kaur, R., Lee, J., Kim, K. H., & Kwon, E. E. (2020). Prospects of biopolymer technology as an alternative option for non-degradable plastics and sustainable management of plastic wastes. *Journal of Cleaner Production*, *258*, 120536.

Kasayanond, A., Umam, R., & Jermsittiparsert, K. (2019). Environmental sustainability and its growth in Malaysia by elaborating the green economy and environmental efficiency. *International Journal of Energy Economics and Policy*, *9*(5), 465–473.

Kopnina, H. (2017). Working with human nature to achieve sustainability: exploring constraints and opportunities. *Journal of Cleaner Production*, *148*, 751–759.

Kuhn, P., Sémeril, D., Matt, D., Chetcuti, M. J., & Lutz, P. (2007). Structure-reactivity relationships in SHOP-type complexes: tunable catalysts for the oligomerisation and polymerisation of ethylene. *Dalton Transactions*, (5), 515–528.

Kumar, S. (2011). Composting of municipal solid waste. *Critical Reviews in Biotechnology*, *31*(2), 112–136.

MacArthur, E. (2017). Beyond plastic waste. *Science*, *358*(6365), 843–843.

Magni, S., Bonasoro, F., Della Torre, C., Parenti, C. C., Maggioni, D., & Binelli, A. (2020). Plastics and biodegradable plastics: ecotoxicity comparison between polyvinylchloride and Mater-Bi(r) micro-debris in a freshwater biological model. *Science of the Total Environment*, *720*, 137602.

Montini, M. (2020). Designing law for sustainability. In V. Mauerhofer et al. (eds.), *Sustainability and Law* (pp. 33–48). Springer.

Moshood, T. D., Nawanir, G., Mahmud, F., Mohamad, F., Ahmad, M. H., & AbdulGhani, A. (2022). Biodegradable plastic applications towards sustainability: a recent innovations in the green product. *Cleaner Engineering and Technology*, 100404.

Muthusamy, M. S., & Pramasivam, S. (2019). Bioplastics-an eco-friendly alternative to petrochemical plastics. *Current World Environment*, *14*(1), 49.

Napper, I. E., & Thompson, R. C. (2019). Environmental deterioration of biodegradable, oxo-biodegradable, compostable, and conventional plastic carrier bags in the sea, soil, and open-air over a 3-year period. *Environmental Science & Technology*, *53*(9), 4775–4783.

Narancic, T., & O'Connor, K. E. (2019). Plastic waste as a global challenge: are biodegradable plastics the answer to the plastic waste problem? *Microbiology*, *165*(2), 129–137.

Narancic, T., Cerrone, F., Beagan, N., & O'Connor, K. E. (2020). Recent advances in bioplastics: application and biodegradation. *Polymers*, *12*(4), 920.

Nazareth, M., Marques, M. R., Leite, M. C., & Castro, Í. B. (2019). Commercial plastics claiming biodegradable status: is this also accurate for marine environments. *Journal of Hazardous Materials*, *366*, 714–722.

Pei, L., Schmidt, M., & Wei, W. (2011). Conversion of biomass into bioplastics and their potential environmental impacts. *Biotechnology of Biopolymers*, *3*, 57–74.

Pico, Y., & Barceló, D. (2019). Analysis and prevention of microplastics pollution in water: current perspectives and future directions. *ACS Omega*, *4*(4), 6709–6719.

Ramadhan, M. O., & Handayani, M. N. (2020, December). The potential of food waste as bioplastic material to promote environmental sustainability: a review. In *IOP Conference Series: Materials Science and Engineering* (Vol. 980, No. 1, p. 012082). IOP Publishing.

Ramesh Kumar, S., Shaiju, P., & O'Connor, K. E. (2020). Bio-based and biodegradable polymers-State-of-the-art, challenges and emerging trends. *Current Opinion in Green and Sustainable Chemistry*, *21*, 75–81.

Roundtable, O. (1994). Part 1 – The imperative of sustainable production and consumption'. Retrieved February, 17, 2019.

Rujnić-Sokele, M., & Pilipović, A. (2017). Challenges and opportunities of biodegradable plastics: a mini review. *Waste Management & Research*, *35*(2), 132–140.

Saleh, T. A., & Gupta, V. K. (2016). *Nanomaterial and Polymer Membranes: Synthesis, Characterization, and Applications*. Elsevier.

Scaffaro, R., Botta, L., Maio, A., Mistretta, M. C., & La Mantia, F. P. (2016). Effect of graphenenanoplatelets on the physical and antimicrobial properties of biopolymer-based nanocomposites. *Materials*, *9*(5), 351.

Sedaghat, S. (2018). Preparation of shitosan/montmorillonite (MMt) nanocomposite as a drug delivery carrier of podophyllotoxin. *Asian Journal of Applied Sciences*, *6*(2).

Shen, M., Song, B., Zeng, G., Zhang, Y., Huang, W., Wen, X., & Tang, W. (2020). Are biodegradable plastics a promising solution to solve the global plastic pollution? *Environmental Pollution*, *263*, 114469.

Shen, M., Zeng, G., Zhang, Y., Wen, X., Song, B., & Tang, W. (2019). Can biotechnology strategies effectively manage environmental (micro) plastics? *Science of the Total Environment*, *697*, 134200.

Sintim, H. Y., Bandopadhyay, S., English, M. E., Bary, A. I., DeBruyn, J. M., Schaeffer, S. M., & Flury, M. (2019). Impacts of biodegradable plastic mulches on soil health. *Agriculture, Ecosystems & Environment*, *273*, 36–49.

Song, J. H., Murphy, R. J., Narayan, R., & Davies, G. B. H. (2009). Biodegradable and compostable alternatives to conventional plastics. *Philosophical Transactions of the Royal Society B: Biological Sciences*, *364*(1526), 2127–2139.

Spierling, S., Knüpffer, E., Behnsen, H., Mudersbach, M., Krieg, H., Springer, S., & Endres, H. J. (2018). Bio-based plastics – a review of environmental, social and economic impact assessments. *Journal of Cleaner Production*, *185*, 476–491.

Steven, S., Octiano, I., & Mardiyati, Y. (2020, September). Cladophora algae cellulose and starch based bio-composite as an alternative for environmentally friendly packaging material. In *AIP Conference Proceedings* (Vol. 2262, No. 1, p. 040006). AIP Publishing LLC.

Tabasi, R. Y., & Ajji, A. (2015). Selective degradation of biodegradable blends in simulated laboratory composting. *Polymer Degradation and Stability*, *120*, 435–442.

Tate, W. L., & Bals, L. (2018). Achieving shared triple bottom line (TBL) value creation: toward a social resource-based view (SRBV) of the firm. *Journal of Business Ethics*, *152*(3), 803–826.

Thakur, S. (2016). *Sodium Alginate, Xanthan Gum Biopolymer Composites: Synthesis, Characterisation and Application in Organic Dye Removal from Water*. University of Johannesburg (South Africa).

Thakur, S., & Arotiba, O. A. (2018). Synthesis, swelling and adsorption studies of a pH-responsive sodium alginate-poly (acrylic acid) superabsorbent hydrogel. *Polymer Bulletin*, *75*(10), 4587–4606.

Thakur, S., Govender, P. P., Mamo, M. A., Tamulevicius, S., & Thakur, V. K. (2017a). Recent progress in gelatin hydrogel nanocomposites for water purification and beyond. *Vacuum, 146*, 396–408.

Thakur, S., Govender, P. P., Mamo, M. A., Tamulevicius, S., Mishra, Y. K., & Thakur, V. K. (2017b). Progress in lignin hydrogels and nanocomposites for water purification: future perspectives. *Vacuum, 146*, 342–355.

Thomas, N. L., Clarke, J., McLauchlin, A. R., & Patrick, S. G. (2012, August). Oxodegradable plastics: degradation, environmental impact and recycling. In *Proceedings of the Institution of Civil Engineers-Waste and Resource Management* (Vol. 165, No. 3, pp. 133–140). ICE Publishing.

Tokiwa, Y., Calabia, B. P., Ugwu, C. U., & Aiba, S. (2009). Biodegradability of plastics. *International Journal of Molecular Sciences, 10*(9), 3722–3742.

Urbina, L., Eceiza, A., Gabilondo, N., Corcuera, M. Á., & Retegi, A. (2020). Tailoring the in situ conformation of bacterial cellulose-graphene oxide spherical nanocarriers. *International Journal of Biological Macromolecules, 163*, 1249–1260.

Van den Oever, M., Molenveld, K., van der Zee, M., & Bos, H. (2017). *Bio-Based and Biodegradable Plastics: Facts and Figures: Focus on Food Packaging in the Netherlands* (No. 1722). Wageningen Food & Biobased Research.

Vieira, S., Castelli, S., Falconi, M., Takarada, J., Fiorillo, G., Buzzetti, F., & Desideri, A. (2015). Role of 13-(di) phenylalkylberberine derivatives in the modulation of the activity of human topoisomerase IB. *International Journal of Biological Macromolecules, 77*, 68–75.

Wang, B., Li, Y., Wu, N., & Lan, C. Q. (2008). CO2 bio-mitigation using microalgae. *Applied Microbiology and Biotechnology, 79*(5), 707–718.

Wang, E., Cao, H., Zhou, Z., & Wang, X. (2020). Biodegradable plastics from carbon dioxide: opportunities and challenges. *Scientia Sinica Chimica, 50*(7), 847–856.

Wang, J., Tan, Z., Peng, J., Qiu, Q., & Li, M. (2016). The behaviors of microplastics in the marine environment. *Marine Environmental Research, 113*, 7–17.

WCED, S. W. S. (1987). World commission on environment and development. *Our Common Future, 17*(1), 1–91.

White, L., & Noble, B. F. (2013). Strategic environmental assessment for sustainability: a review of a decade of academic research. *Environmental Impact Assessment Review, 42*, 60–66.

Wu, F., Misra, M., & Mohanty, A. K. (2020). Tailoring the toughness of sustainable polymer blends from biodegradable plastics via morphology transition observed by atomic force microscopy. *Polymer Degradation and Stability, 173*, 109066.

Xu, J., Jiang, X., & Wu, Z. (2016). A sustainable performance assessment framework for plastic film supply chain management from a Chinese perspective. *Sustainability, 8*(10), 1042.

Yates, M. R., & Barlow, C. Y. (2013). Life cycle assessments of biodegradable, commercial biopolymers – a critical review. *Resources, Conservation and Recycling, 78*, 54–66.

5 Microbial Production of Biopolymers
Recent Advancements and Their Applications

K. Mohanrasu, R. Guru Raj Rao, G. Siva Prakash, Kunyu Zhang, Dong PoSong, Sudhakar Muniyasamy, Thavamani Palanisami, Panneerselvan Logeshwaran and A. Arun

5.1 INTRODUCTION

Polymers derived from petrochemical industries are used to manufacture most plastic products. Owing to their resilience and malleability, these petrochemical-based polymers are employed in various industrial applications, particularly in packaging, medical, transportation, and agriculture (Lambert, 2015). Modern petrochemical plastics have many environmental problems, particularly in waste management and the toxic compounds released into the environment in due process. The plastic material has endured for thousands of years and is difficult to break down into more specific, less toxic components by the environmental elements. In addition to the rising use of plastic, the reliance on crude oil has also increased. Toxic compounds and microplastics are leftover during recycling, which also poses a more significant threat to the environment. Microplastics are easily entrapped into the human and animal food chains, which eventually end up in their biological system. Plastics are susceptible to solar radiation, which produces greenhouse gases (GHGs) such as methane and ethylene. According to Lambert (2015), plastic is widely used in various industries such as 40.1% of packaging, 20.4% of construction, 5.6%–7.0% of automotive electrical and electronic equipment, and 26.9% of agricultural equipment.

Nowadays, society strives for sustainable development, leading to the search for plastic substitutes. The production of biodegradable materials from biological resources is known as "biopolymers" or "biobased polymers", and it is a potential solution for waste management (Lambert, 2015). Biopolymers with at least one monomer derived from renewable sources will mitigate environmental issues (J. W. Lee et al., 2011; Vroman & Tighzert, 2009). Polyhydroxyalkanoates, PLA, cellulose, xanthan, dextran, pullulan, starch and cellulose are examples of biopolymers (Lambert, 2015). Biopolymers are not a new notion; bio-cellulose was created in the 1850s.

DOI: 10.1201/9781003304142-5

In the 20th century, Henry Ford Motor Company experimented with several biopolymers (soy proteins) as an alternative for automotive parts production (Grujić et al., 2017). The oil crisis in the 1970s sparked the importance of biopolymers in the USA. In the 1980s, biodegradable films, sheets, and moulding materials were accessible. Due to the increased plastic pollution and climate change, biopolymers are considered an important alternative to synthetic polymers. Biopolymers have properties similar to petroleum-based polymers, and there are many different forms of biopolymers on the market, including PLA, polyethylene (PE), starch, chitosan, pectin, collagen, polytrimethylene terephthalate, polyp-phenylene succinate (PBS), gelatin, caseins, zein, natural waxes, and microbial-based polymers like polyhydroxyalkanoates. Various biomaterials, including those mentioned above, have been packaging a wide range of food products since long ago. As a result of their mechanical and chemical properties, these biopolymers are gaining traction in food packaging. Global bioplastic manufacturing capacity will rise substantially from around 2.41 million tons in 2021 to 7.59 million tons in 2026 (European Bioplastics, 2021). The high demand for biopolymers is utilised in various applications like packaging services, including both rigid (cosmetics packaging of creams and lipsticks as well as beverage bottles) and flexible (single-use packaging plastics), consumer goods, fibres, agriculture and horticulture, automotive and transport, coating and adhesives, building and construction, consumer electronics and others. Polyhydroxyalkanoates (PHAs) and polylactide (PLA) are the most extensively produced biodegradable polymers for various industrial purposes.

5.2 INTRODUCTION: HISTORY IN THE DEVELOPMENT OF POLY HYDROXYL ALKANOATES (PHAS)

Petrochemical-based polymers are one of the most significant discoveries in the history of humanity and have been used in numerous industries like packaging, transport, automobile parts, agriculture and surgical and medical devices. The overproduction of petrochemical-based plastics has engendered huge complications for terrestrial and marine environments. The plastic is disposed of in two worst modes: environmental dumping and incineration. Dumping plastic from landfills into the domain has been persistent for several years. That does not degrade and remains in the environment for a consistent time, causing land and marine pollution.

Similarly, incineration of plastic waste also releases toxins into the atmosphere, causing hazardous threats to the environment (Nehra et al., 2017). Recently, modern society has recognised plastic as a significant environmental pollutant, and the search for natural alternatives for synthetic polymers has reached an all-time high. Bioplastic has recently been popular for various applications as an alternative to synthetic polymers obtained from diverse plants or microbial sources (Mohanrasu et al., 2021, Rasu and Arun, 2017; Muniyasamy et al., 2019; Muniyasamy and Dada, 2021).

Compared to all biopolymers, microbial-based biopolymers such as polyhydroxyalkanoates (PHAs) are one of the promising alternatives for petroleum-derived polymers owing to their biodegradability and biocompatibility (Kulkarni et al., 2010). PHAs (polyesters) are energy-storage compounds synthesized by various microbes (Gram-negative and Gram-positive bacteria) under nutrient-stress conditions (excessive carbon and limitation of nitrogen or phosphorus) as accumulated granules inside the cytoplasm. PHAs contain various side chains of hydroxyalkanoates (HAs), and around 150 types of (R)-3-hydroxy fatty acids were identified in the PHA family (Roy & Visakh, 2014). Polyhydroxybutyrate (PHB) was first identified in 1926 by Maurice Lemoige from *Bacillus megaterium* bacterium; it is one of the well-studied groups of PHAs with 80% of polymer in total cell dry weight (Keshavarz & Roy, 2010; Yu & Stahl, 2008). Maurice identified lipid-like insoluble material, which was reported in a French journal; moreover, between 1923 and 1951, he published 27 articles reporting a wide variety of microorganisms producing PHB (Chodak, 2008). According to Wilkinson Macre, increased carbon sources (glucose) with nitrogen sources lead to increased formation of granules in the *B. megaterium* (Alves et al., 2017). Subsequently, various microorganisms have been reported to produce P(3HB), such as *Azotobacter*, *Chromatium*, *Hydrogenomonas*, and *Pseudomonas,* between 1959 and 1973 (Braunegg et al., 1998). Wallen and Rohwedder (1974) reported activated sludge as a carbon source for PHA synthesis, followed by numerous waste sources as carbon sources for PHA synthesis. In the 1990s, *Alcaligenes eutrophus* was reported to accumulate 75% of PHB in total cell dry weight that was later renamed *Ralstonia eutropha,* which is one of the potential bacteria for higher PHB production using different carbon sources (B. S. Kim et al., 1994; Wallen & Rohwedder, 1974).

5.3 STRUCTURES AND CLASSIFICATION OF PHA

PHA polymers are structurally classified into three types based on the number of carbon atoms in the monomeric unit with small chain length (*scl*) PHA polymer that ranges from 3 to 5 (C3–C5) carbon atoms, medium chain length (*mcl*) – PHAs ranges from 6 to 14 (C6–C14) carbon atoms and long-chain length (*lcl*) PHAs which poses greater than 14 (>C14) carbon atoms (Pradhan et al., 2020). The *scl* (3–5 carbons) and *mcl* (6–14 carbons) are blended to form copolymers *scl*-co-*mcl* (3–14 carbons); similarly, the *lcl*-PHAs blending was also studied (Steinbuchel and Valentin, 1995; Steinbüchel and Hein, 2001). Short-chain length PHAs polymer includes poly(3-hydroxybutyrate) P(3HB), poly(4-hydroxybutyrate) P(4HB) and poly(3-hydroxyvalerate) P(3HV) or the copolymer P(3HB-co-3HV), Medium-chain lengths PHAs includes 3-hydroxyhexanoate (3HHx), 3-hydroxyheptanoate (3HHp), 3-hydroxyoctanoate (3HO), and 3-hydroxydecanoate (3HDD) (Anderson et al., 1992).

The PHA synthase substrate specificity is the main reason for the varying carbon length of 3HAs. According to Khanna and Srivastava (2005), *A. eutrophus* PHA synthase can polymerise 3–5 carbon atoms of 3HAs, but *Pseudomonas oleovorans* PHA synthase polymerises 6–14 carbon atoms of 3HAs. Both short-chain and medium-chain monomeric units are linked to form hybrid polymers like poly(3-hydroxybutyrate-co-3-hydroxyhexanoate); these polymer formations are caused by the biosynthetic enzyme's stereospecificity with R-configuration in all monomers. The monomeric units have various functional groups like carboxyl, esterified carboxyl, cyano, epoxy, halogen and hydroxyl (Castilho et al., 2009; Ciesielski et al., 2015). Among the PHAs, *scl* groups with four carbon atoms homopolymer (Poly 3-hydroxybutyrate) are industrially first recognised owing to their high melting point of 180°C together with high molecular weight and crystallinity. The *scl* groups homopolymers are highly crystalline and thus easily breakable. In contrast, the copolymers have reduced the brittleness and crystallinity (20%–40%), but using a medium chain length, PHAs with their copolymers do not break easily (Anderson & Dawes, 1990).

Among the PHAs, PHB is a well-characterised polymer with a highly crystalline nature because of stereoregularity; moreover, PHB is insoluble in water and resistant to hydrolytic degradation (Anjum et al., 2016). PHB has good thermoplastic properties with poor mechanical properties such as low tensile strength and Young's modulus with low O_2 permeability than petroleum-based polymers (Sudesh et al., 2000). The densities of amorphous and crystalline PHBs are 1.18 and 1.26 g/ cm^3, respectively, with the molecular weight in the range of 10,000–3,000,000 Da from wild-type bacteria (Anjum et al., 2016). Poly(4-hydroxybutyrate) P(4HB) properties (thermoplastic and tensile strength) are like synthetic polyethylene, which has extreme elastic properties. The copolymer produced from combining any two or more PHB such as 3-hydroxyvalerate, 3-hydroxyhexanoate, 3-hydroxypropionate and 4-hydroxybutyrate improves the material properties like crystallinity, melting point, stiffness and toughness. The P (3HB-co-3HV) is the most investigated copolymer with decreased stiffness, lower melting temperature, and lower crystallinity, but it is flexible and more challenging (Anjum et al., 2016). On the other hand, terpolymer has better desirable properties than copolymers like P (3HB-co-3HV-co-3HHx).

5.4 BIOSYNTHESIS OF PHAS

PHAs are biodegradable polymers produced by several (above 300) gram-positive and gram-negative bacteria inside the cytoplasm under excessive carbon and limited nitrogen or phosphorus conditions. The PHB granules are made inside the cytoplasm; about 0.2 ± 0.5 mm in diameter can be visualised by staining dyes such as Sudan Black B, oxazine dye Nile Blue A/Nile red under phase contrast light microscope due to the high refractivity index (Anjum et al., 2016). The synthesis of PHA is a complex process regulated by many genes that encode a wide range of enzymes that are directly or indirectly involved in PHA synthesis (Laycock et al., 2013). Numerous bacteria are isolated from marine and terrestrial environments for higher PHA-producing capability, and several strains showed enhanced PHA production from marine and terrestrial environments (Mohanrasu and Arun, 2017; Pradhan et al., 2020; Mohanrasu et al., 2021).

PHA production involved in eight pathways is summarised in *Ralstonia eutropha* as a model organism. The PHA production pathway I involved three enzymes: β-Keto thiolase, NADPH-dependent acetoacetyl-CoA reductase and PHA synthase. These enzymes are encoded by phaA, phaB and phaC genes (Laycock et al., 2013). Pathway II is involved in fatty acid uptake, β oxidation from acyl-CoA, 3-hydroxyacyl-CoA, and PHA production by synthase catalysis. Pathway III involves 3-hydroxyacyl-ACP-CoA transferase, malonyl-CoA-ACP transacylase (FabD) encoded by PhaG and substrates converted 3-hydroxyacyl-ACP to 3-hydroxyacyl-CoA and PHA. Pathway IV involves NADPH-dependent acetoacetyl-CoA reductase to oxidise

(S)-(+)-3-hydroxybutyryl-CoA. The remaining pathways are involved in copolymer production; P(4HB) production is done by pathways V and VII in *Clostridium kluyveri* and *A. hydrophila* 4AK4 (Laycock et al., 2013).

5.5 SUBSTRATES FOR PHAS PRODUCTION

The main problem associated with PHB production is higher production costs. The selection of carbon sources and strains enhances PHB production and is cost-effective. The choice of appropriate carbon sources for PHA production, such as simple sugars or industrial, agricultural, and food waste, has proved to be successful in enhancing PHA production (Table 5.1). In the initial stages of PHA production, simple sugars are mainly used as carbon sources for PHA biosynthesis compared to waste sources. The structure of synthesized PHAs is strongly impacted by various carbon sources and different microbes used; this structural composition influences the applications of PHA. The *Pseudomonas* species producing PHA has other functional groups (phenyl, phenoxy, halogens, branched alkyls, olefin and esters) based on the substrates utilised (Pradhan et al., 2020). Kim et al. (2011) analysed 36 different carboxylic acids containing carbon substrates for PHA production employing *Pseudomonas putida* KCTC 2407 to obtain an appropriate functional group into the PHA chain to enhance the physical properties (Pradhan et al., 2020). The carbohydrates are categorised into three types such as monosaccharides, disaccharides, and polysaccharides; monosaccharides are simple sugars that bacteria can efficiently utilise for PHA production. The disaccharides and polysaccharides cannot be directly used; instead, they should be hydrolysed and converted into monosaccharides for PHA production by bacteria (Pradhan et al., 2020).

Several studies reported utilising glucose as a carbon source for PHA production. Even though many researchers have examined different carbon sources (arabinose, glucose, glycerol, lactose, lactic acid, mannitol, sodium acetate, starch and sucrose) for PHA production, most found glucose to be the optimum carbon source for maximum PHA production of 5.61 g/L by *Bacillus megaterium* (Mohanrasu et al., 2020). Arun et al. (2009) isolated 42 different stains from various marine environments and screened their PHB-producing capabilities. Other carbon sources (arabinose, glucose, glycerol, lactose, lactic acid, mannitol, sodium acetate, starch and sucrose) have been screened for best PHB production; compared to other carbon sources, glucose showed higher (4.223 g/L) PHB production. Obruca (2015) reported better PHA production using sucrose directly as a carbon source using *R. eutropha*. In another study, pure glycerol was found to be the best carbon source for PHA production by recombinant *E. coli* and *Shimwellia blattae* strains with 63% and 30.7% PHA content, respectively (Nikel et al., 2010; Sato et al., 2015).

However, in recent years, waste from different agricultural and industrial sources has been utilised as a potential medium to decrease PHA production costs. For many regions globally, waste sources are explored as a possible factor for converting value-added products. In these studies, various waste sources are used for PHA production, such as lignocellulosic biomasses from wheat straw lignocellulosic hydrolysates (Cesário et al., 2014), oil-palm biomass (Hassan et al., 2013), hydrolysates

TABLE 5.1
Production of PHAs by Bacterial Strains Using a Wide Range of Feedstocks

PHAs Type	Carbon Source	Microbial Strains	Production Scale	PHAs (g/L)	References
Poly(3-hydroxybutyrate) (PHB)	Glucose and glycerol	*C. necator* DSM 545 and *Burkholderia sacchari*	Fed-Batch culture	44.25 and 4.48	Rodríguez-Contreras et al. (2015)
P(3HB)	Crude glycerol, rapeseed waste	*Cupriavidus necator* DSM 545	Fed-batch	10.9	García et al. (2013)
P(3HB)		*C. necator* DSM 428	Batch	0.342	Dhangdhariya et al. (2015)
P(3HB)	Glucose	*Micrococcus luteus* (KY494862)	Shake flask	4.632	Mohanrasu et al. (2018)
P(3HB-co-3HV)	Crude glycerol	*Haloferax mediterranei* DSM1411	Fed-batch	15.2	Hermann-Krauss et al. (2013)
P(3HB-co-3HV)	Synthetic wastewater	*Hydrogenophaga palleronii* NBRC102513	Batch	1.01	Venkateswar Reddy et al. (2016)
P(3HB)	Glucose	*Vibrio* sp. (MK4)	Batch fermentation	4.223	Arun et al. (2009)
P(3HB)	Glucose	*Micrococcus luteus* (KY494862)	Batch fermentation	5.61	Mohanrasu et al. (2020)
P(3HB-co-4HB)	Waste corn-steep liquor, gluconate	*Halomonas bluephagenesis* TD01	Fed-batch	66.6	Ye et al. (2018)
P(3HB-co-3HHx)	Crude glycerol	*Ralstonia eutropha* Re2133	Batch	0.57	Bhatia et al. (2018)
P(3HB)	Methanol	*Methylobacterium extorquens* DSMZ 1340	Batch	9.5	Mokhtari-Hosseini et al. (2009)
PHBV	Pre-treated vinasse	*Haloferax mediterranei*	Shake-flasks	19.7	Bhattacharyya et al. (2012)
P(3HB)	Commercial glycerol	*C. necator* DSM 545	Fed-batch	51.2	Cavalheiro et al. (2009)
P(3HB)	Waste glycerol	*C. necator* DSM 545	Fed-batch	38.1	Cavalheiro et al. (2009)
P(3HB)	Glucose	*Bacillus cereus* (KR809374)	Fed-batch	19.15	Dinesh et al. (2020)
P(3HB)	Glucose	*Micrococcus luteus*	Fed-batch	12.18	Mohanrasu et al. (2021)

of spruce sawdust paper waste, wood chips (D.-H. Kim et al., 2011), paddy straw (Sandhya et al., 2013), corn cob, corn stalks, eucalyptus, pine, sugarcane bagasse, sorghum straw, oat straw, barley hull and grasses (Kucera et al., 2017).

Apart from lignocellulosic biomasses, extensive research is done on different agricultural and food wastes such as waste rapeseed oil (Stanislav Obruca et al., 2010), jatropha oil (Ng et al., 2011), coprah oil (Simon-Colin et al., 2008), linseed oil (Bassas et al., 2008), corn oil (Chaudhry et al., 2011), municipal waste waters (Chua et al., 2003), waste sesame oil (Arun et al., 2006), waste potato starch (Haas et al., 2008), waste cooking oil (Haba et al., 2007), waste frying oil (Verlinden et al., 2011), palm oil mill effluents (Wu et al., 2009), brewery waste effluents (Liu et al., 2011), kraft mill waste water (Pozo et al., 2011), biodiesel waste water (Dobroth et al., 2011), alpechin medium (waste water from olive oil mil) (Ntaikou et al., 2009), food processing waste effluents (Reddy et al., 2017) and potato processing waste for PHA production (Rusendi & Sheppard, 1995).

5.6 THE CURRENT SCENARIO FOR INDUSTRIAL-LEVEL PHA PRODUCTION

Among different types of PHA, only four different varieties are gaining industrial importance such as poly[(*R*)-3-hydroxybutyrate] (PHB), poly[(*R*)-3-hydroxybutyrate-*co*-4-hydroxybutyrate] (P3HB4HB), poly[(*R*)-3-hydroxybutyrate-*co*-(*R*)-3-hydroxyvalerate] (PHBV) and poly[(*R*)-3-hydroxybutyrate-*co*-(*R*)-3-hydroxyhexanoate] (PHBHHx) (Chen, 2010). PHA production depends upon several processes such as isolation of the best possible strains from wild or recombinant sources, optimisation of numerous parameters influencing the production, scaling up the output to industrial-level fermentation, biomass drying, extraction, and purification of PHA. Currently, 24 companies are known to produce PHAs globally, but some of them have stopped PHA production due to the rise in the cost of production (Chen, 2010).

5.6.1 Metabolix

Metabolix company has produced PHA for the longest time aimed at commercial applications. This company has more than 500 patents for diverse industrial applications. PHA is made in the name of Mirel, which rapidly reached 50,000 tons production capacity a construction facility (Chen, 2010).

5.6.2 Biopol

Biopol is an Imperial Chemical Industries (ICI), UK, the company that sold its patents to Zeneca, which was brought by Monsanto and currently produces in the name of Metabolix. Biopol mainly had a wide range of products like packaging materials, medical, surgical pins, disposable knives and forks, disposable cups, shampoo bottles and disposable razors using copolymers poly(3-hydroxybutyrate-co-3-hydroxyvalerate) (Anjum et al., 2016).

5.6.3 Nodax

Nodax also produces a copolymer of 3-hydroxybutyrate with medium chain length monomers, including 3-hydroxy hexanoate, 3-hydroxy octanoate and 3-hydroxydecanoate. Nodax has films, latex and fibres or non-wovens.

5.6.4 Biogreen

Biogreen is a Japanese-based company (Mitsubishi Gas Chemicals) that produces P (3HB) from methanol. Biogreen produced various industrially essential materials such as cosmetics containers, packaging, shampoo bottles, cups, milk cartons, diapers, feminine hygiene products, sanitary towels, food additives, sanitary towels, mulch films, herbicides, insecticides, bacterial inoculants, cardiovascular stents and heart valves, vascular grafts, surgical sutures, dusting powders, wound dressings, nerve conduits, bone plates and osteosynthetic materials (Anjum et al., 2016).

5.7 BIODEGRADABILITY AND BIOCOMPATIBILITY OF PHAS

A microorganism degrades complex macromolecules through the action of their enzymes into simpler small molecules, known as biodegradation. Biodegradation (aerobically or anaerobically) is every biopolymer's most desired and essential feature. Moreover, the degraded product should not produce harmful compounds or toxins in the environment. PHAs can be degraded entirely under aerobic and anaerobic conditions; during degradation, they break down into carbon dioxide and methane in soil, lakes, sewage, and marine environments by microorganisms. The incredible feature of PHA is that when disposed of the material environments, the diverse microbes prevailing in the habitat secrete depolymerase enzyme, which hydrolyses the PHA into water-soluble oligomers and monomers that can be efficiently utilised as nutrients for the growth of the microbes (Schneider et al., 2010).

The PHA can be degraded in a wide range of environmental habitats like marine and terrestrial ecosystems; the PHA degradation is influenced by several external environmental factors such as pH, temperature, a load of microbes in the particular environment, humidity and molecular weight of the polymer (Khanna & Srivastava, 2006). Apart from external factors, PHA characteristics such as stereospecificity, crystallinity, molecular weight, melting temperature, and monomeric composition of PHAs also strongly impact PHA degradation (Jendrossek et al., 1996). Several researchers have explored the potential degradation of PHA in terrestrial and marine environments; in the terrestrial environment, 85% of the PHA can be degraded within seven weeks. In aquatic conditions, it requires 254 days to degrade PHAs, but the only limitation is that the temperature should not exceed 60°C (Anjum et al., 2016). The bacterial family such as Micromonosporaceae, Streptosporangiaceae and Streptomycetaceae, Pseudonocardiaceae can degrade P (3HB) in the environment (Anjum et al., 2016).

5.8 PHAS AND THEIR APPLICATIONS IN DIFFERENT SECTORS

The research towards PHA productions has spotlighted the potential owing to its valuable properties in diverse applications such as biomedical, packaging materials, industrial, and agricultural sectors biomedical, packaging materials, industrial, and agricultural applications. The PHA characteristics, including solubility in water, optically active nature, and crystalline and piezoelectricity, make it an attractive material. The significant attributes of PHA are non-toxic, biodegradable, and biocompatible materials compared to other synthetic polymers such as polypropylene (PP) and polyethylene (PE) (Nehra et al., 2017).

5.8.1 Industrial Sector

PHA is used mainly as an alternative to synthetic polymers for numerous industrial applications. The US-based company produced blends of P(3HB) and P(3HO); moreover, the FDA has approved these polymers' usage for food packaging sectors (Rai & Roy, 2011). PHAs from *Aeromonas hydrophila* were produced on an industrial scale by a collaboration of KAIST (Korea) and Procter & Gamble (P&G, USA), where the company made various critical industrial products such as flexible packaging, flushable, non-wovens, thermoformed articles, binders, synthetic paper and medical devices (Rai & Roy, 2011). A German company produced P(3HB) biomer by using *Alcaligene latus* to make innovative materials like pens and combs (Chen & Wu, 2005). BIOPOL® also manufactured a copolymer of P(3HB-co-3HV) for packaging, fishing nets, paper boards, electrical appliances, ropes, preparation of coat papers and storing containers like shampoo razors and motor oil (Pradhan et al., 2020). The copolymers *mcl*-PHA and P(3HB) synthesised by the Nodax™ company were used for the films and foam production (Philip et al., 2007).

In general, plastics are used in food sectors to extend their shelf life by avoiding contact with microorganisms, thus reducing the risk of food spoilage, restricting the moisture and light and maintaining the food texture, retaining the flavours of food by preventing connection with any of the following substances such as oxygen, carbon dioxide, water vapour and volatile compounds. On the contrary, in recent years, PHAs have been an alternative to synthetic polymers for food packaging applications because of their qualities like biodegradability, environmentally friendly and renewable nature. The perishable foods (raw meat, cured meat and meat salad) packaging is done by PHAs to reduce municipal solid wastes. It is also used in various food sectors such as fresh fruit packaging, dairy product packaging, medicine packaging and fragile item packaging (Mudenur et al., 2019).

5.8.2 Medical Sector

Being of bacterial origin, PHAs are extensively used in the medical fields, such as scaffolds in the form of screws, pins, cartilage repair, PHB ultrafine fibres for skin regeneration, bone tissue regeneration, articular cartilage, cardiovascular patch grafting and meniscus repair devices (Ray & Kalia, 2017). Moreover, PHAs are also utilised in stem cell growth, as a vehicle for targeted drug delivery, and as raw material for various moulded products such as syringes, sutures, disposable needles, surgical gloves and gowns (Ray & Kalia, 2017). The targeted drug delivery system with a controlled optimal delivery rate and dose in a specific targeted site is the fascinating application of PHA in biomedical fields. As the polymer is biocompatible and degradable, it is utilised as a drug delivery system both orally and intravenously.

5.8.3 Agricultural Sector

In agricultural fields, PHA nanocomposites are modified like plastic mulch to be used over the surface of the land to prevent weed growth, protect the vital ingredients in soil like nutrients, and escape the soil through evaporation. These types of plastic mulch are exceptionally eco-friendly, produced from renewable sources and reduce labour costs (Ivanov et al., 2015). The PHB used in plant growth-promoting rhizobacteria formulations protects the bacteria from extreme stress conditions like high or low temperatures, desiccation, chemical pesticides and differences in pH during transportation and storage (Benrebah et al., 2007). The PHA acts as transporting materials due to its properties such as lack of toxicity, chemical and physical uniformity, high water-holding capacity, and eco-friendly material used in powder and liquid formulations (Stephens & Rask, 2000). On the other hand, PHAs are also used as a carrier for insecticides that will release the insecticide into the soil while facing specific environmental conditions (Holmes, 1985; Philip et al., 2007)

5.9 LIMITATIONS AND RECENT ADVANCES IN PHAS PRODUCTION AND MARKETING

1. The most significant problem associated with PHA production is the high production cost compared to petroleum-based polymers. Exploiting new application fields should reduce the cost problems.
2. The PHA production costs are 15 times higher than the petroleum-based polymer, which could be credited to using carbon sources such as arabinose, glucose, glycerol, lactose, lactic acid, mannitol, and sodium acetate, starch, and sucrose as the cost of these chemicals are not stable.
3. Therefore, PHA production costs can be considerably lowered using industrial byproducts such as wastewaters, agriculture feedstocks, waste plant oils and kitchen wastes.
4. Another significant limiting factor of PHA production is the extraction process which consumes around 50% of the total production cost and requires various chemical digestion and solvent extraction; moreover, this leads to environmental problems with socio-economic losses.
5. The bioextraction techniques (genetic and metabolic engineering) can be a fantastic alternative for extraction by solvents (hazardous to the environment). These techniques include mealworm systems, bacteriophage-mediated lysis and predatory bacteria.

6. Industrial-scale PHA production with selected appropriate strains should be based on fast-growing capabilities, higher PHA production ability, a wide range of substrate utilisation, salt concentration, temperature and ease of genetic modification. Finally, strain should not produce toxins for various industrial applications.
7. The PHAs blended with different HA monomers such as poly(4-hydroxybutyrate) P(4HB), poly(3-hydroxybutyrate-co-3-hydroxyvalerate) P(3HB-co-3HV), P(3HB-co-3HV-co-3HHx), and blend with natural polymers like poly(3-hydroxybutyrate) (PHB)/chitin blends, poly(3-hydroxybutyrate-co-hydroxyvalerate) (PHBV)/chitosan, poly(3-hydroxybutyrate)/cellulose acetate butyrate PHB/CAB, poly(hydroxybutyrate)/cellulose propionate (PHB/CP), PHB/hydroxyethyl cellulose acetate (HECA) and different PHA nanocomposites have been investigated in various studies for maintaining the outstanding properties and wide industrial application.
8. Despite the existing hurdles, the strategies mentioned above would undoubtedly improve the PHA's production and will positively impact various industrial applications, which would eventually play a vital role in different outfronts shortly.

5.10 BACTERIAL CELLULOSE (BC)

In 1886, bacterial cellulose gained attention due to its unique properties such as biodegradability, biocompatibility, high purity, good chemical stability, higher mechanical strength, crystalline network structure, and better water-holding capacity (Santos et al., 2015). Cellulose is the most abundant, sustainable, and widely used polysaccharide on the planet, and it is one of the most commonly used polysaccharides (Ullah et al., 2016). In terms of long-term development, bacterial cells can produce BC for a broad range of industrial products. It is a semicrystalline polysaccharide composed of linear homopolymers of D-glucopyranose residues ($C_6H_{11}O_5$) linked together by b-(1 → 3)-glycosidic bonds. BC is produced by various bacteria, including *Acetobacter*, *Pseudomonas*, *Sarcina*, and *Gluconacetobacter* (Pandit & Kumar, 2021).

Several parameters have been optimised to maximise BC yield, regardless of whether the cultivation is agitated or static. pH, dissolved oxygen, and temperature are examples of these parameters (K.-Y. Lee et al., 2014). On an industrial scale, BC manufacturing necessitates a continuous or semicontinuous process, low-cost raw ingredients, and little byproduct generation. It also necessitates a high carbon conversion rate to cellulose and a high speed of BC synthesis (Çakar et al., 2014). BC's exceptional qualities make it ideal for various applications, including transparent food packaging, water treatment, battery separators, adsorbents, tissue engineering scaffolds, artificial blood vessels, and electric conductors or magnetic materials. BC must overcome some technical challenges before reaching an industrial scale. Its hydrophobicity, limited solubility, and relatively high cost are among them. Improving cultivation methods through genetic engineering approaches aids in problem resolution. The primary goals of growing BC studies are modifying BC residences and yields, reducing manufacturing expenses, and selecting appropriate business fabrication methods.

5.10.1 Polylactic Acid (PLA)

In recent years, there has been a significant increase in the production of commercial microbial products. Due to the growing global energy and environmental challenges, researchers have been developing green production methods for nearly decades. PLA has gotten increasing attention due to its eco-friendly approaches and use in packaging, pharmaceuticals, and cosmetics. The ring-opening polymerisation (ROP) and direct polycondensation routes are used to convert lactic acid monomer into PLA (Sin et al., 2013a). Lactic acid (LA) comprises three carbon a-hydroxycarboxylic acids found in two stereoisomers: D-LA and L-LA. PLA also has significant market worth because of its biodegradable and biocompatible qualities. LA is produced from a range of carbon sources and then turned into PLA. LA can be synthesised in two ways: chemically or by fermentation.

Compared to chemical synthesis, the fermentative method of PLA production has several advantages, including optically pure D- or L-LA, low energy consumption, waste materials as substrates, and low production temperatures with higher LA-producing microbes (Abdel-Rahman et al., 2011). A catalytic ROP technique converts the dimers of L-lactide, D-lactide, and meso-lactide to a high-molecular-weight polyester (Sin et al., 2013b). The crystallisation rate, melting point, and mechanical properties are based on the stereochemical composition of PLA (Drumright et al., 2000). PLA demand is increasing due to its transparency, durability, and good mechanical strength compared to other biodegradable polymers. The potential biodegradable and biocompatible properties result in significant market expansion for PLA. In 2019, PLA produced around 190,000 tons worldwide 2019, and a multi-million-ton production will be made in the upcoming decade (Jem & Tan, 2020). According to Djukić-Vuković et al. (2019), 3141 patents were filed (LA and PLA) from 2009 to 2018, which showed that 30.1% (947) are associated with human requirements, particularly foods (502), veterinary science and hygiene (432), and medicine.

LA is synthesised primarily through fermentation using a wide range of materials (both synthetic and waste) and various microbes such as bacteria, fungi, and yeast. LA can produce a wide range of microorganisms, including algae and cyanobacteria (*Hydrodictyon reticulum*, *Scenedesmus obliquus*, *Nannochlorum* sp. 26A4, and *Synechococcus elongates*), bacteria (*Bacillus coagulans*, *Lactococcus*, *Lactobacillus plantarum*, *Carnobacterium*, *Lactococcus lactis*, *Tetragenococcus*, *Pediococcus*, and *Vagococcus*), fungi (*Rhizopus oryzae*, Ceriporiopsis subvermispora, *Mucor* and *Monilia* sp) and yeast (*Saccharomyces*, *Zygosacchromyces*, and *Kluyveromyces*) (Djukić-Vuković et al., 2019).

The cost of raw materials is one of the most critical elements in economically manufacturing LA. Various renewable alternative materials, such as whey, yoghurt, glycerol, food industry wastes, starchy biomass, lignocellulosic biomass, algal biomass, and agricultural byproducts, are currently underway to reduce the production costs (Abdel-Rahman et al., 2013). Fermentation methods differ depending on higher yields [batch, fed-batch, semicontinuous (repeated batch), and continuous fermentation]. Each has its own set of advantages and disadvantages, such as ease of operation, low contamination risk, time-saving processes, and high productivity, as well as drawbacks such as lower yield, inhibition of end products and incomplete utilisation of carbon sources (Abdel-Rahman et al., 2013). Before starting the fermentation process, we ought to sort out the optimum parameters for fermentation, such as substrate, organic loading rate, carbon/nitrogen ratio, pH, temperature, inoculum concentration, and inoculum age.

Many major companies have joined the PLA sector, including Toyota, Mitsubishi, Mitsui, and Dow, and some others have intentions to launch, such as Uhde, Galactic and COFCO (Jem et al., 2010). Many governments are actively promoting PLA manufacturing as part of an attempt to develop biobased and biodegradable polymers to reduce pollution. PLA production capacity has expanded from less than 100,000 tons to more than 1 million tons in the last 15 years (Jem et al., 2010). The commercial scale of PLA production is lagging due to higher costs compared to synthetic plastics. Several countries, such as France, Korea, Italy and Taiwan, have increased attempts to ban or restrict the use of traditional plastics applications and tax reductions for biobased plastics to remove PLA's economic obstacle (Jem et al., 2010).

To reduce PLA production costs, a fully integrated ROP capacity should exceed 50,000 tons per year. The cost of unit production is high in a small plant, and the competitive edge is low, but the risk of entry increases in a large field (Jem et al., 2010). Apart from the PLA polymerisation part, the industry should focus on upstream LA and lactide production to minimise the economic scale of production. PLA polymerisation plants can be smaller than 50,000 tons per year if the lactide plant is decoupled from the polymerisation plant and still be competitive (Jem et al., 2010). The synthetic plastic manufacturing and incineration processes are emitting CO_2 and are a potential threat to the environment in the form of global warming. Considering PLA's biocompatibility and bio responsibility, it can be used in various biomedical applications, including sutures, implants, fracture fixation, and drug delivery. A sustainable alternative to petrochemical-derived products has emerged due to PLA's biodegradable characteristics and the nature of its feedstock, which makes it an environmentally sustainable choice. The capacity of a fully integrated ROP PLA production facility would reach 50,000 tons per year to lower PLA production costs. In a small unit, the cost of unit production is high, and the competitive advantage is minimal, but in a large one, the risk of entry is significant (Jem et al., 2010). Apart from PLA polymerisation, the industry should concentrate on upstream LA and lactide synthesis to reduce manufacturing costs. If the lactide plant is separated from the polymerisation plant, PLA polymerisation plants can be less than 50,000 tons per year and still be competitive (Jem et al., 2010). The manufacturing and incineration of synthetic plastics emit CO_2, threatening the environment through global warming. Due to its biocompatibility and responsibility, PLA can be employed in various biomedical applications, including sutures, implants, fracture repair, and drug distribution. PLA has developed as a sustainable alternative to petrochemical-derived products due to its biodegradable properties and the nature of its feedstock, making it an environmentally friendly option.

5.11 CHALLENGES AND OPPORTUNITIES

The novel coronavirus (SARS-CoV-2), or COVID-19, has created a pandemic situation around the globe since December 2019 and has significantly affected people from all walks of life. COVID-19 has dramatically increased human transmission anxiety and resulted in the extensive use of single-use plastics in packed foods, groceries and disposable utensils (Mittal et al., 2022). According to the WHO (2020), 89 million face masks, 76 million glove pairs, and 1.6 million pairs of protective goggles are in need every month to meet the rising global demand, and this vast surge is due to the organisation proposed mandates that hospitals should have attire to strictly. During the pandemic's peak, Wuhan (a city in China) alone generated about 240 tons of medical waste per day during the outbreak, almost six times more than it had produced previously (Singh et al., 2020).

The increased plastic waste from the pandemic has posed an increased threat to the globe; thus, biodegradable plastics are a hot topic today. The excellent property of a biodegradable polymer is biodegradation. The biodegradable polymers are converted into CO_2 (aerobic conditions) and CH_4 (anaerobic conditions) in natural environments with the help of natural microflora. The amount of biodegradation (per cent biodegradation) and production (CO_2 or $CO_2 + CH_4$) should be measured. The biodegradable polymer should be 100% biodegradable and not release any harmful toxic materials during the biodegradation process.

The biodegradation process is ensured by various standards methods such as ASTM D6954-04, AS 4736-2006 (American Society for Testing and Materials), ASTM D5209-92, ASTM D5338-98, ASTM D5526-94, ASTM D5951-96, ASTM D5988-03, ASTM D6002-96, ASTM D6340-98, ASTM D6400-99, ASTM D6691-01, ASTM D6692-01, ASTM D7081-05, DIN V 54900-2, EN 13432:2000, EN 14045:2003 (EN 14046:2003, EN 14047:2002, EN 14048:2002, EN 14806:2005, ISO 14851:1999 (International Organization for Standardization), ISO 14852:1999, ISO 14855:1999, ISO 14593:1999, ISO 15314:2004, ISO 16929:2002, ISO 17556:2003, ISO 20200:2004, BS 8472, BS EN ISO 14855-2:2009 (British Standards) and BS EN 13432:2000) (Ammala et al., 2011).

Biopolymer degradation is influenced by various environmental factors such as the nature of the biopolymer, moisture, temperature and pH. In addition, the biodegradation rate is affected by the size of the polymer, crystallinity, copolymer composition and molecular weight (Kale et al., 2007). The biodegradation rate is positively altered by water and moisture; increased moisture and water content result in higher microbial growth and faster degradation (Thakur et al., 2018).

Currently, various countries have taken steps for policies related to bioplastic production. The Malaysian government took the new initiative for Bioeconomy Initiative Malaysia, an initiative to accelerate biotechnology commercialisation (Moshood et al., 2021). This project includes industrial biotechnology, biobased chemical, medical equipment, medical biotechnology, biopharmaceutical, agricultural biotechnology and vaccine manufacturing (Moshood et al., 2021). In China, the National Development and Reform Committee have allocated funds to biodegradable plastics. The biopolymer research undertaken by different Chinese institutes such as the University of Tsinghua, the Chinese Academy of Sciences' Institutes of Physics and Chemistry, and Sichuan University mainly focused on PLA production (Moshood et al., 2021). The Japanese government took the initiative to reduce fossil fuel usage and promote biotechnology biomass to mitigate global warming. Toyota plans to shift 20% of its plastics to bio-sourced plastics by 2015 for its cars (Moshood et al., 2021).

The Korean government reduced the dependability on crude oil through the Industrial Biotechnological Promotion Plan in 2012 (Moshood et al., 2021). Korean-based companies such as LGHausys, SK Chemical, and Hyundai Motors are substituting their petroleum-based material with toxic biopolymers (Moshood et al., 2021). The Korean Bioplastics Association supports the bioplastics supply and biomass authentication. Thailand has over 4000 plastics industry companies, and the nation has a strong biomass industry (Moshood et al., 2021). In Thailand, the bioplastic industry has been formed as a strategic industrial sector for boosting economic growth and sustainability. In the European Green Agreement, there has been a commitment by European countries to transform their linear economies into balanced circular economies by 2050 (Moshood et al., 2021).

The bioplastics industry in India is still in its infancy, and very few companies are involved in the production of bioplastics. In India, nowadays, various aspects such as government funds for bioplastics, Environmental awareness projects, and easy access to feedstock are promoting the growth of bioplastics industries (Mittal et al., 2022). An initial move has been taken, which is banning single-use plastics in various hill stations in India, and scientists are working towards an alternative for bioplastics. Biotechnology-based companies such as Biogreen, Envigreen, 2M Biotech, Plastobags, Truegreen and Ecolife have started bioplastics production in India. United States Department of Agriculture (USDA) has a Bio Preferred program, which initiated the purchase of biobased products and reduced reliance on petroleum-based products (USDA, 2020). The major bioplastics manufacturers have filed U.S. patent (230) for the application of (polyhydroxyalkanoate (PHA), PLA and starch) between 1993 and 2012, such as Novamont, Metabolix, Cargill (Nature Works), Cereplast, Kimberly Clark, Biotec and Nissei (Byun and Kim, 2014).

According to National Research Strategy Bioeconomy 2030: Our Route towards a Biobased Economy is done by Germany, which moves towards the bioplastics alternative for petroleum-based products (BMBF, 2011). The Japanese government announced two programmes, the Biotechnology Strategic Scheme and the Biomass Nippon, whose objective is to promote renewable sources utilisation (Taniguchi, 2018). In Spain, several organisations are promoting bioplastics, such as the Spanish Technological Biomass Platform (BIOPLAT), the Spanish Sustainability Observatory (OSE), the Platform of Biotechnological Markets, and the National Food & Agriculture Forestry Innovation and Research Program to promote linear economy to circular economy (Garrido et al., 2021). There is no official policy on bioplastics in Italy other than banning the distribution of traditional plastic carrier bags in favour of biodegradable bags that have been in effect (Garrido et al., 2021). Although the Canadian government lacks an overarching bioeconomy strategy, it has a series of diverse and often competing policy frameworks and visions (Birch, 2016).

5.12 SUMMARY AND OUTLOOK: A PERSONAL VIEWPOINT

Globally, most of the population is unaware of bioplastics, yet many believe biodegradable plastics are better for the environment than "traditional" plastics. In the designated disposal environment, microorganisms break down the bioplastics safely and efficiently via a microbial food chain. Bioplastic waste disposal must be managed carefully using various methods, such as mechanical or chemical recycling, and it might be better from an economic and environmental standpoint. Biodegradable polymers are frequently promoted in the media. Plastic derived from plants, eco-friendly plastic, and plastic without CO_2 footprint are all gaining significant attention. However, attempting to make this a reality will be challenging.

Our current issues are likely to be remedied by bioplastics, yet such a strategy would be exorbitant and have significant ramifications for our economy and working-class people. Bioplastics have been lauded for their contribution to sustainable living, although most have been regarded from an eco-friendly perspective. Pepsico, for example, has been acclaimed for producing PET from 100% biobased material that is biobased but not biodegradable. Few nations have rules and regulations, a disadvantage in bioplastics manufacturing. Carbon tax increases and stricter government regulations, as well as public opinion, are supported. However, the function of political law will grow in importance.

ACKNOWLEDGEMENTS

The authors thank the Technology Mission Division (Energy, Water and All Others), Department of Science and Technology, Ministry of Science and Technology, Government of India (Reference Number: DST/TMD/IC-MAP/2K20/02 Project Title: DST-IIT Hyderabad Integrated Clean Energy Material Acceleration Platform on Bioenergy and Hydrogen); Department of Science and Technology-Science and Engineering Research Board (DST-SERB-No. SB/YS/LS-47/2013), India; Department of Science and Technology-Science and Engineering Research Board (DST-SERB-No. SB/YS/LS-47/2013), India; Department of Science and Technology-Promotion of University Research and Scientific Excellence (DST-PURSE) [DST letter No. SR/PURSE phase 2/38 (G); dated: February 21, 2017], India; RUSA – Phase 2.0 grant [Letter No. F.24-51/2014-U, Policy (TN Multi-Gen), Dept. of Edn. Govt. of India; dated: October 09, 2018]; Scheme for Promotion of Academic and Research Collaboration (SPARC) (No. SPARC/2018–2019/P485/SL; dated: March 15, 2019) and University Science Instrumentation Centre (USIC), Alagappa University, Karaikudi, Tamil Nadu, India.

REFERENCES

Abdel-Rahman, M. A., Tashiro, Y., & Sonomoto, K. (2011). Lactic acid production from lignocellulose-derived sugars using lactic acid bacteria: Overview and limits. *Journal of Biotechnology*, *156*(4), 286–301. https://doi.org/10.1016/j.jbiotec.2011.06.017

Abdel-Rahman, M. A., Tashiro, Y., & Sonomoto, K. (2013). Recent advances in lactic acid production by microbial fermentation processes. *Biotechnology Advances*, *31*(6), 877–902. https://doi.org/10.1016/j.biotechadv.2013.04.002

Alves, M. I., Macagnan, K. L., Rodrigues, A. A., de Assis, D. A., Torres, M. M., de Oliveira, P. D., Furlan, L., Vendruscolo, C. T., & Moreira, A. da S. (2017). Poly(3-hydroxybutyrate)-P(3HB): Review of production process technology. *Industrial Biotechnology*, *13*(4), 192–208. https://doi.org/10.1089/ind.2017.0013

Ammala, A., Bateman, S., Dean, K., Petinakis, E., Sangwan, P., Wong, S., Yuan, Q., Yu, L., Patrick, C., & Leong, K. H. (2011). An overview of degradable and biodegradable polyolefins. Progress *in Polymer Science*, *36*(8), 1015–1049. https://doi.org/10.1016/j.progpolymsci.2010.12.002

Anderson, A. J., & Dawes, E. A. (1990). Occurrence, metabolism, metabolic role, and industrial uses of bacterial polyhydroxyalkanoates. *Microbiological Reviews*, *54*(4), 450–472. https://doi.org/0146-0749/90/040450-23$02.00/0

Anderson, A. J., Williams, D. R., Taidi, B., Dawes, E. A., & Ewing, D. F. (1992). Studies on copolyester synthesis by Rhodococcus ruber and factors influencing the molecular mass of polyhydroxybutyrate accumulated by Methylobacterium extorquens and Alcaligenes eutrophus. *FEMS Microbiology Letters*, *103*(2–4), 93–101. https://doi.org/10.1111/j.1574-6968.1992.tb05826.x

Anjum, A., Zuber, M., Zia, K. M., Noreen, A., Anjum, M. N., & Tabasum, S. (2016). Microbial production of polyhydroxyalkanoates (PHAs) and its copolymers: A review of recent advancements. *International Journal of Biological Macromolecules*, *89*, 161–174. https://doi.org/10.1016/j.ijbiomac.2016.04.069

Arun, A., Murrugappan, R., Ravindran, A. D. D., Veeramanikandan, V., & Balaji, S. (2006). Utilization of various industrial wastes for the production of poly-β-hydroxy butyrate (PHB) by Alcaligenes eutrophus. *African Journal of Biotechnology*, *5*(17), 1524–1527. https://doi.org/10.5897/AJB06.283

Arun, A., Arthi, R., Shanmugabalaji, V., & Eyini, M. (2009). Microbial production of poly-β-hydroxybutyrate by marine microbes isolated from various marine environments. *Bioresource Technology*, *100*(7), 2320–2323. https://doi.org/10.1016/j.biortech.2008.08.037

Bassas, M., Marqués, A. M., & Manresa, A. (2008). Study of the crosslinking reaction (natural and UV induced) in polyunsaturated PHA from linseed oil. *Biochemical Engineering Journal*, *40*(2), 275–283. https://doi.org/10.1016/j.bej.2007.12.022

Benrebah, F., Prevost, D., Yezza, A., & Tyagi, R. (2007). Agro-industrial waste materials and wastewater sludge for rhizobial inoculant production: A review. *Bioresource Technology*, *98*(18), 3535–3546. https://doi.org/10.1016/j.biortech.2006.11.066

Bhatia, S. K., Kim, J.-H., Kim, M.-S., Kim, J., Hong, J. W., Hong, Y. G., Kim, H.-J., Jeon, J.-M., Kim, S.-H., Ahn, J., Lee, H., & Yang, Y.-H. (2018). Production of (3-hydroxybutyrate-co-3-hydroxyhexanoate) copolymer from coffee waste oil using engineered Ralstonia eutropha. *Bioprocess and Biosystems Engineering*, *41*(2), 229–235. https://doi.org/10.1007/s00449-017-1861-4

Bhattacharyya, A., Pramanik, A., Maji, S., Haldar, S., Mukhopadhyay, U., & Mukherjee, J. (2012). Utilization of vinasse for production of poly-3-(hydroxybutyrate-co-hydroxyvalerate) by Haloferax mediterranei. *AMB Express*. https://doi.org/10.1186/2191-0855-2-34

Birch, K. (2016). Emergent imaginaries and fragmented policy frameworks in the Canadian bio-economy. *Sustainability*, *8*(10), 1007. https://doi.org/10.3390/su8101007

Braunegg, G., Lefebvre, G., & Genser, K. F. (1998). Polyhydroxyalkanoates, biopolyesters from renewable resources: Physiological and engineering aspects. *Journal of Biotechnology*, *65*(2-3), 127–161. https://doi.org/10.1016/S0168-1656(98)00126-6

Byun, Y., & Kim, Y. T. (2014). Utilization of Bioplastics for Food Packaging Industry. In *Innovations in Food Packaging* (pp. 369–390). Elsevier. https://doi.org/10.1016/B978-0-12-394601-0.00015-1

Çakar, F., Özer, I., Aytekin, A. Ö., & Şahin, F. (2014). Improvement production of bacterial cellulose by semi-continuous process in molasses medium. *Carbohydrate Polymers*, *106*, 7–13. https://doi.org/10.1016/j.carbpol.2014.01.103

Castilho, L. R., Mitchell, D. A., & Freire, D. M. G. (2009). Production of polyhydroxyalkanoates (PHAs) from waste materials and by-products by submerged and solid-state fermentation. *Bioresource Technology*, *100*(23), 5996–6009. https://doi.org/10.1016/j.biortech.2009.03.088

Cavalheiro, J. M. B. T., de Almeida, M. C. M. D., Grandfils, C., & da Fonseca, M. M. R. (2009). Poly(3-hydroxybutyrate) production by *Cupriavidus necator* using waste glycerol. *Process Biochemistry*, *44*(5), 509–515. https://doi.org/10.1016/j.procbio.2009.01.008

Cesário, M. T., Raposo, R. S., de Almeida, M. C. M. D., van Keulen, F., Ferreira, B. S., & da Fonseca, M. M. R. (2014). Enhanced bioproduction of poly-3-hydroxybutyrate from wheat straw lignocellulosic hydrolysates. *New Biotechnology*, *31*(1), 104–113. https://doi.org/10.1016/j.nbt.2013.10.004

Chaudhry, W. N., Jamil, N., Ali, I., Ayaz, M. H., & Hasnain, S. (2011). Screening for polyhydroxyalkanoate (PHA)-producing bacterial strains and comparison of PHA production from various inexpensive carbon sources. *Annals of Microbiology*, *61*(3), 623–629. https://doi.org/10.1007/s13213-010-0181-6

Chen, G.-Q. (2010). *Industrial Production of PHA* (Vol. 14, pp. 121–132). https://doi.org/10.1007/978-3-642-03287-5_6

Chen, G.-Q., & Wu, Q. (2005). The application of polyhydroxyalkanoates as tissue engineering materials. *Biomaterials*, *26*(33), 6565–6578. https://doi.org/10.1016/j.biomaterials.2005.04.036

Chodak, I. (2008). Polyhydroxyalkanoates: Origin, properties and applications. In *Monomers, Polymers and Composites from Renewable Resources* (pp. 451–477). Elsevier. https://doi.org/10.1016/B978-0-08-045316-3.00022-3

Chua, A. S., Takabatake, H., Satoh, H., & Mino, T. (2003). Production of polyhydroxyalkanoates (PHA) by activated sludge treating municipal wastewater: effect of pH, sludge retention time (SRT), and acetate concentration in influent. *Water Research*, *37*(15), 3602–3611. https://doi.org/10.1016/S0043-1354(03)00252-5

Ciesielski, S., Mozejko, J., & Pisutpaisal, N. (2015). Plant oils as promising substrates for polyhydroxyalkanoates production. *Journal of Cleaner Production*. https://doi.org/10.1016/j.jclepro.2014.09.040

Dhangdhariya, J. H., Dubey, S., Trivedi, H. B., Pancha, I., Bhatt, J. K., Dave, B. P., & Mishra, S. (2015). Polyhydroxyalkanoate from marine Bacillus megaterium using CSMCRI's Dry Sea Mix as a novel growth medium. *International Journal of Biological Macromolecules*. https://doi.org/10.1016/j.ijbiomac.2015.02.009

Dinesh, G. H., Nguyen, D. D., Ravindran, B., Chang, S. W., Vo, D.-V. N., Bach, Q.-V., Tran, H. N., Basu, M. J., Mohanrasu, K., Murugan, R. S., Swetha, T. A., Sivapraksh, G., Selvaraj, A., & Arun, A. (2020). Simultaneous biohydrogen (H2) and bioplastic (poly-β-hydroxybutyrate-PHB) productions under dark, photo, and subsequent dark and photo fermentation utilizing various wastes. *International Journal of Hydrogen Energy*, *45*(10), 5840–5853. https://doi.org/10.1016/j.ijhydene.2019.09.036

Djukić-Vuković, A., Mladenović, D., Ivanović, J., Pejin, J., & Mojović, L. (2019). Towards sustainability of lactic acid and poly-lactic acid polymers production. *Renewable and Sustainable Energy Reviews*, *108*, 238–252. https://doi.org/10.1016/j.rser.2019.03.050

Dobroth, Z. T., Hu, S., Coats, E. R., & McDonald, A. G. (2011). Polyhydroxybutyrate synthesis on biodiesel wastewater using mixed microbial consortia. *Bioresource Technology*, *102*(3), 3352–3359. https://doi.org/10.1016/j.biortech.2010.11.053

Drumright, R. E., Gruber, P. R., & Henton, D. E. (2000). Polylactic acid technology. *Advanced Materials*, *12*(23), 1841–1846. https://doi.org/10.1002/1521-4095(200012)12:23<1841::AID-ADMA1841>3.0.CO;2-E

European Bioplastics, & Nova-Institute. (2020). Bioplastics market development update 2020. *European Bioplastics Org.*, *2021*, 2020-2021.

Federal Ministry of Education and Research (BMBF). (2011). *National Research Strategy BioEconomy 2030 Our Route towards a Biobased Economy*. https://knowledge4policy.ec.europa.eu/sites/default/files/bioeconomy_2030_germany.pdf

García, I. L., López, J. A., Dorado, M. P., Kopsahelis, N., Alexandri, M., Papanikolaou, S., Villar, M. A., & Koutinas, A. A. (2013). Evaluation of by-products from the biodiesel industry as fermentation feedstock for poly(3-hydroxybutyrate-co-3-hydroxyvalerate) production by *Cupriavidus necator*. *Bioresource Technology*, *130*, 16–22. https://doi.org/10.1016/j.biortech.2012.11.088

Garrido, R., Cabeza, L. F., & Falguera, V. (2021). An overview of bioplastic research on its relation to national policies. *Sustainability*, *13*(14), 7848. https://doi.org/10.3390/su13147848

Grujić, R., Vujadinović, D., & Savanović, D. (2017). Biopolymers as Food Packaging Materials. In *Advances in Applications of Industrial Biomaterials* (pp. 139–160). Springer International Publishing. https://doi.org/10.1007/978-3-319-62767-0_8

Haas, R., Jin, B., & Zepf, F. T. (2008). Production of poly(3-hydroxybutyrate) from waste potato starch. *Bioscience, Biotechnology, and Biochemistry*, *72*(1), 253–256. https://doi.org/10.1271/bbb.70503

Haba, E., Vidal-Mas, J., Bassas, M., Espuny, M. J., Llorens, J., & Manresa, A. (2007). Poly 3-(hydroxyalkanoates) produced from oily substrates by *Pseudomonas aeruginosa* 47T2 (NCBIM 40044): Effect of nutrients and incubation temperature on polymer composition. *Biochemical Engineering Journal*, *35*(2), 99–106. https://doi.org/10.1016/j.bej.2006.11.021

Hassan, M. A., Yee, L.-N., Yee, P. L., Ariffin, H., Raha, A. R., Shirai, Y., & Sudesh, K. (2013). Sustainable production of polyhydroxyalkanoates from renewable oil-palm biomass. *Biomass and Bioenergy*, *50*, 1–9. https://doi.org/10.1016/j.biombioe.2012.10.014

Hermann-Krauss, C., Koller, M., Muhr, A., Fasl, H., Stelzer, F., & Braunegg, G. (2013). Archaeal production of polyhydroxyalkanoate (PHA) Co- and terpolyesters from biodiesel industry-derived by-products. *Archaea*, 2013, 1–10. https://doi.org/10.1155/2013/129268

Holmes, P. A. (1985). Application of PHB: A microbially produced biodegradable thermoplastic. *Physics in Technology*, *32*(16), 32–36.

Ivanov, V., Stabnikov, V., Ahmed, Z., Dobrenko, S., & Saliuk, A. (2015). Production and applications of crude polyhydroxyalkanoate-containing bioplastic from the organic fraction of municipal solid waste. *International Journal of Environmental Science and Technology*, *12*(2), 725–738. https://doi.org/10.1007/s13762-014-0505-3

Jem, K. J., & Tan, B. (2020). The development and challenges of poly (lactic acid) and poly (glycolic acid). *Advanced Industrial and Engineering Polymer Research*, *3*(2), 60–70. https://doi.org/10.1016/j.aiepr.2020.01.002

Jem, K. J., van der Pol, J. F., & de Vos, S. (2010). Microbial lactic acid, its polymer poly(lactic acid), and their industrial applications. In *Plastics from Bacteria*. Springer Nature (pp. 323–346). https://doi.org/10.1007/978-3-642-03287-5_13

Jendrossek, D., Schirmer, A., & Schlegel, H. G. (1996). Biodegradation of polyhydroxyalkanoic acids. *Applied Microbiology and Biotechnology*, *46*(5–6), 451–463. https://doi.org/10.1007/s002530050844

Kale, G., Kijchavengkul, T., Auras, R., Rubino, M., Selke, S. E., & Singh, S. P. (2007). Compostability of bioplastic packaging materials: An overview. *Macromolecular Bioscience*, *7*(3), 255–277. https://doi.org/10.1002/mabi.200600168

Keshavarz, T., & Roy, I. (2010). Polyhydroxyalkanoates: bioplastics with a green agenda. *Current Opinion in Microbiology*. *13*(3), 321–326. https://doi.org/10.1016/j.mib.2010.02.006

Khanna, S., & Srivastava, A. K. (2005). Recent advances in microbial polyhydroxyalkanoates. *Process Biochemistry*, *40*(2), 607–619. https://doi.org/10.1016/j.procbio.2004.01.053

Khanna, S., & Srivastava, A. K. (2006). Computer simulated fed-batch cultivation for over production of PHB: A comparison of simultaneous and alternate feeding of carbon and nitrogen. *Biochemical Engineering Journal*, *27*(3), 197–203. https://doi.org/10.1016/j.bej.2005.08.006

Kim, B. S., Lee, S. C., Lee, S. Y., Chang, H. N., Chang, Y. K., & Woo, S. I. (1994). Production of poly(3-hydroxybutyric acid) by fed-batch culture of *Alcaligenes eutrophus* with glucose concentration control. *Biotechnology and Bioengineering*, *43*(9), 892–898. https://doi.org/10.1002/bit.260430908

Kim, D.-H., Kim, S.-H., Kim, H.-W., Kim, M.-S., & Shin, H.-S. (2011). Sewage sludge addition to food waste synergistically enhances hydrogen fermentation performance. *Bioresource Technology*, *102*(18), 8501–8506. https://doi.org/10.1016/j.biortech.2011.04.089

Kucera, D., Benesova, P., Ladicky, P., Pekar, M., Sedlacek, P., & Obruca, S. (2017). Production of polyhydroxyalkanoates using hydrolyzates of spruce sawdust: Comparison of hydrolyzates detoxification by application of overliming, active carbon, and lignite. *Bioengineering*, *4*(4), 53. https://doi.org/10.3390/bioengineering4020053

Kulkarni, S. O., Kanekar, P. P., Nilegaonkar, S. S., Sarnaik, S. S., & Jog, J. P. (2010). Production and characterization of a biodegradable poly (hydroxybutyrate-co-hydroxyvalerate) (PHB-co-PHV) copolymer by moderately haloalkalitolerant *Halomonas campisalis* MCM B-1027 isolated from Lonar Lake, India. *Bioresource Technology*. https://doi.org/10.1016/j.biortech.2010.07.089

Lambert, S. (2015). Biopolymers and their application as biodegradable plastics. In *Microbial Factories* (pp. 1–9). Springer India. https://doi.org/10.1007/978-81-322-2595-9_1

Laycock, B., Halley, P., Pratt, S., Werker, A., & Lant, P. (2013). The chemomechanical properties of microbial polyhydroxyalkanoates. Progress *in Polymer Science*, *38*(3–4), 536–583. https://doi.org/10.1016/j.progpolymsci.2012.06.003

Lee, J. W., Kim, H. U., Choi, S., Yi, J., & Lee, S. Y. (2011). Microbial production of building block chemicals and polymers. *Current Opinion in Biotechnology*, *22*(6), 758–767. https://doi.org/10.1016/j.copbio.2011.02.011

Lee, K.-Y., Buldum, G., Mantalaris, A., & Bismarck, A. (2014). More than meets the eye in bacterial cellulose: Biosynthesis, bioprocessing, and applications in advanced fiber composites. *Macromolecular Bioscience*, *14*(1), 10–32. https://doi.org/10.1002/mabi.201300298

Liu, H., Parker, J., & Sun, W. (2011). Estimating rates at which books are mis-shelved. *CHANCE*, *24*(1), 36–43. https://doi.org/10.1080/09332480.2011.10739850

Mittal, M., Mittal, D., & Aggarwal, N. K. (2022). Plastic accumulation during COVID-19: Call for another pandemic; bioplastic a step towards this challenge? *Environmental Science and Pollution Research*, *29*(8), 11039–11053. https://doi.org/10.1007/s11356-021-17792-w

Rasu, K. M., & Arun, A. (2017). Exploring biodegradable polymer production from marine microbes. In *Biodegradable Polymers: Recent Developments and New Perspectives* (pp. 33–64). IAPC Publishing.

Mohanrasu, K., Premnath, N., Siva Prakash, G., Sudhakar, M., Boobalan, T., & Arun, A. (2018). Exploring multi potential uses of marine bacteria; an integrated approach for PHB production, PAHs and polyethylene biodegradation. *Journal of Photochemistry and Photobiology B: Biology*, *185*, 55–65. https://doi.org/10.1016/J.JPHOTOBIOL.2018.05.014

Mohanrasu, K., Rao, R. G. R., Dinesh, G. H., Zhang, K., Prakash, G. S., Song, D.-P., Muniyasamy, S., Pugazhendhi, A., Jeyakanthan, J., & Arun, A. (2020). Optimization of media components and culture conditions for polyhydroxyalkanoates production by *Bacillus megaterium*. *Fuel*, *271*, 117522. https://doi.org/10.1016/j.fuel.2020.117522

Mohanrasu, K., Guru Raj Rao, R., Dinesh, G. H., Zhang, K., Sudhakar, M., Pugazhendhi, A., Jeyakanthan, J., Ponnuchamy, K., Govarthanan, M., & Arun, A. (2021). Production and characterization of biodegradable polyhydroxybutyrate by Micrococcus luteus isolated from marine environment. *International Journal of Biological Macromolecules*, *186*, 125–134. https://doi.org/10.1016/J.IJBIOMAC.2021.07.029

Mokhtari-Hosseini, Z. B., Vasheghani-Farahani, E., Heidarzadeh-Vazifekhoran, A., Shojaosadati, S. A., Karimzadeh, R., & Khosravi-Darani, K. (2009). Statistical media optimization for growth and PHB production from methanol by a methylotrophic bacterium. *Bioresource Technology*. https://doi.org/10.1016/j.biortech.2008.11.024

Moshood, T., Nawanir, G., Mahmud, F., Mohamad, F., Ahmad, M., & Abdul Ghani, A. (2021). Expanding policy for biodegradable plastic products and market dynamics of bio-based plastics: Challenges and opportunities. *Sustainability*, *13*(11), 6170. https://doi.org/10.3390/su13116170

Mudenur, C., Mondal, K., Singh, U., & Katiyar, V. (2019). Production of polyhydroxyalkanoates and its potential applications. In *Advances in Sustainable Polymers* (pp. 131–164). Springer Singapore. https://doi.org/10.1007/978-981-32-9804-0_7

Muniyasamy, S., & Dada, O. E. (2021). Recycling of plastics and composites materials and degradation technologies for bioplastics and biocomposites. In *Waste Management in the Fashion and Textile Industries* (pp. 311–333). Elsevier. https://doi.org/10.1016/B978-0-12-818758-6.00017-X

Muniyasamy, S., Muniyasamy, S., Mohanrasu, K., Mohanrasu, K., Gada, A., Gada, A., Mokhena, T. C., Mokhena, T. C., Mtibe, A., Boobalan, T., Paul, V., & Arun, A. (2019). Biobased biodegradable polymers for ecological applications: A move towards manufacturing sustainable biodegradable plastic products. In *Integrating Green Chemistry and Sustainable Engineering* (pp. 215–253). John Wiley & Sons, Inc. https://doi.org/10.1002/9781119509868.ch8

Nehra, K., Jamdagni, P., & Lathwal, P. (2017). Bioplastics: A sustainable approach toward healthier environment. In *Plant Biotechnology: Recent Advancements and Developments* (pp. 297–314). Springer Singapore. https://doi.org/10.1007/978-981-10-4732-9_15

Ng, K.-S., Wong, Y.-M., Tsuge, T., & Sudesh, K. (2011). Biosynthesis and characterization of poly(3-hydroxybutyrate-co-3-hydroxyvalerate) and poly(3-hydroxybutyrate-co-3-hydroxyhexanoate) copolymers using jatropha oil as the main carbon source. *Process Biochemistry*, *46*(8), 1572–1578. https://doi.org/10.1016/j.procbio.2011.04.012

Nikel, P. I., Giordano, A. M., de Almeida, A., Godoy, M. S., & Pettinari, M. J. (2010). Elimination of d-lactate synthesis increases poly(3-hydroxybutyrate) and ethanol synthesis from glycerol and affects cofactor distribution in recombinant Escherichia coli. *Applied and Environmental Microbiology*, *76*(22), 7400–7406. https://doi.org/10.1128/AEM.02067-10

Ntaikou, I., Kourmentza, C., Koutrouli, E. C., Stamatelatou, K., Zampraka, A., Kornaros, M., & Lyberatos, G. (2009). Exploitation of olive oil mill wastewater for combined biohydrogen and biopolymers production. *Bioresource Technology*, *100*(15), 3724–3730. https://doi.org/10.1016/j.biortech.2008.12.001

Obruca, S. (2015). Use of lignocellulosic materials for PHA production. *Chemical and Biochemical Engineering Quarterly*, *29*(2), 135–144. https://doi.org/10.15255/CABEQ.2014.2253

Obruca, Stanislav, Marova, I., Snajdar, O., Mravcova, L., & Svoboda, Z. (2010). Production of poly(3-hydroxybutyrate-co-3-hydroxyvalerate) by Cupriavidus necator from waste rapeseed oil using propanol as a precursor of 3-hydroxyvalerate. *Biotechnology Letters*, *32*(12), 1925–1932. https://doi.org/10.1007/s10529-010-0376-8

Pandit, A., & Kumar, R. (2021). A review on production, characterization and application of bacterial cellulose and its biocomposites. *Journal of Polymers and the Environment*, *29*(9), 2738–2755. https://doi.org/10.1007/s10924-021-02079-5

Philip, S., Keshavarz, T., & Roy, I. (2007). Polyhydroxyalkanoates: Biodegradable polymers with a range of applications. *Journal of Chemical Technology & Biotechnology*, *82*(3), 233–247. https://doi.org/10.1002/jctb.1667

Pozo, G., Villamar, A. C., Martínez, M., & Vidal, G. (2011). Polyhydroxyalkanoates (PHA) biosynthesis from kraft mill wastewaters: Biomass origin and C:N relationship influence. *Water Science and Technology*, *63*(3), 449–455. https://doi.org/10.2166/wst.2011.242

Pradhan, S., Dikshit, P. K., & Moholkar, V. S. (2020). Production, Characterization, and Applications of Biodegradable Polymer: Polyhydroxyalkanoates. In *Advances in Sustainable Polymers* (pp. 51–94). https://doi.org/10.1007/978-981-15-1251-3_4

Rai, R., & Roy, I. (2011). Chapter 3. Polyhydroxyalkanoates: The emerging new green polymers of choice. In *A Handbook of Applied Biopolymer Technology. Synthesis, Degradation and Applications* (pp. 79–101). Royal Society of Chemistry, London. https://doi.org/10.1039/9781849733458-00079

Rasu, K. M., & Arun, A. (2017). Exploring biodegradable polymer production from marine microbes. In *Biodegradable Polymers: Recent Developments and New Perspectives* (pp. 33–64). IAPC Publishing. https://doi.org/10.5599/obp.14.7

Ray, S., & Kalia, V. C. (2017). Biomedical applications of polyhydroxyalkanoates. *Indian Journal of Microbiology*, *57*(3), 261–269. https://doi.org/10.1007/s12088-017-0651-7

Reddy, M. V., Yajima, Y., Choi, D., & Chang, Y.-C. (2017). Biodegradation of toxic organic compounds using a newly isolated *Bacillus sp.* CYR2. *Biotechnology and Bioprocess Engineering*, *22*(3), 339–346. https://doi.org/10.1007/s12257-017-0117-0

Rodríguez-Contreras, A., Koller, M., Miranda-de Sousa Dias, M., Calafell-Monfort, M., Braunegg, G., & Marqués-Calvo, M. S. (2015). Influence of glycerol on poly(3-hydroxybutyrate) production by *Cupriavidus necator* and *Burkholderia sacchari*. *Biochemical Engineering Journal*, *94*, 50–57. https://doi.org/10.1016/j.bej.2014.11.007

Roy, I., & Visakh, P. M. (Eds.). (2014). *Polyhydroxyalkanoate (PHA) based Blends, Composites and Nanocomposites*. Royal Society of Chemistry. https://doi.org/10.1039/9781782622314

Rusendi, D., & Sheppard, J. D. (1995). Hydrolysis of potato processing waste for the production of poly-β-hydroxybutyrate. *Bioresource Technology*, *54*(2), 191–196. https://doi.org/10.1016/0960-8524(95)00124-7

Sandhya, M., Aravind, J., & Kanmani, P. (2013). Production of polyhydroxyalkanoates from Ralstonia eutropha using paddy straw as cheap substrate. *International Journal of Environmental Science and Technology*, *10*(1), 47–54. https://doi.org/10.1007/s13762-012-0070-6

Santos, S. M., Carbajo, J. M., Quintana, E., Ibarra, D., Gomez, N., Ladero, M., Eugenio, M. E., & Villar, J. C. (2015). Characterization of purified bacterial cellulose focused on its use on paper restoration. *Carbohydrate Polymers*, *116*, 173–181. https://doi.org/10.1016/j.carbpol.2014.03.064

Sato, S., Maruyama, H., Fujiki, T., & Matsumoto, K. (2015). Regulation of 3-hydroxyhexanoate composition in PHBH synthesized by recombinant Cupriavidus necator H16 from plant oil by using butyrate as a co-substrate. *Journal of Bioscience and Bioengineering*, *120*(3), 246–251. https://doi.org/10.1016/j.jbiosc.2015.01.016

Schneider, A. L. S., Silva, D. D., Garcia, M. C. F., Grigull, V. H., Mazur, L. P., Furlan, S. A., Aragão, G. F., & Pezzin, A. P. T. (2010). Biodegradation of poly(3-hydroxybutyrate) produced from *Cupriavidus necator* with different concentrations of oleic acid as nutritional supplement. *Journal of Polymers and the Environment*, *18*(3), 401–406. https://doi.org/10.1007/s10924-010-0184-1

Simon-Colin, C., Raguénès, G., Crassous, P., Moppert, X., & Guezennec, J. (2008). A novel mcl-PHA produced on coprah oil by Pseudomonas guezennei biovar. Tikehau, isolated from a "kopara" mat of French Polynesia. *International Journal of Biological Macromolecules*, *43*(2), 176–181. https://doi.org/10.1016/j.ijbiomac.2008.04.011

Sin, L. T., Rahmat, A. R., & Rahman, W. A. W. A. (2013a). Synthesis and Production of Poly(Lactic Acid). In *Polylactic Acid* (pp. 71–107). Elsevier. https://doi.org/10.1016/B978-1-4377-4459-0.00002-0

Sin, L. T., Rahmat, A. R., & Rahman, W. A. W. A. (2013b). Thermal properties of poly(lactic acid). In *Polylactic Acid* (pp. 109–141). Elsevier. https://doi.org/10.1016/B978-1-4377-4459-0.00003-2

Singh, N., Tang, Y., Zhang, Z., & Zheng, C. (2020). COVID-19 waste management: Effective and successful measures in Wuhan, China. *Resources, Conservation and Recycling*, *163*, 105071. https://doi.org/10.1016/j.resconrec.2020.105071

Steinbüchel, A., & Hein, S. (2001). *Biochemical and Molecular Basis of Microbial Synthesis of Polyhydroxyalkanoates in Microorganisms* (pp. 81–123). https://doi.org/10.1007/3-540-40021-4_3

Steinbuchel, A., & Valentin, H. E. (1995). Diversity of bacterial polyhydroxyalkanoic acids. *FEMS Microbiology Letters*, *128*(3), 219–228. https://doi.org/10.1111/j.1574-6968.1995.tb07528.x

Stephens, J. H.., & Rask, H.. (2000). Inoculant production and formulation. *Field Crops Research*, *65*(2–3), 249–258. https://doi.org/10.1016/S0378-4290(99)00090-8

Sudesh, K., Abe, H., & Doi, Y. (2000). Synthesis, structure and properties of polyhydroxyalkanoates: Biological polyesters. *Progress in Polymer Science* (*Oxford*). https://doi.org/10.1016/S0079-6700(00)00035-6

Taniguchi, R. (2018). Resource mobilization by "Strange Bedfellows": A case study of "Biomass Nippon Strategy." *2018 Portland International Conference on Management of Engineering and Technology* (*PICMET*), 1–9. https://doi.org/10.23919/PICMET.2018.8481884

Thakur, S., Chaudhary, J., Sharma, B., Verma, A., Tamulevicius, S., & Thakur, V. K. (2018). Sustainability of bioplastics: Opportunities and challenges. *Current Opinion in Green and Sustainable Chemistry*, *13*, 68–75. https://doi.org/10.1016/j.cogsc.2018.04.013

Ullah, H., Wahid, F., Santos, H. A., & Khan, T. (2016). Advances in biomedical and pharmaceutical applications of functional bacterial cellulose-based nanocomposites. *Carbohydrate Polymers*, *150*, 330–352. https://doi.org/10.1016/j.carbpol.2016.05.029

USDA. (2020). *USDA Biopreferred*. USDA.

Venkateswar Reddy, M., Mawatari, Y., Yajima, Y., Satoh, K., Venkata Mohan, S., & Chang, Y.-C. (2016). Production of poly-3-hydroxybutyrate (P3HB) and poly(3-hydroxybutyrate-co-3-hydroxyvalerate) P(3HB-co-3HV) from synthetic wastewater using Hydrogenophaga palleronii. *Bioresource Technology*, *215*, 155–162. https://doi.org/10.1016/j.biortech.2016.03.025

Verlinden, R. A., Hill, D. J., Kenward, M. A., Williams, C. D., Piotrowska-Seget, Z., & Radecka, I. K. (2011). Production of polyhydroxyalkanoates from waste frying oil by Cupriavidus necator. *AMB Express*, *1*(1), 11. https://doi.org/10.1186/2191-0855-1-11

Vroman, I., & Tighzert, L. (2009). Biodegradable polymers. *Materials*, *2*(2), 307–344. https://doi.org/10.3390/ma2020307

Wallen, L. L., & Rohwedder, W. K. (1974). Poly-.beta.-hydroxyalkanoate from activated sludge. *Environmental Science & Technology*, *8*(6), 576–579. https://doi.org/10.1021/es60091a007

WHO. (2020). *Shortage of personal protective equipment endangering health workers worldwide, 3 March 2020a*, Geneva

Wu, T. Y., Mohammad, A. W., Jahim, J. M., & Anuar, N. (2009). A holistic approach to managing palm oil mill effluent (POME): Biotechnological advances in the sustainable reuse of POME. *Biotechnology Advances*, *27*(1), 40–52. https://doi.org/10.1016/j.biotechadv.2008.08.005

Ye, J., Huang, W., Wang, D., Chen, F., Yin, J., Li, T., Zhang, H., & Chen, G.-Q. (2018). Pilot Scale-up of poly(3-hydroxybutyrate-co-4-hydroxybutyrate) production by *Halomonas bluephagenesis* via cell growth adapted optimization process. *Biotechnology Journal*, *13*(5), 1800074. https://doi.org/10.1002/biot.201800074

Yu, J., & Stahl, H. (2008). Microbial utilization and biopolyester synthesis of bagasse hydrolysates. *Bioresource Technology*. https://doi.org/10.1016/j.biortech.2008.03.071

6 Agro-based Bioplastic Production and Its Application

R. Nithya, K. Mohanrasu and A. Arun

6.1 INTRODUCTION

Plastic is an invulnerable material that can be recast into different forms and utilized in a wide range of applications due to its lightweight, hygienic, and corrosion-free. Day-to-day lifeand the survivability of human beings without plastic is an unimaginable part of the modern world (Umesh et al., 2018). The attractive qualities of plastic are high heat combustion, cost-effectiveness, ease of production, transparency, softness, flexibility, and increased availability in the local community. Plastics play a vital role as innovative materials for use in engineered tissues, absorbable sutures, prosthetics, and other medical applications (Andrady and Neal, 2009).

The necessity and usage of plastics are not only dependent on their mechanical and thermal properties;they aremainly familiar with stability and durability (Rivard et al., 1995). Initially, the first man-made plastic is Parkesine, which was derived from the cellulose biological compound and it can be moulded to any shape (Shah et al., 2021). Later, the natural materials were replaced by chemical polymers (synthetic polymers) which end up in waste all around the environment. Synthetic polymers like polyamides, nylon, polystyrene,teflon, polyethylene terephthalate and polyethylene are derived from petroleumbyproducts. The consumption and user-friendly nature of synthetic plastic results in the depletion of fossil fuels, difficulty in solid waste management and harm to nature (Sharma et al., 2017).

Plastic usage is directly proportional to the accumulation of waste on land and in the oceans (de Moura, et al., 2017). Most types of plastic are not biodegradable but photodegradable, and they slowly break down into small fragments known as microplastics. Due to high UV irradiation and abrasion by waves, the degradation of the microplastic process is much slower in the ocean due to cooler temperatures and reduced UV exposure. Accordingly, plastic is made up of various toxic chemicals which cause serious environmental threats to modern society and result in soil, air, water and also obstruct the groundwater movement (Silva et al., 2014).

The main disadvantage of plastic waste may usually be in the form of film and contain harmful heavy metals, which, when mixed up with water or rainwater, result in crumbling soil fertility. Hydrocarbon-based plastic waste will have a comparatively high calorific value when used for incineration or boiler, but at lower temperatures, the burning of plastics may release deadly poisonous chemical gases into the

DOI: 10.1201/9781003304142-6

air, including dioxins, which are corrupting to human beings. Using the reprocessing line, Plastic waste can also be used to produce new plastic-based products (Saiki and De Brito, 2012). A serious environmental concern is created mainly by the usage of plastic bags, since they last in surroundings for up to 1000 years and do not decompose. Accumulation or disposal of a large quantity of plastic bags ends up in pollution of land, water bodies, and air, destroying the biosphere and directly or indirectly affecting living organisms (Raghatate Atul, 2012).

Plastic particles and their chemical additives cause environmental and health hazards to living organisms in the world. Generally, plastic debris entershumans directly or indirectly through parasites as vectors and causes particle toxicity and chemical toxicity (Vethaak and Leslie, 2016). Using the animal model studies, the researchers proved that microplastic traverse into the cell membranes, the blood-brain barrier, and even the human placenta results in oxidative stress, lung and gut injury, cell damage, and inflammation (Kershaw and Rochman, 2015; Gasperi et al., 2015; McCormick et al., 2014; Galloway, 2015).

Geyer et al., (2017) published the first global data on the production, usage and end-of-life of plastics. He clearly declared that current extensive and expanding use of plastic as same is continue, roughly 12,000 Mt of plastic waste will be accumulated in the natural environment by 2050. Considering the environmental and health issues caused by plastic and its Greenhouse gas emissions pays the way to find alternative ways towards a circular economy, which helps to decrease the non-renewable resources production throughreuse and recycling of the materials (Rosenboom et al., 2022). The increasing concern aboutsynthetic plastics issues has made scientistsattracted towards "green" plastics such as bioplastics which are produced from renewable resources (Siracusa and Blanco, 2020). To avoid the nondegradable petroleum product, sustainable, significant and inexpensive methods is needed to produce the best-quality of biobased renewable materials (Bohmert-Tatarev et al., 2011).

Bioplastics are vouched as a noteworthy alternative to plastics because their production is not dependent on fossil fuels and is cost-effective. In 2020, European Bioplastics declared that according to the development of the bioplastics market, nearly 1% of bioplastics, more than 368 million tons of plastic produced every year. Due to increases in the demand for the use of plastics, the bioplastics market is growing and diversifying all over the world. In the same year, the report declared that the capacity of bioplastics globally is expected to increase roughly from 2.11 to 2.87 million tons in 2025. It is the best substitute for every conventional plastic material.The commercialization bioplastics such as polyethylenefuranoate (PEF), polypropylene-based (PP), polyhydroxyalkanoates (PHAs), and polylactic acid (PLA) are expected to grow in their manufacturing process over the next few years (Shen et al., 2009).

Biopolymers are synthesised from biomass feedstock or a combination of sustainable, eco-friendly substances, and it has been reported that the manufacturing capacity of bioplastics has increased from 0.36 Mt in 2007 to 3.45 Mt in 2020 (Agustin et al., 2014). It has the capability to diminish the landfills, enhance the degradation and eliminate the toxins which create air pollution. The significant essential for the production of bioplastics is mainly dependent on the usage of economical raw materials and methodology adapted for the production. Bioplastic products are mainly produced from various sustainable feedstocks, which are available in plenty and are called agro-waste products (Nee and Othman, 2022). Agro-based polymers are one of the best and most well-known polymers.

Agricultural waste is normally incinerated or dumped, which causesenvironmental problems. It contains polysaccharidesand carbohydrates, so it gives alternative ideas to use as a starting material to produce biobased polymers. The incorporation of agricultural residues into the polymer is highly innovative because it has high strength, stiffness, availability, recyclability, low cost, environmental friendliness, no toxicity, lower pollution emissions and low density of natural fibres (Pirayesh et al., 2013; Borri, et al., 2013; Reixach et al., 2013; Hamza et al., 2013; Battegazzore et al., 2014).Therefore, ignored wastes like agricultural residues should be put in the attention and purposely studied as substrates for sustainable energy.

6.2 CLASSIFICATION OF PRODUCTION OF BIOBASED MATRIX

Biobasedmatricesare ecologically benign,biocompatible, and biodegradable. The origin of biopolymerisgrouped into four classifications, as shown in Figure 6.1. The classification is based on the synthesis of polymers from chemicals and bacteria, as well as blends with polymers and from biomass or natural resources (Satyanarayana et al., 2009).

6.3 NEED OF RECYCLING OF AGRICULTURAL WASTE

Agricultural wastes areunwanted materials produced entirely from agricultural operations directly related to the growing of crops or the primary purpose of making a profit by rising of animals or by agro-industries. Plant waste from agriculture poses a high negative impact problem on the environment due to their landfills, soil pollution, and incineration process. Agricultural waste is often disposed of in open trash

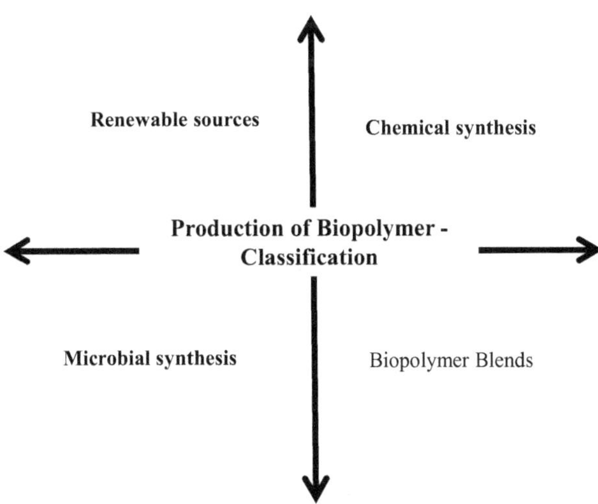

FIGURE 6.1 Classification of production of biopolymer.

dumps, where it can contribute to severe soil contamination and increase the dangers already present because stored garbage can emit toxic gases. The great strategy to control environmental pollution is only by the management of agricultural waste (Mostafa et al., 2018).

Because the cost of recovery and collecting for reuse is more than the lucrative value of various forms of waste as the end products from the agro-industries, are discarded, which has a negative impact on the environment (Panesar and Kaur, 2015). Nearly, 90 million tons of oil equivalents of agricultural waste were resulting from enormous supply chains (Gontard, et al., 2018). Only a small portion of the massive waste accumulation is used in the manufacturing of animal feeds, manure, and other value-added products, and the remaining tons of waste are causing pollution and environmental hazards. However, the successful utilization of waste could be a good remedy for the development of degradable products, if appropriate biotechnological interventions are available (Panesa and Kaur, 2015).

The agro-wasteisrich in micro and macronutrients especially starch and cellulose, which can be extracted and moulded into new forms. Using this principle, the reinforcement of agricultural residues into biopolymer is presently an interestingtopic in the research field due to their compatibility, recyclability, lowcost, and energy consumption (Borri et al., 2013; Reixach et al., 2013; Hamza et al., 2013; Battegazzore et al., 2014). Moreover, it is rich in fermentable sugars, which microbes can directly utilize and convert into high-value components (Ali and Zulkali, 2011). One example of byproducts is organic acids such as lactic acid, which are obtained from industrial and agricultural waste as substrates.

6.4 AGRO-WASTE INTO BIOPLASTICS

The general process of producing bioplastics from agro-waste materials is pretreatment, extraction and characterizing of materials (Hagemann and D'Amico, 2009), as shown in Figure 6.2.

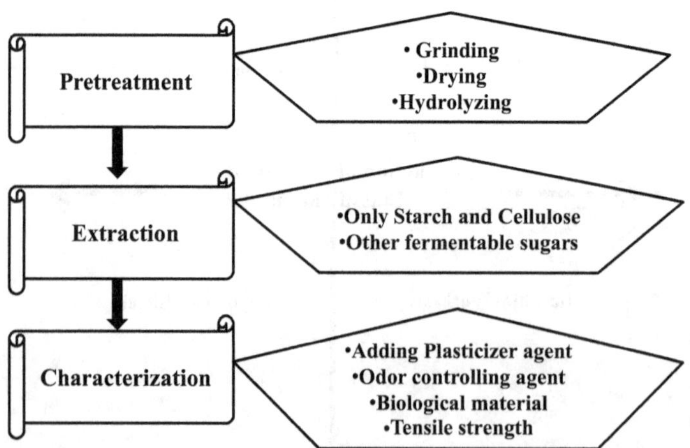

FIGURE 6.2 Steps involved in conversion of agro-waste into bioplastic.

6.5 CRITERIA FOR THE SELECTION OF AGRO-WASTE RESIDUES

The use of agro-wastes in the production of biobased matrix is primarily due to the large quantity of starting materials/precursors available, cost-effectiveness, andavailability. India and China are the two most powerful economies with the potential to increase the production of fruit and vegetable-based biopolymers (Sharma et al., 2020). Figure 6.3 shows the important criteria for selectingagro-waste for bioplastic production.

The lignocellulose fibres are rich in waste materials like pineapple, sisal, jute and curaua (Satyanarayana et al., 2009), and the end products are determined by extraction method. The mechanical strength of produced biomaterials is increased by the large quantities of cellulose, so selecting the precursors with cellulose is important (Maraveas, 2020). For example, corn and stalks are rich in cellulose, which is the major waste generated from agriculture.

6.6 STAGES OF BIODEGRADATION

Biodegradation is defined as the degradation of materials in the presence of biological activity. Biodegradation occurs in three stages: (i) biodeterioration – the physical, chemical, and mechanical properties of the polymer are altered by microorganisms; (ii) biofragmentation – polymers degrade into oligomers and monomers as a result of microbial activity, and (iii) assimilation – it is a continues fragmentation process. Fragmented carbon polymers are converted into carbon dioxide, water, and biomass (Lucas et al., 2008).

Degradability of synthetic polymers is impossible because of their chemical structure, complexity, chain of polymers, and crystallinity structure. Polymers with shorter chains, less complexity, and amorphous structure are typically easily degradable by microorganisms, and environmental factors such as pH, temperature, and moisture play an important role in biodegradation (Massardier-Nageotte et al., 2006; Kale et al., 2007).

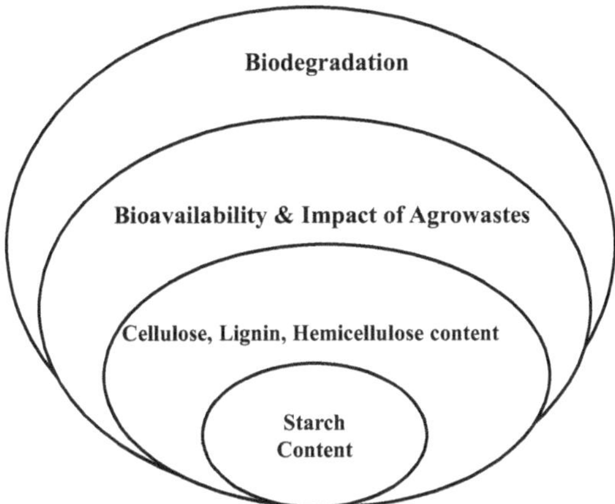

FIGURE 6.3 Criteria for the selection of suitable agro-waste residues.

6.7 PRECURSOR FROM AGRICULTURAL WASTES

6.7.1 Cellulose and Its Sources

Cellulose is the most abundant biomass material on the planet, and it is well-known as natural fibres or regenerated fibres obtained from cotton, jute and rayon (Iwata, 2015). Cellulose is one of the important sources for bioplastic production, but it does not have a plasticity characteristic. The derivatives of cellulose are cellulose nanocrystals (Arrieta et al., 2015), nanofiber cellulose (Siró and Plackett, 2010), and acetate butyrate (Grunert and Winter, 2002) are used directly in the production of bioplastics. Even though, cellulose lacks plasticity nature, itspresence in bioplastics preparation will increase the physical properties of polymers and as fillers. Biocomposites are natural filler-based materials that have been successfully produced and marketed. The familiar choice of filler in thermoplastic materials is cellulose (natural fillers), which isinexpensiveand obtained from renewable resources like coconut shells, palm kernel shells, oil palm empty fruit bunch, and rice husk (Chun et al., 2012; Salmah et al., 2013; Chun et al., 2013).

6.7.2 Cellulose from Cocoa Pod Husk

Cocoa is an important agricultural crop in several tropical countries (Vriesmann et al., 2012), and Cocoa Pod Husk is a waste generated from cocoa pod net weight ranging from 52% to 76% (Koay et al., 2013). Lubis et al. (2018) revealed that bioplastics were produced from starch of jack fruit seed reinforced with cellulose of cocoa pod husk. In this method, glycerol acts as a plasticizer. Using alkali treatment, bleaching and hydrochloric acid hydrolysis, microcrystalline cellulose was produced. It showed 74% crystallinity was confirmed in XRD, and rod shape (5–10 μm length) with adiameter of 11.635 nm was proved by SEM.

Azmin and Nor (2020) proved that cellulose and fibre were extracted from cocoa pod husk and sugarcane bagasse and incorporated together to form bioplastic materials. Their observations showed that the ratio of cellulose to fibre is 75:25,carriesgood water absorption and plays the best role in food packaging. The produced materials proved to bethebest antimicrobial by reducing the growth of mould and preservingthe food for a longer time by preventing the moisture transfer between the environment and food. Due to the presence of hydrophilic nature of bioplastic derived from cellulose, decreases the water vapour barrier, which leads to brittleness and reduced mechanical properties.

6.7.3 Cellulose from Oil Palm Empty Fruit Bunch

Accumulation of palm fruit waste after crude oil extraction is 28.65 million Mt each year in the land. It is eradicated by burning or dumping in the land, resulting in air pollution. Palm fruit waste contains more cellulose (40.37%) than hemicelluloses (20.06%) and lignin (23.89%) (Isroi et al.,, 2017). Due to the high content of cellulose in palm fruit waste, it is used as a good source of cellulose derivatives.The cellulose from the waste is isolated and bleached to remove lignin using sodium hydroxide and sodium hypochlorite methods, and the cellulose content is increased by about 97%.

Oxidation of cellulose helps to decrease the crystallinity and improve the carboxyl side chain of the components. Finally, for bioplastics production the purified cellulose is mixed with glycerol (plasticizer) and cassava starch as composites. It clearly ends up in the production of bioplasticsheets using casting methodology (Cifriadi et al., 2017).

6.7.4 Cellulose from Pineapple Peel

Carboxymethyl cellulose (CMC)hasahigh degree of solubility and is used as a resource for fabricating bioplastics, extracted from agricultural wastes. Pineapple is apopular edible fruit and is produced as a solid waste during processing. The accumulated wastes are rich in cellulose andcan be extracted by refluxing with acidic or alkali solution (Xu et al., 2009). Using etherification, CMC is produced from pineapple peel cellulose. Chumee and Khemmakama (2014) proved the CMC extraction by etherification and refluxing of pineapple peel powder.

6.7.5 Cutin and Its Sources

The outermost membrane of plant cuticles is composed of cutin, which protects the plant epidermis of aerial organs. The size and thickness of plant cuticles range between 0.02 and 200 m (Nawrath, 2006). Pure cutin and its monomers are rich in biopolyester, non-toxic, biodegradable, waterproof, UV-blocking and amorphous. It also plays an important role in the protection against pathogensand abiotic factors and also in avoiding water loss from internal tissues. These features increase the interest in considering cutin as a realistic, raw material in food-packaging material instead of conventional plastic (Zhang and Uyama, 2016; Heredia-Guerrero etal., 2017).

The main source of cutin biomass is tomato pomace. Due to the low price and more harvesting, tomatoes are used for multiple purposes, particularly atthe industry level. In the tomato processing industry, the skin of tomato (tomato pomace) accumulation is more because of its high production of juice, paste, pureeand sauce. About 20wt% of polyester is present in tomato pomace, which is considered agood raw material for bioplastics production. Using the melt polycondensation method, from the tomato pomace unsaturated and polyhydroxylated fatty acids were converted to free-standing films in the presence of catalyst Tin(II) 2-ethylhexanoate (Heredia-Guerrero etal., 2019). Benítez et al. (2018) produced unsaturated fatty acids without a catalyst. Finally, inexpensive resource for the fabrication of bioplastics has been synthesized from tomato pomace agro-waste, which has the features of amorphous, hydrophobic nature, insoluble, infusible, and thermally stable.

6.7.6 Starch and Its Sources

Starch is a naturally occurring polymer that is one of the major carbohydrates. It is derived from agricultural resources such as corn, wheat, rice, tapioca, amaranth, cashewnuts, and potatoes in the form of granules (Ren et al., 2009).After extraction, it is white in colour, tasteless, odourless, and undissolved in cold water or alcohol, easily available with low-cost, non-toxic and biologically absorbable (Sanyang et al., 2018). It is also renewable, biodegradable, considered as avalid choice of sustainable resources and capability to reveal thermoplastic properties.

In plant tissue, each granule of starch contains polysaccharides such as amylopectin (70%–80%) found in million molecules which are accompanied by amylose (20%–30%).The structure of both polysaccharides has D-glucose units in$^4 C_1$ conformation. The length of the amylose chains ranges from 10^2 to 10^4 glucose units, and amylopectin varies from 10^4 to 10^5 glucose units (Ponstein, 1990). Starch is semicrystalline in nature due to its 20%–40% crystallinity (Mentzer et al., 1984), and the presence of hydrophilic propertiesmakes itincompatiblewith all hydrophobic polymers. Due to these properties, starch plays an important role as an alternative to replace thermoplastic.

Native starch cannot be used directly as traditional plastic because ithas limited high water affinity, brittleness, limited long-term stability caused by water absorption, ageing caused retro gradation, poor mechanical properties and bad processability. Thermoplastic starch is produced by using disruption, plasticization and the application of mechanical and thermal treatment (Mathew and Dufresne, 2002; VanderBurgt et al., 1996). Blending plasticized starch with some other degradable polymer is also thebest solution to produce cost-effective materials (Gaspar et al., 2005).

Usage of starch as thermoplastic in polymer technology is based on two forms:starch-filled plastics and structural starch modifications. In starch-filled plastics, the microbial uptake of the starch leads to increased porosity by increasing the surface area. In structural starch modifications, using heat and mechanical methods, the native starch forms are modified, and methods are listed (Bastioli et al., 2013). The significance of starch-based biodegradable plastics is that they decrease the cost and usage of synthetic plastics, which enhance the biodegradability of the products (Wool et al., 2000).

Plasticizers play an important role in the blending of starch into plastic, and they are a low molecular weight substancethat reduces the hydrogen bonding of the chains, which helps to increase the flexibilityand processability of polymers. Hydrogen bond-forming abilities aresignificant between plasticizers and starch molecules, which arethe main properties of starch-based bioplastics. The physical characteristics of the processed starch depend on the plasticizers quantity and its types by controlling destructurization and depolymerization of the final materials (Averous, 2004).Various plasticizers such as water, amide, sugars, glycol, quaternary amine, urea, glycerol and sorbitol are used to prepare thermoplastic starch blends (Shi et al., 2007).

In 2020, it was reported that starch-based plastics havethe largest production capacity of about 1.3 Mt, and the remaining depends on the production of polylactic acid (PLA) and polyhydroxyalkanoates (PHAs) (Shen et al., 2009).

6.7.6.1 Starch from Potato Waste

Potato is the world's fourth major crop next to rice, wheat and maize (Leo et al., 2008). Inthe world, India stands in third place as the largest potato producer after China and the Russian Federation, with nearly 41,483 thousand tons of potato produced

(FAOSTAT, 2016). Every year, nearly millions of tons of skin waste are generated in the potato chip industry and for disposal,a significant amount of money is required for transportation (Rogols et al., 2003). Composting or using it as feed for animals is a conventional management strategy for discarding peel waste to avoid pollutants (Nelson, 2010). The waste contains large quantities of starch, cellulose, hemicelluloses, carbohydrates and proteins, which lead to thedevelopment of biodegradable polymers.Due to the presence of excess starch in potato peels, it is a potential application for the development of biopolymer film (Kang and Min, 2010).

Xie et al. (2020) proved that the combination of potato peel with bacterial cellulose (BC) and curcumin leads to the formation of active potato peel films, and these bond interactions strengthened the mechanical properties of the films, water vapour permeability and the light transparency. During the time of packaging, curcumin present in the films inhibits the lipid oxidation of fresh pork at the time of storage. He clearly reported that potato films with BC will have potential in packaging companiesand are the best solution for replacing thermoplastic polymers.

Degradability of potato peel bioplastics was determined by Arikan and Bilgen (2019). They concluded that using potato peels, waste bioplastics can be produced, and it will be completely degraded within 28 days compared with commercial bioplastics. Using ultrasound treatment, biopolymer film development was done by incorporating potato peel waste and sweet lime pomace. It clearly reported that an increase in the ultrasound treatment gives the best properties of biopolymer film such as weight loss, hardness and microbial load reduction(Borah et al., 2017).

6.7.6.2 Starch from Jackfruit Seed

In Asia, the most tropical popular fruit is jack fruit, which contains about 100–500 seeds in single fruit and is enriched with protein and starch. Mukprasirt and Sajjaanantakul (2004) compared the cotyledon starch with modified starch, and it clearly revealed that seed starch isrich in amyloseand protein contents, requiresless temperature for gelatinizationand has high stability compared with modified starches. They concluded seed starch can replace the modified starch.

Starch is an important precursor in the production of biodegradable films due to its renewable and inexpensive sources. About 8%–15% of jackfruit seed is made up of starch, and it is used in bioplastic production. From the seeds, starch was extracted using grinding and deposition methods. After deposition, the starch-rich suspension was filtered using Whatman filter paper and dried at 70°C. The dried starch powder was reinforced with cellulose as filler using plasticizer (glycerol), and finally, bioplastics wereoptimized with different ratios. This research reflects the usage of jackfruit seeds as starch source to produce bioplastics (Lubis et al., 2018).

6.8 ROLE OF MICROBES IN AGRO-BASED PLASTIC PRODUCTION

6.8.1 Polyhydroxyalkanoates (PHA)

PHAs are important biopolymers that are produced by the bacteria and stored inside the cell in the form of inclusion bodies as reservoirs (Shrivastavet al., 2010).Several studies declared that various PHAs are produced by different bacterial strains, and due to their biodegradable properties,they arean alternative to petroleum-based plastics (López-Cuellar et al., 2011). The composition of monomers of PHAs is directly related to the available carbon source and to the bacteria species. The selection of raw materials to produce PHAs is difficult, since it depends on the cost-effectiveness. It can be solved only by the use of industrial byproducts like farming, domestic waste materials, fats and cellulose-rich wastes (Rasu and Arun, 2017). The best PHAs can be produced by agro-industrial waste as substrates since it is rich in all needed compounds.

In everyday industrial life, the food industry generates waste palm oil, which is the best feedstock for the production of polyhydroxylalkonates (PHA). In the world of biopolymer production, using waste palm oil will decrease the costs and the adoption of PHA. The waste palm oil contains fatty acid-long chains as a carbon source, and during the fermentation by bacteria,it is converted into structurally different PHA with 6–14 carbon atoms. Using saponified waste palm oil as a carbon source, *Pseudomonas* sp. G101 was able to produce an increased amount of biopolymer (PHA) with the best thermal and physical properties (Możejko and Ciesielski, 2013).

Vega-Castro et al.(2016) produced and characterized the PHA from solid waste of pineapple and fermented by the bacteria *Ralsthonia eutropha*. By optimizing different fermentation conditions, PHAs wereproduced and using FTIR, GCMS, and NMR, the produced compounds were characterized. They proved at pH 9, 60 hr fermentation, the best PHAs were obtained, and it is auseful alternative for processing of waste generated in agro-industries.

6.8.2 Lactic Acid

Lactic acid is the basic chemicalused to produce PLA, and the most commonly biodegradable plasticused today. LA can preventthe growth of many pathogenic microorganisms, and it is used as flavour, acidulant, and preservative in food. It also acts as a scraping solution in high quantities and an important ingredient in skin care creams designed to remove dead cells from the surface of the skin (Komesu et al., 2017; Smith, 1996). LA is essential in the production of poly (lactic acid) (PLA) and poly (lactic-co-glycolic acid) (PLGA), the most common eco-friendly bioplastics at present (Anderson and Shive, 1997; Tyler et al., 2016).

Using biotechnological techniques, lactic acid is produced from agricultural residues and isanimportant componentin producing biocompatible polymers. The produced polymers are used in high-end applications, but their production and recovery costs are extremely expensive.The exact method to produce LA is fermentation by bacteria, fungi and yeast (Gao et al., 2011). *Carnobacterium, Enterococcus, Lactobacillus, Lactococcus, Leuconostoc, Pediococcus, Streptococcus, Tetragenococcus, and Vagococcus* are the lactic acid-producing bacteria. The LA producer's (LA bacterias) characteristics are not similar toeach other since their morphological, physiological and biochemical features are different in nature, and each should be considered for fermentation on exact raw materials. The metabolic engineering techniques and optimization of LA production improvement aregoing on (Upadhyaya et al., 2014).

Cui et al. (2011) showed that using mixed culture system, glucose was converted into LA by *L. rhamnosus* through the Embden–Meyerhof–Parnas (EMP) pathway,

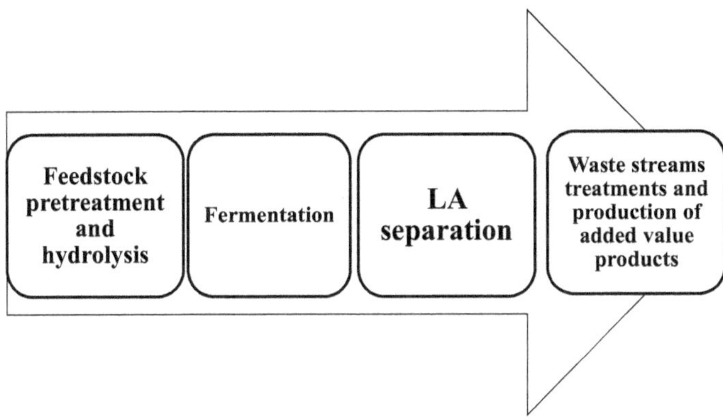

FIGURE 6.4 Stages in lactic acid production.

and at the same time, xylose from the cornstoverwas converted to LA by *L. brevis* strain. The foremost company, Nature Works LLC, USA, stands as a leading manufacturer of PLA and PLGA from agro-residues like corn starch, sugar beet, and sugarcane as raw materials. The lab-scale production of PLA resin is dependent on corbion, and there is extensive research being conducted around the world to derive the LA using second-generation materials such as bagasse, corn stover, and wheat straw as sources to produce the PLA. Reports on the processing of PLA obtained by polymerization of LA from waste and byproducts of agriculture are scarce. Both academia and bioplastic producers are urged to conduct more intensive research in the domain of lactide production methods by using second-generation feedstock for its polymerization in order to demonstrate the feasibility of PLA production from specific agricultural substrates.Generally, four major stages may possiblybe distinguished in fermentative LA production (Figure 6.4).

6.8.2.1 Feedstock Pretreatment and Hydrolysis

The valorization process of LA production directly depends on the substrate chemical composition. The primary precursor of agricultural residues is solid, and in the secondary, either solid or liquid. During processing, the storage time, unwanted bacterial growth, and deterioration are analysed based on the water content present in the substrate. The complex carbohydrates are hydrolysed in the pretreatment, which helps the microbes to access the simple compounds in substrate to produce the LA. Because some substrates are high in readily available monosaccharides, microorganisms can be used without any pretreatment (Ohkouchi and Inoue, 2006; Venus et al., 2018; Lu et al., 2010).

In LA production, particular microbes hydrolysed the carbohydrates into fermentable sugars. The production of fermentable sugars from carbohydrates is called saccharification. The enzymes play an important role in the hydrolysis of starch through gelatinization, liquefaction, and saccharification. The optimization of hydrolysis for different substrates is necessary to find the enzyme selection, temperature, mixing conditions, etc. (Panesar and Kaur, 2015).

6.8.2.2 Fermentation

The LA production mainly depends on the pH, and it varies from 5.0 and 7.0, and it will prevent the inhibition of the cell growth (Wang et al., 2015). Nearly 90% of LA was produced using $Ca(OH)_2$ as neutralizing agent, and calcium lactate was formed. The liberation of LA from calcium lactate was achieved by using H_2SO_4, and finally,a significant amount of gypsum ($CaSO_4$) was yielded (Yang et al., 2013). The produced gypsum results in global warming (Pal et al., 2009; Groot and Borén, 2010), and presently so, many research works are being carriedout to convert it into PLA, which is planned to be usedin medical applications (Murariu and Dubois, 2016). Batch process is easy to operate with low contamination, which is the single-step mode. But the substrate and product inhibition will be more (Abdel-Rahman et al., 2013; Cubas-Cano et al., 2018).In fed-batch mode, inhibition of substrate is low, and microorganisms are maintained in a prolonged lag phase, which helps to increase the production of LA (Shi et al., 2012). Effective fermentation depends on the feeding of sterile media into the fermentor exponentially or constantly or intermittently in exact time intervals.

In LA fermentation, the highest productivity can be reached in the continuous culture method at steady-state condition. In this method, due to the constant rate and fast growth of microbes, LA is produced as primary metabolite. Compared with other fermentation, conventional continuous fermentation results in higher productivity. The production rate is higherdue to outcome product inhibition, more dilution rate or shorter residence time. The only longer-term yield of LA depends on constant fermentation, especially by reducing or omitting the time essential for inoculum preparation, which allows microorganisms to grow faster (Abdel-Rahman et al., 2013; Chang et al., 1994).

The cassava is an important crop and basic food in many countries and due to industrial application, more quantity of bagasse was accumulated in the environment. The cassava bagasse contains 52% starch and also fibrous-rich residues. Rojan et al.(2005) clearly proved thatcassava waste was sacrificed and fermented by *L.* casei, which leads to the conversion of lactic acid production in a single step.

6.8.2.3 Recovery of LA

The quality and cost of LA depend on the recovery step, especially separation and purification. The cost of recovery consumes nearly half of the total production cost (Wasewar, 2005). Usage of further complex substrates with different compositions also creates a complication in the final product of LA, and theseparameters also play an important role in the purity and cost of LA. The separation of high-purity LA is achieved by the electrodialysis method. During the separation, frequent membrane blocking and deionization of broth areobserved, and so this technique ends up with a high-cost process (Komesu et al., 2017). Reactive extraction has been praised for high recovery and simple scale (Krzyżaniak et al., 2013). Good separation yields, reasonable costs and shorter process times havebeen reached by ion exchange chromatography than anyother downstream processes (Bishai et al., 2015), and high purity is achieved by anion and cation chromatography.

6.8.3 Chitosan

A linear cationic polysaccharide with a high molecular weight is chitosan. It is made up of randomly dispersed N-acetyl-D-glucosamine (acetylated unit) and -(1,4)-linked D-glucosamine (deacetylated unit) (Zargar et al., 2015; Islam et al., 2017).

6.8.3.1 Agro-wastes as Medium to Produce Chitosan

Leite et al. (2015) produced chitosan from sugarcane bagasse and corn steep liquor as substrate and fermentation by *Syncephalastrum racemosum.* Using sugarcane bagasse as carbon source and corn steep liquor as nitrogen source, the fungus was grown, and chitosan was extracted using alkali acid method. The viscosity and molecular weight were determined by vibrational spectroscopy, and the chitosan yield was about 23.5 mg/g. It clearly addressed that agricultural waste from the industries can be used to prepare a cost-effective medium to generate the biopolymer through fungal fermentation.

Instead of being utilized as filler in other materials, the chitosan-based substance is typically applied in biomedical and food applications. A catheter, a small tube that is inserted into a patient's body to cure diseases or perform surgery, can be coated with chitosan-based film in the field of biomedical applications. Chitosan is a substance that is placed on the outside of PET or polyethylene (PE) catheters because of its anti-adhesive and antibacterial properties (Lo et al., 2014; Muzzarelli, 1983). But before coating the catheter surface, chitosan must be heparinized to improve its blood compatibility and overcome its poor solubility. In application, the chitosan-heparin-coated catheter can therefore display high compatibility with the fluid and tissue around it in addition to thrombo-resistant qualities.

Chitosan is frequently used as a packaging material in the food industry (Nunes et al., 2018). There are a number of chitosan-based film types that have been researched for use in food packaging, including chitosan/polysaccharides-based films, chitosan/protein-based films, and chitosan/extracts-based films (Wang et al., 2018). Using grafting and cross-linking, chitosan is functionalized to get a perfect mechanical, antioxidant and antibacterial action. Liu et al.(2018) proved chitosan functionalization by grafting using 4-hexyloxyphenol, which has the ability to increase the radical scavenging capacity,that is, antioxidant activity.

6.9 OTHER AGRICULTURAL WASTES

In the current scenario of the world facing food scarcity, the research world is finding an alternative source to manufacture bioplastics,that is, plan to use the non-edible portion. The fruits skin like orange peel and banana peel, andvegetables like potato peel; cassava peels are used as a precursor to produce biopolymers. The most modern trend in the production of bioplastic films today is from polysaccharide residue feedstock, which is in great demand (Shah et al., 2021).

Pomegranate peel is one of the rich sources of bioactive compounds (Gumienna et al., 2016), which contains lignin 5.7%, hemicelluloses, cellulose 26.2% and pectin 27%. The polysaccharides were extracted and converted into simple sugars using the acid hydrolysis method, which is present in the pomegranate peel. Finally, it breaks down into hemicelluloses, celluloses and lignin compounds, and further, it acts as the primary source for the production of biopolymer carbohydrates present in the orange peel, and it can be used for the production of bioplastics. The accumulation of unprocessed peel createsmuchenvironmental pollution, and so conversion is very important in mutual way (Chozhavendhan et al., 2020).

6.10 APPLICATION OF AGRO-BASED BIOPLASTICS

Packaging of food, construction and farming are the places of important applications of agro-based bioplastics. This is due to the impact of quality properties like mechanical, chemical and physical of the bioplastics. Biopolymers with significant tensile strength/Young modulus are required for high-strength applications in construction and agriculture (Figure 6.5).

In packaging applications, the material's flexibility is one of the important factors. For example, tamarind fruit fibre hasa tensile strength of approximately 1360 MPa, resulting in the best mechanical characteristics ofbiopolymers produced from it and being replaceable to synthetic carbon (Binoj et al., 2017). In construction, biopolymer usage is enhanced by nourishing materials like carbon nanotubes, nanofibers and fillers from agro-wastes (Nagarajan et al., 2020). Agricultural shade nets and mulching films are produced using biobased polymers comprised of cellulose, starch, polyhydroxyalkanoates (PHA), bio-polyethylene and PLA(Briassoulis&Giannoulis, 2018; Mukherjee et al., 2019). Because commercial pesticides are so hazardous, using fewer of them has both ecological and financial advantages. Additionally, shade nets have better mechanical qualities than conventional low-density polyethylene films. The nets also assist in filtering UV light, which is detrimental to plant growth. These nets' tensile strength, mesh sizes, surface colour, and chemical makeup all have an impact on their commercial use. High-tensile strength shade nets have a longer usable life and can endure weather-related dangers including strong winds, sunshine, and hail (Mukherjee et al., 2019).

Due to the sustainability of biopolymers, it is used in food-packaging materials, and it is further enhanced using surface modification techniques by nano-fibrillated cellulose at different concentrations (Ilyas et al., 2019). Photobleaching was used to change the biopolymer structure of corn starch and blueberry powder, which helped develop intelligent foodpackaging solutions. The biofilms' luminance values

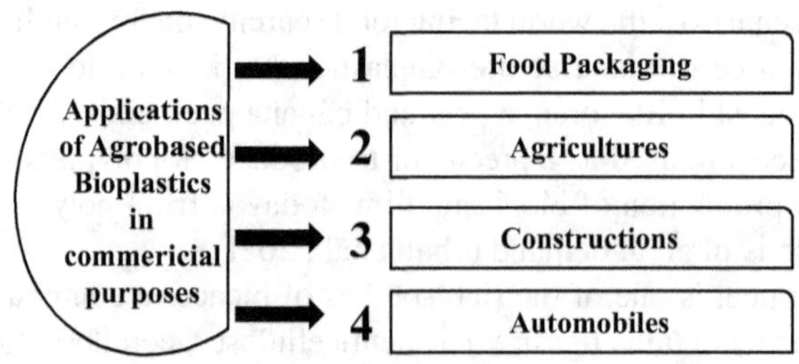

FIGURE 6.5 Application of agro-based bioplastics.

(surface colour), which were reduced by photobleaching, made them excellent colorimetric markers for the deterioration of packed food. In situations with acidic and basic pH levels, the biofilms, respectively, turned blue and red. Fermentation and an increase in pH are characteristics of food that havebeen packaged poorly. The human eye can distinguish the colour variations. Unfortunately, because they are unrelated to the shelf life of foods that contain various biomolecules including proteins, lipids, salt, and sugar, the intelligent pH detection findings are inconclusive (Luchese et al., 2017).

Biobased polymers' usable lives are influenced by how much UV radiation they are exposed to radiation that causes photo-oxidation heat that causes thermal deterioration, and the possibility of dissolution,both mechanical strength in water and high strength applications. There are numerous choices for end-of-life care for biodegradable materials, such as residential and commercial breakdown enzymatic depolymerization, catalytic recycling, chemical recycling, mechanical recycling, compostingandanaerobic digestion (Vroman & Tighzert, 2009).

6.11 CONCLUSION AND FUTURE PERSPECTIVES

The dependence on non-renewable resources has raised economic and environmental concerns, which have fuelled the development of the biopolymer sector over the years. The creation of novel bioplastics is one way to support the circular economy, which aims to reduce waste and resource consumption over time. Every nation needs the agro-industry because it is the cornerstone of economic growth. Trash that will pollute the environment has been produced during the processing and production processes in the agroindustrial sector.These agricultural wastes have emerged as enticing substitute sources for the creation of high-value products.As science and technology advanced, the uses of agricultural wastes in many industries, including the bioplastics industry, were identified.Future studies should concentrate on the characteristics of bioplastics made from agricultural waste. Future research on bioplastics nanocomposites and fibre-based composites is crucial to overcoming the inferior performance of bioplastics. The government should also play a part by boosting incentives for the creation and use of biopolymers from agricultural waste. The function of public-private partnerships should be addressed more carefully for the benefit of the bioplastics industry and the community as a whole.This chapter clearly revealed the state-of-the-art of agro-waste-based bioplastics and theircontributionto thebest economic growth of society. They also enhance a green environment as the agricultural wastes can be renewed into novel value-added products.

ACKNOWLEDGEMENTS

Authors acknowledge the financial support provided by the Technology Mission Division (Energy, Water andAll others), Department of Science andTechnology, Ministry of Science and Technology, Government of India (Reference Number: DST/TMD/IC-MAP/2K20/02 Project Title: DST-IIT Hyderabad Integrated Clean Energy Material Acceleration Platform on Bioenergy and Hydrogen); Department of Science and Technology-Science and Engineering Research Board (DST-SERB-No.SB/YS/

LS-47/2013), India; Department of Science and Technology-Science and Engineering Research Board (DST-SERB-No.SB/YS/LS-47/2013), India; Department of Science and Technology-Promotion of University Research and Scientific Excellence (DST-PURSE) [DST letter No.SR/PURSE phase 2/38 (G);dated: February 21, 2017], India; RUSA – Phase 2.0 grant [Letter No. F.24–51/2014-U, Policy (TN Multi-Gen), Department of Education, Government of India;dated:October 09, 2018]; Scheme for Promotion of Academic and Research Collaboration (SPARC) (No. SPARC/2018–2019/P485/SL; dated: March 15, 2019) and University Science Instrumentation Centre (USIC), Alagappa University, Karaikudi, Tamil Nadu, India.

REFERENCES

Abdel-Rahman, M. A., Tashiro, Y., & Sonomoto, K. (2013). Recent advances in lactic acid production by microbial fermentation processes. *Biotechnology Advances*, *31*(6), 877–902.

Agustin, M. B., Ahmmad, B., Alonzo, S. M. M., & Patriana, F. M. (2014). Bioplastic based on starch and cellulose nanocrystals from rice straw. *Journal of Reinforced Plastics and Composites*, *33*(24), 2205–2213.

Ali, H. K. Q., & Zulkali, M. M. D. (2011). Utilization of agro-residual ligno-cellulosic substances by using solid state fermentation: A review. *Hrvatski časopis za prehrambenu tehnologiju, biotehnologiju i nutricionizam*, *6*(1–2), 5–12.

Anderson, J. M., & Shive, M. S. (1997). Biodegradation and biocompatibility of PLA and PLGA microspheres. *Advanced Drug Delivery Reviews*, *28*(1), 5–24.

Andrady, A. L., & Neal, M. A. (2009). Applications and societal benefits of plastics. *Philosophical Transactions of the Royal Society B: Biological Sciences*, *364*(1526), 1977–1984.

Arikan, E. B., & Bilgen, H. D. (2019). Production of bioplastic from potato peel waste and investigation of its biodegradability. *International Advanced Researches and Engineering Journal*, *3*(2), 93–97.

Arrieta, M. P., Fortunati, E., Dominici, F., López, J., & Kenny, J. M. (2015). Bionanocomposite films based on plasticized PLA-PHB/cellulose nanocrystal blends. *Carbohydrate Polymers*, *121*, 265–275.

Averous, L. (2004). Biodegradable multiphase systems based on plasticized starch: A review. *Journal of Macromolecular Science, Part C: Polymer Reviews*, *44*(3), 231–274.

Azmin, S. N. H. M., & Nor, M. S. M. (2020). Development and characterization of food packaging bioplastic film from cocoa pod husk cellulose incorporated with sugarcane bagasse fibre. *Journal of Bioresources and Bioproducts*, *5*(4), 248–255.

Bastioli, C., Magistrali, P., & Garcia, S. G. (2013). Starch. In *Bio-Based Plastics: Materials and Applications* (pp. 9–33). Wiley.

Battegazzore, D., Bocchini, S., Alongi, J., & Frache, A. (2014). Rice husk as bio-source of silica: preparation and characterization of PLA-silica bio-composites. *RSC Advances*, *4*(97), 54703–54712.

Benítez, J. J., Castillo, P. M., Del Río, J. C., León-Camacho, M., Domínguez, E., Heredia, A., & Heredia-Guerrero, J. A. (2018). Valorization of tomato processing by-products: Fatty acid extraction and production of bio-based materials. *Materials*, *11*(11), 2211.

Binoj, J. S., Raj, R. E., & Daniel, B. S. S. (2017). Comprehensive characterization of industrially discarded fruit fiber, Tamarindus indica L. as a potential eco-friendly bio-reinforcement for polymer composite. *Journal of Cleaner Production*, *142*, 1321–1331.

Bishai, M., De, S., Adhikari, B., & Banerjee, R. (2015). A platform technology of recovery of lactic acid from a fermentation broth of novel substrate Zizyphus oenophlia. *3 Biotech*, *5*(4), 455–463.

Bohmert-Tatarev, K., McAvoy, S., Daughtry, S., Peoples, O. P., & Snell, K. D. (2011). High levels of bioplastic are produced in fertile transplastomic tobacco plants engineered with a synthetic operon for the production of polyhydroxybutyrate. *Plant Physiology*, *155*(4), 1690–1708.

Borah, P. P., Das, P., & Badwaik, L. S. (2017). Ultrasound treated potato peel and sweet lime pomace based biopolymer film development. *Ultrasonics Sonochemistry*, *36*, 11–19.

Borri, A., Corradi, M., & Speranzini, E. (2013). Reinforcement of wood with natural fibers. *Composites Part B: Engineering*, *53*, 1–8.

Briassoulis, D., & Giannoulis, A. (2018). Evaluation of the functionality of bio-based plastic mulching films. *Polymer Testing*, *67*, 99–109.

Chang, H. N., Yoo, I. K., & Kim, B. S. (1994). High density cell culture by membrane-based cell recycle. *Biotechnology Advances*, *12*(3), 467–487.

Chozhavendhan, S., Singh, M. V. P., Fransila, B., Kumar, R. P., & Devi, G. K. (2020). A review on influencing parameters of biodiesel production and purification processes. *Current Research in Green and Sustainable Chemistry*, *1*, 1–6.

Chumee, J., & Khemmakama, P. (2014). Carboxymethyl cellulose from pineapple peel: Useful green bioplastic. In *Advanced Materials Research* (Vol. 979, pp. 366–369). Trans Tech Publications Ltd.

Chun, K. S., Husseinsyah, S., & Osman, H. (2012). Mechanical and thermal properties of coconut shell powder filled polylactic acid biocomposites: effects of the filler content and silane coupling agent. *Journal of Polymer Research*, *19*(5), 1–8.

Chun, K. S., Husseinsyah, S., & Osman, H. (2013). Properties of coconut shell powder-filled polylactic acid ecocomposites: Effect of maleic acid. *Polymer Engineering & Science*, *53*(5), 1109–1116.

Cifriadi, A., Panji, T., Wibowo, N. A., & Syamsu, K. (2017, May). Bioplastic production from cellulose of oil palm empty fruit bunch. In *IOP Conference Series: Earth and Environmental Science* (Vol. 65, No. 1, p. 012011). IOP Publishing.

Cubas-Cano, E., González-Fernández, C., Ballesteros, M., & Tomás-Pejó, E. (2018). Biotechnological advances in lactic acid production by lactic acid bacteria: lignocellulose as novel substrate. *Biofuels, Bioproducts and Biorefining*, *12*(2), 290–303.

Cui, F., Li, Y., & Wan, C. (2011). Lactic acid production from corn stover using mixed cultures of *Lactobacillus rhamnosus* and *Lactobacillus brevis*. *Bioresource Technology*, *102*(2), 1831–1836.

de Moura, I. G., de Sá, A. V., Abreu, A. S. L. M., & Machado, A. V. A. (2017). Bioplastics from agro-wastes for food packaging applications. In *Food Packaging* (pp. 223–263). Academic Press.

FAOSTAT. (2016). Food and Agriculture Organization of the United Nations Cropping Database. https://faostat.fao.org/site/339/default.aspx2012 (Accessed 5 August 2016).

Galloway, T. S. (2015). Micro-and nano-plastics and human health. In *Marine Anthropogenic Litter* (pp. 343–366). Springer.

Gao, C., Ma, C., & Xu, P. (2011). Biotechnological routes based on lactic acid production from biomass. *Biotechnology Advances*, *29*(6), 930–939.

Gaspar, M., Benkő, Z., Dogossy, G., Reczey, K., & Czigany, T. (2005). Reducing water absorption in compostable starch-based plastics. *Polymer Degradation and Stability*, *90*(3), 563–569.

Gasperi, J., Dris, R., Rocher, V., & Tassin, B. (2015). Microplastics in the continental area: An emerging challenge. *Welcome to Issue*, *4*, 18.

Geyer, R., Jambeck, J. R., & Law, K. L. (2017). Production, use, and fate of all plastics ever made. *Science Advances*, *3*(7), e1700782.

Gontard, N., Sonesson, U., Birkved, M., Majone, M., Bolzonella, D., Celli, A., & Sebok, A. (2018). A research challenge vision regarding management of agricultural waste in a circular bio-based economy. *Critical Reviews in Environmental Science and Technology*, *48*(6), 614–654.

Groot, W. J., & Borén, T. (2010). Life cycle assessment of the manufacture of lactide and PLA biopolymers from sugarcane in Thailand. *The International Journal of Life Cycle Assessment*, *15*(9), 970–984.

Grunert, M., & Winter, W. T. (2002). Nanocomposites of cellulose acetate butyrate reinforced with cellulose nanocrystals. *Journal of Polymers and the Environment*, *10*(1), 27–30.

Gumienna, M., Szwengiel, A., & Górna, B. (2016). Bioactive components of pomegranate fruit and their transformation by fermentation processes. *European Food Research and Technology*, *242*(5), 631–640.

Hagemann, R. T., & D'Amico, D. (2009). *U.S. Patent Application No. 12/127,855.*

Hamza, S., Saad, H., Charrier, B., Ayed, N., & Charrier-El Bouhtoury, F. (2013). Physico-chemical characterization of Tunisian plant fibers and its utilization as reinforcement for plaster based composites. *Industrial Crops and Products*, *49*, 357–365.

Heredia-Guerrero, J. A., Benítez, J. J., Cataldi, P., Paul, U. C., Contardi, M., Cingolani, R., ... & Athanassiou, A. (2017). All-natural sustainable packaging materials inspired by plant cuticles. *Advanced Sustainable Systems*, *1*(1–2), 1600024.

Heredia-Guerrero, J. A., Caputo, G., Guzman-Puyol, S., Tedeschi, G., Heredia, A., Ceseracciu, L., ... & Athanassiou, A. (2019). Sustainable polycondensation of multifunctional fatty acids from tomato pomace agro-waste catalyzed by tin (II) 2–ethylhexanoate. *Materials Today Sustainability*, *3*, 100004.

Ilyas, R. A., Sapuan, S. M., Ibrahim, R., Abral, H., Ishak, M. R., Zainudin, E. S., ... & Jumaidin, R. (2019). Sugar palm (Arenga pinnata (Wurmb.) Merr) cellulosic fibre hierarchy: A comprehensive approach from macro to nano scale. *Journal of Materials Research and Technology*, *8*(3), 2753–2766.

Islam, S., Bhuiyan, M. A., & Islam, M. N. (2017). Chitin and chitosan: Structure, properties and applications in biomedical engineering. *Journal of Polymers and the Environment*, *25*(3), 854–866.

Isroi, C. A., Panji, T., Wibowo, N. A., & Syamsu, K. (2017). Bioplastic production from cellulose of oil palm empty fruit bunch. *IOP Conference Series: Earth and Environmental Science*, *65*(1).

Iwata, T. (2015). Biodegradable and bio-based polymers: Future prospects of eco-friendly plastics. *Angewandte Chemie International Edition*, *54*(11), 3210–3215.

Kale, G., Kijchavengkul, T., Auras, R., Rubino, M., Selke, S. E., & Singh, S. P. (2007). Compostability of bioplastic packaging materials: An overview. *Macromolecular Bioscience*, *7*(3), 255–277.

Kang, H. J., & Min, S. C. (2010). Potato peel-based biopolymer film development using high-pressure homogenization, irradiation, and ultrasound. *LWT-Food Science and Technology*, *43*(6), 903–909.

Kershaw, P. J., & Rochman, C. M. (2015). Sources, fate and effects of microplastics in the marine environment: Part 2 of a global assessment. *Reports and Studies-IMO/FAO/Unesco-IOC/WMO/IAEA/UN/UNEP Joint Group of Experts on the Scientific Aspects of Marine Environmental Protection (GESAMP) Eng No. 93.*

Koay, S. C., Husseinsyah, S., & Osman, H. (2013). Modified cocoa pod husk-filled polypropylene composites by using methacrylic acid. *BioResources*, *8*(3), 3260–3275.

Komesu, A., de Oliveira, J. A. R., da Silva Martins, L. H., Maciel, M. R. W., & Maciel Filho, R. (2017). Lactic acid production to purification: A review. *BioResources*, *12*(2), 4364–4383.

Krzyżaniak, A., Leeman, M., Vossebeld, F., Visser, T. J., Schuur, B., & de Haan, A. B. (2013). Novel extractants for the recovery of fermentation derived lactic acid. *Separation and Purification Technology*, *111*, 82–89.

Leite, M. V., Stamford, T. C. M., Stamford-Arnaud, T. M., Lima, J. M. N., Silva, A. M., Okada, K., & Campos-Takaki, G. M. (2015). Conversion of agro-industrial wastes to chitosan production by Syncephalastrum racemosum UCP 1302. *International Journal of Applied Research in Natural Products*, *8*(4), 5–11.

Leo, L., Leone, A., Longo, C., Lombardi, D. A., Raimo, F., & Zacheo, G. (2008). Antioxidant compounds and antioxidant activity in "early potatoes". *Journal of Agricultural and Food Chemistry*, *56*(11), 4154–4163.

Liu, N., Ni, S., Ragauskas, A. J., Meng, X., Hao, N., & Fu, Y. (2018). Laccase-mediated functionalization of chitosan with 4-hexyloxyphenol enhances antioxidant and hydrophobic properties of copolymer. *Journal of Biotechnology*, *269*, 8–15.

Lo, J., Lange, D., & Chew, B. H. (2014). Ureteral stents and foley catheters-associated urinary tract infections: the role of coatings and materials in infection prevention. *Antibiotics*, *3*(1), 87–97.

López-Cuellar, M. R., Alba-Flores, J., Rodríguez, J. G., & Pérez-Guevara, F. (2011). Production of polyhydroxyalkanoates (PHAs) with canola oil as carbon source. *International Journal of Biological Macromolecules*, *48*(1), 74–80.

Lu, Z., He, F., Shi, Y., Lu, M., & Yu, L. (2010). Fermentative production of L (+)-lactic acid using hydrolyzed acorn starch, persimmon juice and wheat bran hydrolysate as nutrients. *Bioresource Technology*, *101*(10), 3642–3648.

Lubis, M., Gana, A., Maysarah, S., Ginting, M. H. S., & Harahap, M. B. (2018, February). Production of bioplastic from jackfruit seed starch (*Artocarpus heterophyllus*) reinforced with microcrystalline cellulose from cocoa pod husk (*Theobroma cacao L.*) using glycerol as plasticizer. In *IOP Conference Series: Materials Science and Engineering* (Vol. 309, No. 1, p. 012100). IOP Publishing.

Lucas, N., Bienaime, C., Belloy, C., Queneudec, M., Silvestre, F., & Nava-Saucedo, J. E. (2008). Polymer biodegradation: Mechanisms and estimation techniques-A review. *Chemosphere*, *73*(4), 429–442.

Luchese, C. L., Sperotto, N., Spada, J. C., & Tessaro, I. C. (2017). Effect of blueberry agro-industrial waste addition to corn starch-based films for the production of a pH-indicator film. *International Journal of Biological Macromolecules*, *104*, 11–18.

Maraveas, C. (2020). Production of sustainable and biodegradable polymers from agricultural waste. *Polymers*, *12*(5), 1127.

Massardier-Nageotte, V., Pestre, C., Cruard-Pradet, T., & Bayard, R. (2006). Aerobic and anaerobic biodegradability of polymer films and physico-chemical characterization. *Polymer Degradation and Stability*, *91*(3), 620–627.

Mathew, A. P., & Dufresne, A. (2002). Plasticized waxy maize starch: Effect of polyols and relative humidity on material properties. *Biomacromolecules*, *3*(5), 1101–1108.

McCormick, A., Hoellein, T. J., Mason, S. A., Schluep, J., & Kelly, J. J. (2014). Microplastic is an abundant and distinct microbial habitat in an urban river. *Environmental Science & Technology*, *48*(20), 11863–11871.

Mentzer, M. J., Whistler, R. L., BeMiller, J. N., & Paschall, E. F. (1984). *Starch: Chemistry and Technology*. Academic Press.

Mostafa, N. A., Farag, A. A., Abo-dief, H. M., & Tayeb, A. M. (2018). Production of biodegradable plastic from agricultural wastes. *Arabian Journal of Chemistry*, *11*(4), 546–553.

Możejko, J., & Ciesielski, S. (2013). Saponified waste palm oil as an attractive renewable resource for mcl-polyhydroxyalkanoate synthesis. *Journal of Bioscience and Bioengineering*, *116*(4), 485–492.

Mukherjee, A., Knoch, S., Chouinard, G., Tavares, J. R., & Dumont, M. J. (2019). Use of bio-based polymers in agricultural exclusion nets: A perspective. *Biosystems Engineering*, *180*, 121–145.

Mukprasirt, A., & Sajjaanantakul, K. (2004). Physico-chemical properties of flour and starch from jackfruit seeds (Artocarpus heterophyllus Lam.) compared with modified starches. *International Journal of Food Science & Technology*, *39*(3), 271–276.

Murariu, M., & Dubois, P. (2016). PLA composites: From production to properties. *Advanced Drug Delivery Reviews*, *107*, 17–46.

Muzzarelli, R. A. (1983). Heparin-like substances and blood-compatible polymers obtained from chitin and chitosan. In *Polymers in Medicine* (pp. 359–374). Springer.

Nagarajan, K. J., Balaji, A. N., Basha, K. S., Ramanujam, N. R., & Kumar, R. A. (2020). Effect of agro waste α-cellulosic micro filler on mechanical and thermal behavior of epoxy composites. *International Journal of Biological Macromolecules*, *152*, 327–339.

Nawrath, C. (2006). Unraveling the complex network of cuticular structure and function. *Current Opinion in Plant Biology*, *9*(3), 281–287.

Nee, F. C., & Othman, S. A. (2022). Preparation and characterization of irradiated bioplastic from Cassava peel-A review. In *Journal of Physics*: *Conference Series* (Vol. 2169, No. 1, p. 012041). IOP Publishing.

Nelson, M. L. (2010). Utilization and application of wet potato processing coproducts for finishing cattle. *Journal of Animal Science*, *88*(suppl_13), E133–E142.

Nunes, C., Coimbra, M. A., & Ferreira, P. (2018). Tailoring functional chitosan-based composites for food applications. *The Chemical Record*, *18*(7–8), 1138–1149.

Ohkouchi, Y., & Inoue, Y. (2006). Direct production of L (+)-lactic acid from starch and food wastes using *Lactobacillus manihotivorans* LMG18011. *Bioresource Technology*, *97*(13), 1554–1562.

Pal, P., Sikder, J., Roy, S., & Giorno, L. (2009). Process intensification in lactic acid production: A review of membrane based processes. *Chemical Engineering and Processing*: *Process Intensification*, *48*(11–12), 1549–1559.

Panesar, P. S., & Kaur, S. (2015). Bioutilisation of agro-industrial waste for lactic acid production. *International Journal of Food Science & Technology*, *50*(10), 2143–2151.

Pirayesh, H., Khanjanzadeh, H., & Salari, A. (2013). Effect of using walnut/almond shells on the physical, mechanical properties and formaldehyde emission of particleboard. *Composites Part B*: *Engineering*, *45*(1), 858–863.

Ponstein, A. S. (1990). Starch Synthesis in Potato Tubers. [Thesis fully internal (DIV), University of Groningen].

Raghatate Atul, M. (2012). Use of plastic in a concrete to improve its properties. *International Journal of Advanced Engineering Research and Studies*, *1*(3), 109–111.

Rasu, K. M., & Arun, A. (2017). Exploring biodegradable polymer production from marine microbes. In *Biodegradable Polymers – Recent Developments and New Perspectives* (pp. 33–64). IAPC Publishing.

Reixach, R., Espinach, F. X., Franco-Marquès, E., Ramirez de Cartagena, F., Pellicer, N., Tresserras, J., & Mutjé, P. (2013). Modeling of the tensile moduli of mechanical, thermomechanical, and chemi-thermomechanical pulps from orange tree pruning. *Polymer Composites*, *34*(11), 1840–1846.

Ren, J., Fu, H., Ren, T., & Yuan, W. (2009). Preparation, characterization and properties of binary and ternary blends with thermoplastic starch, poly (lactic acid) and poly (butylene adipate-co-terephthalate). *Carbohydrate Polymers*, *77*(3), 576–582.

Rivard, C., Moens, L., Roberts, K., Brigham, J., & Kelley, S. (1995). Starch esters as biodegradable plastics: Effects of ester group chain length and degree of substitution on anaerobic biodegradation. *Enzyme and Microbial Technology*, *17*(9), 848–852.

Rogols, S., Sirovatka, D. M., & Widmaier, R. G. (2003). *U.S. Patent No. 6,547,867*. Washington, DC: U.S. Patent and Trademark Office.

Rojan, P. J., Nampoothiri, K. M., Nair, A. S., & Pandey, A. (2005). L (+)-lactic acid production using Lactobacillus casei in solid-state fermentation. *Biotechnology Letters*, *27*(21), 1685–1688.

Rosenboom, J. G., Langer, R., & Traverso, G. (2022). Bioplastics for a circular economy. *Nature Reviews Materials*, *7*(2), 117–137.

Saikia, N., & De Brito, J. (2012). Use of plastic waste as aggregate in cement mortar and concrete preparation: A review. *Construction and Building Materials*, *34*, 385–401.

Salmah, H., Koay, S. C., & Hakimah, O. (2013). Surface modification of coconut shell powder filled polylactic acid biocomposites. *Journal of Thermoplastic Composite Materials, 26*(6), 809–819.

Sanyang, M. L., Ilyas, R. A., Sapuan, S. M., & Jumaidin, R. (2018). Sugar palm starch-based composites for packaging applications. In *Bionanocomposites for Packaging Applications* (pp. 125–147). Springer, Cham.

Satyanarayana, K. G., Arizaga, G. G., & Wypych, F. (2009). Biodegradable composites based on lignocellulosic fibers-An overview. Progress *in Polymer Science, 34*(9), 982–1021.

Shah, M., Rajhans, S., Pandya, H. A., & Mankad, A. U. (2021). Bioplastic for future: A review then and now. *World Journal of Advanced Research and Reviews, 9*(2), 056–067.

Sharma, C., Manepalli, P. H., Thatte, A., Thomas, S., Kalarikkal, N., & Alavi, S. (2017). Biodegradable starch/PVOH/laponite RD-based bionanocomposite films coated with graphene oxide: Preparation and performance characterization for food packaging applications. *Colloid and Polymer Science, 295*(9), 1695–1708.

Sharma, P., Gaur, V. K., Kim, S. H., & Pandey, A. (2020). Microbial strategies for bio-transforming food waste into resources. *Bioresource Technology, 299*, 122580.

Shen, L., Haufe, J., & Patel, M. K. (2009). Product overview and market projection of emerging bio-based plastics PRO-BIP 2009. *Report for European Polysaccharide Network of Excellence (EPNOE) and European Bioplastics* (Vol. 243, pp. 1–245).

Shi, R., Liu, Q., Ding, T., Han, Y., Zhang, L., Chen, D., & Tian, W. (2007). Ageing of soft thermoplastic starch with high glycerol content. *Journal of Applied Polymer Science, 103*(1), 574–586.

Shi, Z., Wei, P., Zhu, X., Cai, J., Huang, L., & Xu, Z. (2012). Efficient production of l-lactic acid from hydrolysate of Jerusalem artichoke with immobilized cells of Lactococcus lactis in fibrous bed bioreactors. *Enzyme and Microbial Technology, 51*(5), 263–268.

Shrivastav, A., Mishra, S. K., Shethia, B., Pancha, I., Jain, D., & Mishra, S. (2010). Isolation of promising bacterial strains from soil and marine environment for polyhydroxyalkanoates (PHAs) production utilizing Jatropha biodiesel byproduct. *International Journal of Biological Macromolecules, 47*(2), 283–287.

Silva, R. V., De Brito, J., & Dhir, R. K. (2014). Properties and composition of recycled aggregates from construction and demolition waste suitable for concrete production. *Construction and Building Materials, 65*, 201–217.

Siracusa, V., & Blanco, I. (2020). Bio-Polyethylene (Bio-PE), Bio-Polypropylene (Bio-PP) and Bio-Poly (ethylene terephthalate) (Bio-PET): Recent developments in bio-based polymers analogous to petroleum-derived ones for packaging and engineering applications. *Polymers, 12*(8), 1641.

Siró, I., & Plackett, D. (2010). Microfibrillated cellulose and new nanocomposite materials: A review. *Cellulose, 17*(3), 459–494.

Smith, W. P. (1996). Epidermal and dermal effects of topical lactic acid. *Journal of the American Academy of Dermatology, 35*(3), 388–391.

Tyler, B., Gullotti, D., Mangraviti, A., Utsuki, T., & Brem, H. (2016). Polylactic acid (PLA) controlled delivery carriers for biomedical applications. *Advanced Drug Delivery Reviews, 107*, 163–175.

Umesh, M., Mani, V. M., Thazeem, B., & Preethi, K. (2018). Statistical optimization of process parameters for bioplastic (PHA) production by Bacillus subtilis NCDC0671 using orange peel-based medium. *Iranian Journal of Science and Technology, Transactions A: Science, 42*(4), 1947–1955.

Upadhyaya, B. P., DeVeaux, L. C., & Christopher, L. P. (2014). Metabolic engineering as a tool for enhanced lactic acid production. *Trends in Biotechnology, 32*(12), 637–644.

Van der Burgt, M. C., Van der Woude, M. E., & Janssen, L. P. B. M. (1996). The influence of plasticizer on extruded thermoplastic starch. *Journal of Vinyl and Additive Technology, 2*(2), 170–174.

Vega-Castro, O., Contreras-Calderon, J., León, E., Segura, A., Arias, M., Pérez, L., & Sobral, P. J. (2016). Characterization of a polyhydroxyalkanoate obtained from pineapple peel waste using *Ralsthonia eutropha*. *Journal of Biotechnology*, *231*, 232–238.

Venus, J., Fiore, S., Demichelis, F., & Pleissner, D. (2018). Centralized and decentralized utilization of organic residues for lactic acid production. *Journal of Cleaner Production*, *172*, 778–785.

Vethaak, A. D., & Leslie, H. A. (2016). *Plastic debris is a human health issue.*

Vriesmann, L. C., Teofilo, R. F., & de Oliveira Petkowicz, C. L. (2012). Extraction and characterization of pectin from cacao pod husks (*Theobroma cacao* L.) with citric acid. *LWT*, *49*(1), 108–116.

Vroman, I., & Tighzert, L. (2009). Biodegradable polymers. *Materials*, *2*(2), 307–344.

Wang, Y., Tashiro, Y., & Sonomoto, K. (2015). Fermentative production of lactic acid from renewable materials: Recent achievements, prospects, and limits. *Journal of Bioscience and Bioengineering*, *119*(1), 10–18.

Wang, H., Qian, J., & Ding, F. (2018). Emerging chitosan-based films for food packaging applications. *Journal of Agricultural and Food Chemistry*, *66*(2), 395–413.

Wasewar, K. L. (2005). Separation of lactic acid: Recent advances. *Chemical and Biochemical Engineering Quarterly*, *19*, 159–172.

Wool, R. P., Raghavan, D., Wagner, G. C., & Billieux, S. (2000). Biodegradation dynamics of polymer-starch composites. *Journal of Applied Polymer Science*, *77*(8), 1643–1657.

Xie, Y., Niu, X., Yang, J., Fan, R., Shi, J., Ullah, N., & Chen, L. (2020). Active biodegradable films based on the whole potato peel incorporated with bacterial cellulose and curcumin. *International Journal of Biological Macromolecules*, *150*, 480–491.

Xu, W., Reddy, N., & Yang, Y. (2009). Extraction, characterization and potential applications of cellulose in corn kernels and Distillers' dried grains with solubles (DDGS). *Carbohydrate Polymers*, *76*(4), 521–527.

Yang, S. T., El-Ensashy, H., & Thongchul, N. (Eds.). (2013). *Bioprocessing Technologies in Biorefinery for Sustainable Production of Fuels, Chemicals, and Polymers*. John Wiley & Sons, Inc.

Zargar, V., Asghari, M., & Dashti, A. (2015). A review on chitin and chitosan polymers: structure, chemistry, solubility, derivatives, and applications. *ChemBioEng Reviews*, *2*(3), 204–226.

Zhang, B. X., & Uyama, H. (2016). Biomimic plant cuticle from hyperbranched poly (ricinoleic acid) and cellulose film. *ACS Sustainable Chemistry & Engineering*, *4*(1), 363–369.

7 Biobased Polyesters from Biobased Monomers

Synthesis and Properties

Rakgoshi Lekalakala, Pertunia Mothoa, John Letwaba, Orebotse Joseph Botlhoko and Sudhakar Muniyasamy

7.1 INTRODUCTION

The detrimental environmental impact of fossil-based plastics has driven the need for sustainable alternatives. Biopolymers have emerged as a promising solution due to their renewable nature, biodegradability, and potential for resource circularity. As a result, intensive research efforts have been focused on understanding, optimizing, and improving the synthesis chemistry and manufacturing processes of biopolymers. This chapter explores how the principled restriction of using only bio-sourced monomers not only complicates the technical requirements for synthesis technologies but also presents an opportunity to develop manufacturing technology based on the values of green chemistry and circular economy (Aaliya et al., 2021; Sadasivuni et al., 2020; Devadas et al., 2021).

Biopolymers can be derived from renewable resources such as biomass, agricultural waste, or microorganisms. Unlike fossil-based plastics, which rely on non-renewable fossil fuel reserves, biopolymers offer a sustainable alternative by utilizing resources that can be replenished within a reasonable time frame. This renewable characteristic reduces the carbon footprint associated with their production and use, contributing to the mitigation of climate change. Furthermore, biopolymers exhibit biodegradability due to their elemental composition of primarily carbon, hydrogen, and oxygen, where higher oxygen content relative to conventional plastics increases the bioavailability of biopolymers to microorganisms that can metabolize them (Sarkar et al., 2020; Stoica et al., 2022; Andreeßen and Steinbüchel, 2019). This also makes bio-sourcing of biopolymer monomers from organic matter advantageous since the elemental composition is more likely to be in a similar range from biomass to monomer. This allows the resulting biopolymers to break down naturally in the environment through biological processes, which is particularly crucial in reducing plastic pollution, as traditional plastics can persist in the environment for hundreds of years. Biodegradable biopolymers offer the potential for reduced waste accumulation

DOI: 10.1201/9781003304142-7

and lower environmental impacts, leading to a more sustainable future (Wang et al., 2020; Weng et al., 2013; Yates and Barlow, 2013).

It is in keeping with the principles of sustainability and circularity to use only bio-sourced monomers in biopolymer synthesis to ensure maximum positive environmental impact and build bio economies independent of fossil-based petroleum markets. However, this restriction adds complexity to the technical requirements of synthesis technologies. Bio-sourced monomers often have varying compositions and complex structures, making their polymerization processes more challenging than those of traditional plastics derived from a single source (De Clercq et al., 2017; Liu et al., 2012). Moreover, biopolymer synthesis often requires the development of efficient catalysts and reaction conditions specific to biobased feedstocks for both monomer and biopolymer synthesis. The selection and optimization of these catalysts and conditions are critical to achieving high yields, desired molecular structures, and suitable material properties within the molecular weight of biodegradability. Researchers face the task of understanding and manipulating the intricate chemistry involved in biopolymer synthesis to achieve economically viable and sustainable production processes (Bayu et al., 2018).

Another advantage of using only bio-sourced monomers is that within the available synthesis protocols of catalysis and biological (fermentation, bacterial, and enzymatic) for their synthesis is embodied green chemistry, which aims to minimize the use of hazardous substances, reduce energy consumption, and promote environmentally benign processes. Circular economy principles advocate for resource efficiency, recycling, and waste reduction. Manufacturing technologies for biopolymers can embrace these principles by utilizing renewable energy sources, employing non-toxic catalysts and solvents, and optimizing reaction conditions to minimize waste generation. By focusing on green chemistry and circular economy principles, the development of biopolymer manufacturing technologies can create a synergy between environmental sustainability and economic viability. The advancement of these technologies has the potential to transform industries and reduce reliance on fossil-based plastics, ultimately leading to a more sustainable future (Koutinas et al., 2014; Jha and Kumar, 2019; Fasciotti, 2017; Hernández et al., 2014; Sharma and Mudhoo, 2011).

Figure 7.1 shows the viability of a spectrum of different biomaterials sources that can be converted to biopolymers. The most common route from these sources is to break them down into platform chemicals, which can further be processed to specific monomers catalytically or biologically, which may then be polymerized to get the desired material where the material properties are determined by controlling the polymerization parameters. The most important of these sources is cellulose since it is a non-food source and the most abundant with the most concerted farming effort for paper demand, which is in rapid decline for the selfsame environmental concerns. This avails strong synergies for such companies with vast inventories of standing timber to add biodegradable polymers to their product catalog (Schnepp, 2013; van Beilen, 2008; Yabushita et al., 2014).

Figure 7.2 shows the critical biological pathways in which polysaccharide biomass (including cellulose and starch) can be converted to biopolymers by fermentation of the elementary sugars resulting from the hydrolysis of the biomass. There are several methodologies available for the hydrolysis of saccharide polymers to their sugar

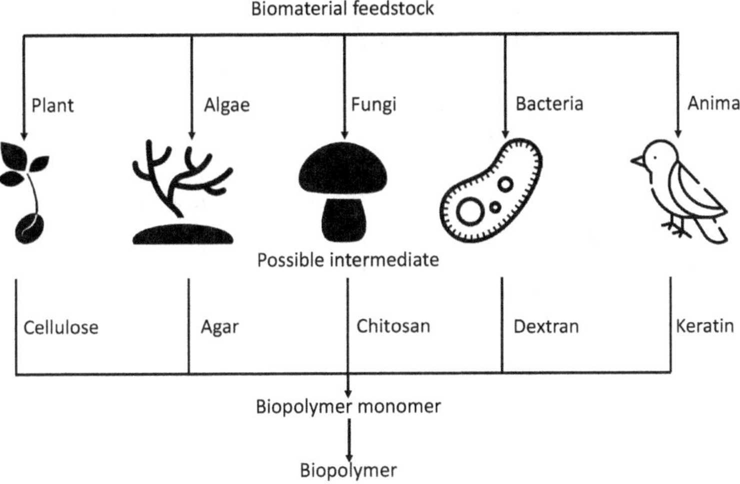

FIGURE 7.1 Viable biomaterial source for biopolymer synthesis (Sarkar et al., 2020).

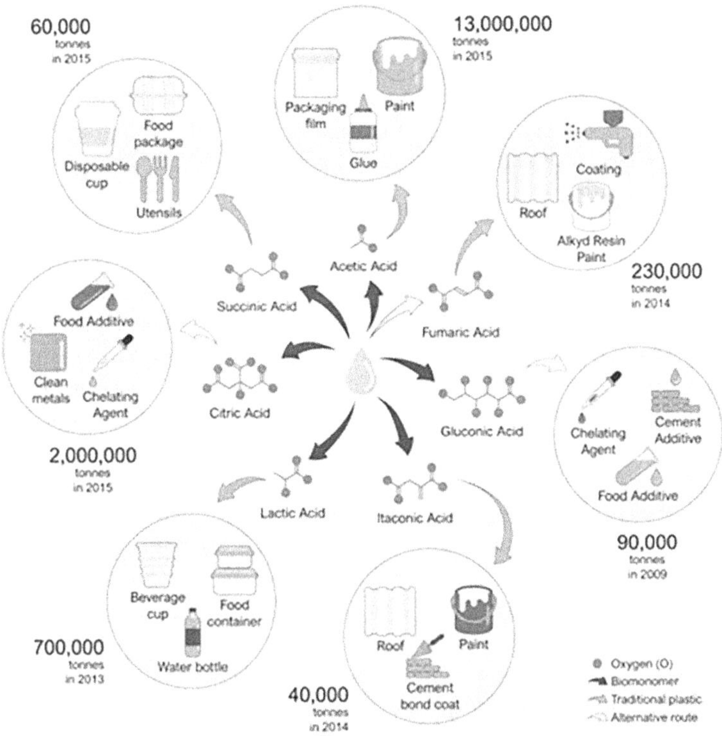

FIGURE 7.2 Possible transformations of organic acids feedstock to biopolymers (Yadav et al., 2018).

monomers including enzymatic, fermentation, bacterial, and catalytic, among those of critical acclaim. The same methodologies can be further employed in downstream process to convert the resulting sugars to specific monomer with subsequent polymerization to yield the desired biopolymer.

The synthesis of biopolymers can be undertaken, encompassing monomer production and polymerization with tailoring of synthesis process protocol to achieve desired properties in the final product according to the chosen application. The monomer synthesis may be the most critical aspect of biopolymer synthesis, given that conventional polycondensation would still be employed for polymerization. Biopolymers are proving to be one of the best advancements in the polymer sector for solving environmental problems with various applications including biomedicine, additive technologies, film, fibers, packaging, automotive, and agriculture (Balla et al., 2021).

Polylactic acid (PLA) is one the most common biopolyester in the market with competitive mechanical strength, biocompatibility, and biodegradability. It is usually synthesized from lactide monomer resulting from lactic acid bacterial fermentation, with properties that depend on the polymerization process parameters and additives. Ring opening polymerization (ROP) of lactide monomer is the standard procedure for making PLA. It involves using one of a variety of catalysts including those based on zinc, tin, lead, and aluminum with initiators, including sec-, -n, and tert-butyl lithium, and solvents, including diphenyl ether, chloroform, and toluene. By controlling the residence time, catalyst type, concentration, and temperature, ROP can produce polymers with different molecular weights and the resulting polymer's L- and D-lactic acid unit ratio and order could both be modified to tailored properties in the final polymer product (Komesu et al., 2017; Maharana et al., 2009).

7.2 MONOMER SYNTHESIS

The synthesis of biopolymer monomers plays a crucial role in achieving efficient and scalable production of these environmentally friendly materials. This section aims to explore the recent advancements in biopolymer monomer synthesis, highlighting key methods, strategies, and challenges involved in their production. Among the most common methods for converting biomass to biopolymer monomers include catalytic chemistry and enzymatic, bacterial, or fungal biological processes. Among these, the most critical may be catalytic conversion due to scalability, flexibility, lower theoretical limit of reaction time, and lower sensitivity to changes in process conditions under the concept of biorefinery. Biological conversion processes have their benefit over catalytic processes with high selectivity for desired monomers according to known bioagent metabolism, low energy requirements, and overall benign process conditions. With greater emphasis on cellulosic feedstock due to its abundance, lack of food application, and established value chain, the most critical process may be catalytic chemistry. In addition to the advantages already mentioned for this process, cellulose has a highly crystalline structure with no solubility in most common solvents and thus requires extensive pretreatment and may not be readily bioavailable to be metabolized by microbes at rates that would be in keeping with the demand for biopolymers (Kong et al., 2008; Yang et al., 2015).

The most crucial parameters for catalytic conversion include catalyst selectivity, conversion rate, and yield. The operating conditions of temperature and pressure will also be relevant for the practicality and commercial viability of the monomer synthesis technology. Chemical synthesis processes tend to critically rely on catalyst parameters mentioned above in addition to the cost of materials, cost of the catalyst synthesis process, catalyst lifespan, recyclability and susceptibility to fouling. Other critical factors for chemical synthesis may include the energy cost for reactor heating and initial CAPEX related to the pressure requirements of the reactor system (Rose & Palkovits, 2011; de Clercq et al., 2017; Li et al., 2018).

The concept of biorefinery does present an attractive value proposition for global demand volumes with catalytic conversion and provided the energy used for the process is renewable. At a low enough levelized cost of energy and provided the cellulosic raw material is farmed at scale in a manner similar to the eucalyptus farming for paper, chemical conversion route via biorefinery concept may be a competitive alternative to petroleum-based polymer markets. This may make such companies as paper pulping ideal candidates for transition into polymer markets while further introducing diversity in the product portfolio. In this manner, the conversion of cellulose may be better streamlined for scalability where one platform chemical is produced and fed into further downstream conversion processes, which may be catalytic or biological in nature based on demand and/or prevailing market conditions (Yang et al., 2016; Tan et al., 2018).

Conversion of cellulose to biopolymer monomers typically requires pretreatment of cellulose feedstock due to its inherent inertness and tightly bound molecular structure with high crystallinity and strong hydrogen bonding (Figure 7.3). This is usually a prerequisite for both catalytic and biological conversion processes in order to avail reaction sites for the catalyst or improve bioavailability. Pretreatment of cellulose improves reaction rate and yield. Physical, biological, physicochemical, and chemical pretreatment methods are known in literature and achieve the objective by reducing crystallinity, degree of lignification, structural heterogeneity, complexity of cell-wall constituents, and molecular weight of cellulose. Figure 7.5 shows various pretreatment methods available where their action works to disrupt the inherent properties of cellulose outlined above (Baruah et al., 2018; Zhao et al., 2012).

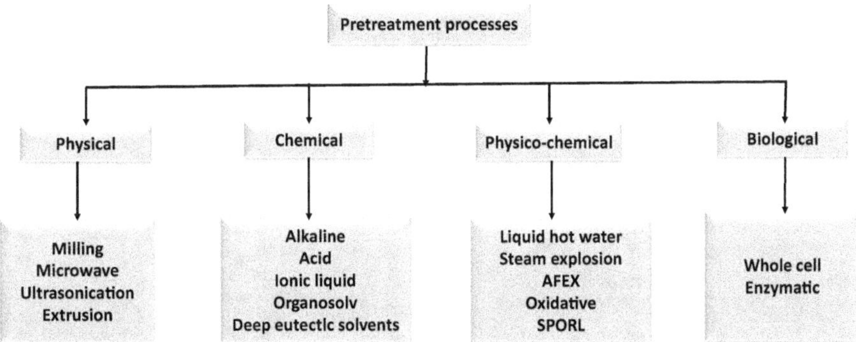

FIGURE 7.3 Pretreatment processes for lignocellulosic biomass redesigned on the basis of information available in (Baruah et al. 2018).

7.2.1 Catalytic Conversion of Biomass

While cellulose is the most abundant non-food biomass available, its low reactivity has fundamentally limited its transformation to other products besides paper manufacturing. Heterogenous catalysts are expected to overcome this drawback with a wide range of reaction conditions, easy recovery and recycling of catalyst, making them attractive for use in biorefineries where platform chemicals can be produced efficiently at scale and further processed to specific monomers in downstream processes with agile flexibility based on demand. Chemical and biological processes may find synergies where, due to the inertness of cellulose, chemical processes can be used for pretreatment and conversion to platform chemicals, which can be used as feedstock in downstream biological processing to individual monomers based on demand (Baruah et al., 2018; Yabushita et al., 2014).

Glucose is one of the more attractive platform chemicals as it can conceptually be transformed into most of the organic acids (lactic, succinic, galactic, adipic, etc.) relevant for biopolymer synthesis or their precursors using the appropriate catalyst, bacteria, fungi, or enzyme. Hydrolysis of cellulose can occur chemically, and several catalysts have been developed for this process with different operating conditions. Another valuable monomer for polymer chemistry which can be derived from glucose is ethylene glycol, which is a ubiquitous diol monomer in the majority of condensation polymers including biodegradable polyesters. Although water at high temperature and pressure can convert cellulose to glucose without any additives, lack of selectivity and further degradation of glucose due to harsh conditions is a challenge that necessitates the use of a catalyst to improve yield, selectivity, and lower operating conditions (Baruah et al., 2018; Yabushita et al., 2014) (Figure 7.4).

Sorbitol, 5-hydroxymethylfurfural (5-HMF), gluconic acid, and others are also platform chemicals that can also be catalytically derived from cellulose. Sorbitol has the advantage over glucose in that it is more thermally stable than glucose due to the aldehyde functional group in its linear structure and, thus, less prone to degradation

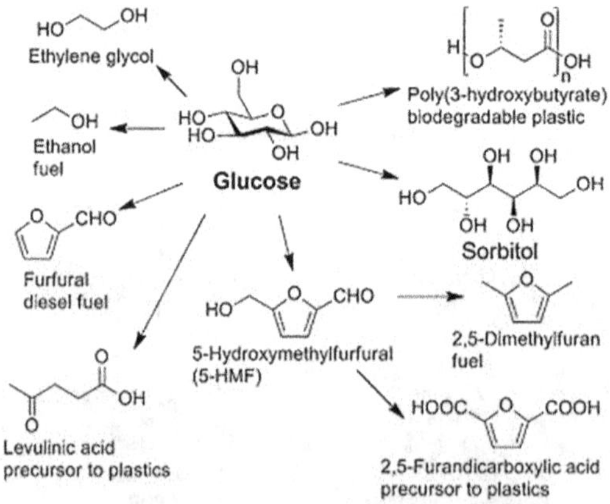

FIGURE 7.4 Glucose hydrolysate from cellulose as a platform chemical for biopolymer monomer synthesis (Yabushita et al., 2014).

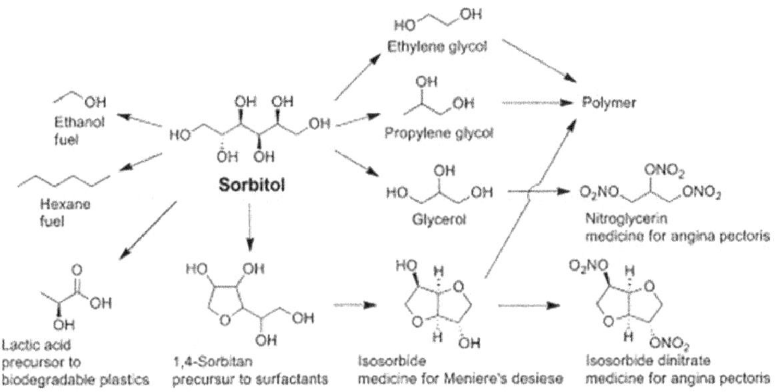

FIGURE 7.5 Cellulose-derived sorbitol as a platform chemical for biopolymer monomer or their precursor (Yabushita et al., 2014).

during the hydrolysis of cellulose. Hydrolytic hydrogenation of cellulose results in sorbitol over several catalysts including those based on ruthenium, platinum, nickel, and palladium. Acetaldehyde and hydrogen cyanide can react to produce lactonitrile, which can then be hydrolyzed to produce lactic acid. Ammonium chloride or ammonium sulfate are the byproducts of hydrolysis, which can be carried out using either hydrochloric acid or sulfuric acid. Two of the most significant chemical producers of lactic acid are the American company Sterling Chemicals Inc. and the Japanese company Musashino. Through catalytic processes, it is feasible to synthesize both racemic and enantiopure lactic acids from different starting materials, such as vinyl acetate or glycerol (Furtwengler & Avérous, 2018; Ochoa-Gómez & Roncal, 2017; Yabushita et al., 2014; Razali & Abdullah, 2017) (Figure 7.5).

7.2.2 Biological Monomer Synthesis

Glucose, sucrose, lactose, and galactose are several of the saccharide sources for fermentation to biopolymer monomers. Genetically engineered bacteria strains like *A. succinogenes, A. succinoproducenes*, and *Mannheimia succinoproducen* are used for a high yield of acid monomer production from lignocellulosic biomass. Genetic and metabolic engineering are key research areas for optimizing the productivity, selectivity, and rate of monomer synthesis. The fermentation process can be carried out aerobically or anaerobically, where anaerobic conditions have been reported to outperform aerobic conditions for some monomers with higher productivity. There may be inhibitors present in lignocellulosic feedstocks, which can occasionally reduce the yield and efficiency of acid monomer synthesis. These inhibitors, which include weak acids, furans, and phenolic compounds, can all be largely eliminated by activated carbon adsorption.

By integrating the two processes—the substrate's enzymatic hydrolysis and its fermentation—into one, known as simultaneous saccharification and fermentation, the generation of monomeric acid from lignocellulosic biomass can be made more

(a) Fermentation and neutralization

$C_6H_{12}O_6 + Ca(OH)_2 \xrightarrow{Fermentation} (2CH_3CHOHCOO\text{-})\ Ca^{2+} + 2H_2O$

(b) (Hydrolysis by H_2SO_4

$(2CH_3CHOHCOO^-)\ Ca^{2+} + H2SO_4 \longrightarrow 2CH_3CHOHCOOH + CaSO_4$

(c) Esterification

$CH_3CHOHCOOH + CH_3OH \longrightarrow CH3CHOHCOOCH_3 + H_2O$

(d) Hydrolysis by H_2O

$CH_3CHOHCOOCH_3 + H2O \longrightarrow CH_3CHOHCOOH + CH_3OH$

FIGURE 7.6 Lactic acid monomer synthesis by fermentation (a–d) (Balla et al., 2021).

efficient. Numerous experiments have been done to use various enzymes and bacteria to break down the lignocellulosic material into organic acids.

While biological monomer synthesis offers several advantages detailed earlier, challenges of low yield, low productivity, high cost of substrate, costly filtration, and difficulty in handling potentially pathogenic materials. Among the important biodegradable polyester is PLA. Its monomer lactic acid is industrially manufactured through the anaerobic fermentation of simple sugars through bacterial fermentation with the use of lactic acid bacteria such as *Lactobacillus casei, Lactococcus lactis Lactobacillus acidophilus, Lactobacillus delbrueckii subsp. bulgaricus (Lactobacillus bulgaricus), Lactobacillus helveticus, Streptococcus salivarius subsp. Thermophilus*. The most often utilized *Lactobacillus* stem, *Lactobacillus helveticus*, produces a racemic blend of LA (Figure 7.6).

Because it is more chemically and economically practical than the chemical pathway, microbial carbohydrate fermentation is the primary method used in the industrial manufacture of lactic acid. Bacterial fermentation generally takes place in batch reactors where water and bacterial cultures are combined to create a broth-like mixture where carbs are transformed into LA. After the broth containing calcium lactate has gone through the fermentation process, it is filtered to remove cells, carbon treated, evaporated, and acidified with sulfuric acid to produce lactic acid and calcium sulfate. Without the use of water, the fermentation process can also occur in a solid-state reactor, although this method is very unstable because temperature cannot be easily monitored and controlled.

Escherichia coli has been thoroughly investigated for its ability to manufacture succinic acid from refined sugars such as glucose, fructose, galactose, lactose, xylose, and sucrose as substrates, while *Actinobacillus succinogenes* and *Anaerobiospirillum succinoproducens* were less frequently utilized for such substrates. Additionally, succinic acid can be generated utilizing *E. coli* from lignocellulosic biomass, such as maize stalk, cane molasses, plant hydrolysates, and softwood hydrolysates. *E. coli* produced a succinic acid concentration of 57.8 g/L using corn stalk enzymatic hydrolysate as a substrate. For the first 12 hours, this process was carried out in anaerobic culture in a stage-aeration fermenter. Figure 7.7 shows various fermentation routes to biopolymer monomers (Balla et al., 2021; De Clercq et al., 2017; Patel et al., 2006).

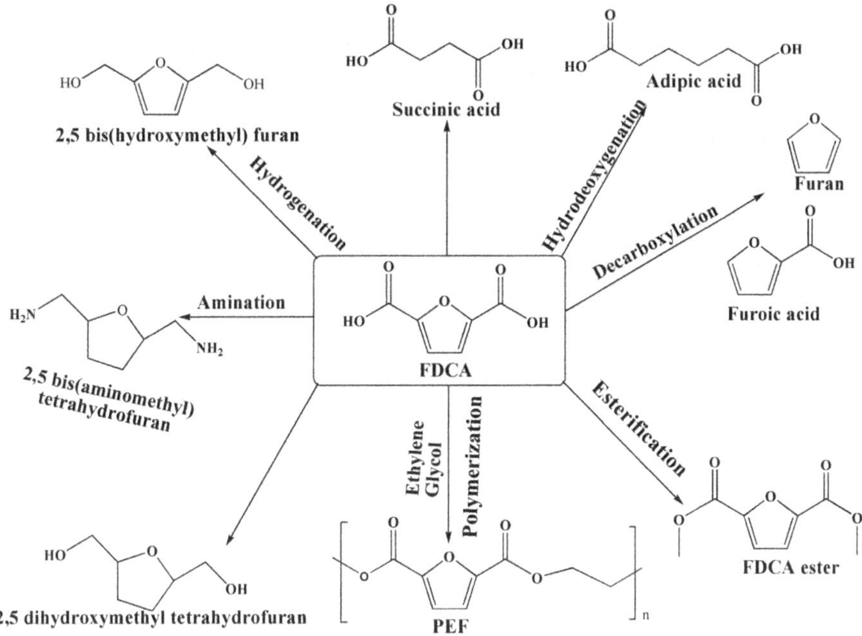

FIGURE 7.7 Conversion of 2,5-furan dicarboxylic acid (FDCA) to various important monomers (Zhang et al., 2015).

7.3 BIOPOLYMER SYNTHESIS AND PROPERTIES

Melt polycondensation as well as ROP techniques are two of the major synthesis technologies for biodegradable polyesters. Catalysts are used during polymerization, and together, process conditions determine the quality and properties of the polymeric output. Another factor during polymerization is the competing reverse reaction of hydrolysis, which decomposes the polymer chains, limiting the achievable molecular. This may be overcome by succeeding steps of solid-state polycondensation and chain extension. Other methods for achieving higher molecular weight include azeotropic dehydrative condensation, direct condensation polymerization, and/or anhydride polymerization, where lack of water condensate limits the reverse reaction of depolymerization. It is generally desirable to optimize for higher molecular weight using the above protocols as this positively correlates with the properties of the resulting polymer including mechanical, barrier, and thermal properties. There is, however, an upper limit above which molecular weight may inhibit the biodegradation rate to a point below set standard requirements (Bayu et al., 2018; Akhtar et al., 2014; Yabushita et al., 2014).

7.3.1 PLA

Three types of PLA are possible since LA is a chiral molecule containing D-type and L-type isomers: poly(L-lactic acid), poly(D-lactic acid), and poly(D,L-lactic acid). Most of the time, lactide ROP is used to create the high-molecular-weight

PLA that is sold commercially. By distilling condensation water with or without a catalyst, direct polycondensation of lactic acid is typically carried out in bulk while vacuum and temperature are gradually raised. Two reaction equilibrium processes are involved in the polycondensation system of LA: (i) the dehydration equilibrium for esterification and (ii) the ring-chain equilibrium involving the depolymerization of PLA into lactide. Water is eliminated as a byproduct during the step-growth polymerization reaction known as direct polycondensation. However, the very viscous reaction mixture makes it difficult to extract water from it, and as a result, direct polycondensation of LA only yields low-molecular-weight PLA (50,000 g/mol) due to the unfavorable reaction equilibrium constant and the low reaction rate. The most often employed catalysts for the polycondensation process of PLA are transition metal compounds (e.g., $SnCl_2$, ZnO, Fe $(Lactate)_3$, and $Ti(OBu)_4$) and Bronsted acids (e.g., H_2SO_4, p-toluenesulfonic acid, CH_3SO_3H, perfluorinated resin, and sulfonic acid) (Figure 7.8).

ROP allows for precise control over polymerization chemistry, leading to variations in polymer properties for specific applications. Commercial production of high molecular weight poly-L-lactic acid (PLLA) involves ROP of L-lactide obtained from the decomposition of low molecular weight PLLA. The ROP process includes the polymerization of L-lactide, followed by depolymerization to lactide, which is then further polymerized to produce high molecular weight PLA. The composition of the resulting polymer can be influenced by the stereochemistry of L-lactide, affecting its properties.

Several companies, such as Cargill Dow, Shimadzu, and Dupont, produce PLA of varying molecular weights using ROP. The ROP of lactide is carried out through different methods like solution polymerization, bulk polymerization, melt polymerization, and suspension polymerization. The specific mechanism of ROP depends on the catalyst used, which can be ionic-, coordination-, or free radical-type. Transition and non-transition metal compounds, including tin, lead, zinc, bismuth, yttrium, iron, aluminum, and magnesium, serve as catalysts for lactide ROP. Tin(II) compounds are commonly used due to their efficiency. However, some tin-based catalysts can be toxic, while others have lower toxicity. Environmentally friendly catalysts, like titanium oxides, have also been developed for lactide polymerization.

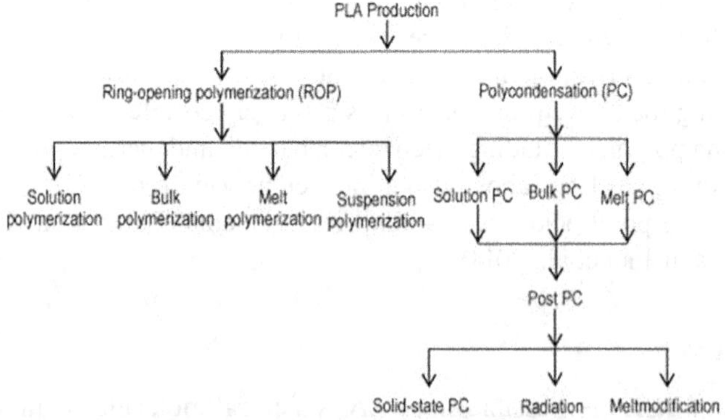

FIGURE 7.8 PLA synthesis methods (Maharana et al., 2009).

The synthesized PLA may contain residual catalysts and monomers, requiring purification and removal. The presence of catalyst in the polymer can range from low to high concentrations, while the removal of residual monomers is essential. However, the purification and isolation process for L-lactide from PLA is expensive, limiting its widespread use as a commodity material. Despite its advantages in producing high molecular weight PLA with high yields, ROP also has certain disadvantages.

Polycondensation of lactic acid (LA) in the presence of catalysts results in PLA and water as byproducts. Binary catalysts, consisting of metal compounds activated with proton acids, are found to be more effective than single-metal-compound catalysts. Direct polycondensation of LA typically yields low molecular weight polymers with substandard mechanical properties. To overcome this, various methods have been developed, such as using organic solvents or multifunctional branching agents like dipentaerythritol, which lead to star-shaped polymers. Melt polycondensation has been introduced as a cost-effective method for obtaining high molecular weight PLLA. Polycondensation can also be performed in the presence of difunctional monomers, generating telechelic prepolymers that can be further reacted to form high molecular weight polymers. These polymers exhibit similar behavior to poly(lactide) homopolymers prepared by ROP.

The properties of PLA, as exemplified in Table 7.1, are influenced by the proportion and allocation of the two stereoisomers of LA within the polymer chains. The thermal, mechanical, and biodegradation properties of PLA depend on these factors. PLA with a high proportion of the l-isomer tends to result in crystalline products, while higher levels of the d-isomer (>15%) lead to amorphous products. Commercial PLLA grades are, on average, semicrystalline, with a high melting point of around 160°C and a glass transition temperature varying from 55°C to 60°C. It is necessary for PLA to have some crystalline fraction as it tends to improve the end product. Properties of PLA may also be affected by residual catalysts, the use of chain extenders and end-capping molecules (Yabushita et al., 2009).

TABLE 7.1
Mechanical Properties of Various Commercially Available PLA Materials (Maharana et al., 2009)

Polymer	*Tg* (°C)	*Tm* (°C)	Tensile Strength (MPa)	Tensile Modulus (MPa)	Flexural Modulus (MPa)	Elongation Yield (%)	Elongation Break (%)
l-PLA (MW: 50,000)	54	170	28	1200	1400	3.7	6
l-PLA (MW: 100,000)	58	159	50	2700	3000	2.6	3.3
l-PLA (MW: 300,000)	59	178	48	3000	3250	1.8	2
d,l-PLA (MW: 20,000)	50	–	n/a	n/a	n/a	n/a	n/a
d,l-PLA (MW: 107,000)	51	–	29	1900	1950	4	6
d,l-PLA (MW: 550,000)	53	–	35	2400	2350	3.5	5

7.3.2 PBAT

Another commercially important biopolymer is PBAT, a biodegradable polymer aliphatic–aromatic co-polyester with exceptional biodegradability bestowed from aliphatic polyesters and effective performance from aromatic polyester segment. When compared to low-density PE, the mechanical characteristics of PBAT are similar and with high promise for a variety of possible applications.

In the polycondensation procedure for PBAT synthesis, catalysts based on organometallic compounds of zinc, tin, and titanium can be used. Three key steps make up the synthesis of PBAT: pre-mixing with catalyst and other additives, pre-polymerization, and final polymerization. Long reaction periods, extreme vacuum, and typically higher temperatures than 190°C are needed to prepare PBAT and encourage condensation processes to get rid of lighter byproducts.

To optimize PBAT crystallization behavior and prevent tack, nucleating agents can be employed in the final polymerization process of PBAT manufacturing. The most appropriate nucleating agents are often inorganic compounds, such as talc, chalk, mica, or silicon oxides. As color stabilizers, phosphorus compounds like phosphoric acid and phosphorous acid can be added during the pre- or post-polymerization processes.

The composition of the monomers and their molecular weight have an impact on the mechanical characteristics of PBAT. On the one hand, Young's modulus is claimed to rise with the amount of terephthalate units while elongation at break declines. On the other side, when the molecular weight grows, the tensile strength rises but the elongation at break falls. The mechanical characteristics of PBAT may be customized depending on the process variables used, such as pressure and reactor temperature, as these factors have an impact on PBAT's molecular weight throughout the reaction (Ferreira et al., 2019).

7.4 CONCLUSION

Biobased polymer become sustainable alternatives to conventional non-biodegradable plastics, where packaging and agricultural products are the major sectors in the bioplastic industry. Modification of biopolymer technology helps to improve biopolymer properties as well as reduce the cost of production and reach wider applications. Biopolymers are increasingly solving the problems of pollution and sustainability. Their performance is, however, still lackluster for some high-end applications. This limits their proliferation and prospects for replacing conventional plastics. The molecular weight of biopolymers is inversely correlated to biodegradability and biodegradation rate. This while being directly proportional to material performance due to higher lower free volume (mechanical; high-strength structural applications, thermal; dimensional stability and retention of properties at above ambient service temperature application, barrier; food packaging films with sufficient protection against the ingress of oxygen from the atmosphere, chemical resistance); provide packaging for hazardous substances where recycling at end-of-life is not viable.

REFERENCES

Aaliya, B., Sunooj, K. V., & Lackner, M. (2021). Biopolymer composites: A review. *International Journal of Biobased Plastics*, *3*(1), 40–84.

Akhtar, J., Idris, A., & Abd. Aziz, R. (2014). Recent advances in production of succinic acid from lignocellulosic biomass. *Applied Microbiology and Biotechnology*, *98*, 987–1000.

Andreeßen, C., & Steinbüchel, A. (2019). Recent developments in non-biodegradable biopolymers: Precursors, production processes, and future perspectives. *Applied Microbiology and Biotechnology*, *103*, 143–157.

Balla, E., Daniilidis, V., Karlioti, G., Kalamas, T., Stefanidou, M., Bikiaris, N. D., … & Bikiaris, D. N. (2021). Poly (lactic Acid): A versatile biobased polymer for the future with multifunctional properties-From monomer synthesis, polymerization techniques and molecular weight increase to PLA applications. *Polymers*, *13*(11), 1822.

Baruah, J., Nath, B. K., Sharma, R., Kumar, S., Deka, R. C., Baruah, D. C., & Kalita, E. (2018). Recent trends in the pretreatment of lignocellulosic biomass for value-added products. *Frontiers in Energy Research*, *6*, 141

Bayu, A., Yoshida, A., Karnjanakom, S., Kusakabe, K., Hao, X., Prakoso, T., … & Guan, G. (2018). Catalytic conversion of biomass derivatives to lactic acid with increased selectivity in an aqueous tin (II) chloride/choline chloride system. *Green Chemistry*, *20*(17), 4112–4119.

De Clercq, R., Dusselier, M., & Sels, B. F. (2017). Heterogeneous catalysis for bio-based polyester monomers from cellulosic biomass: Advances, challenges and prospects. *Green Chemistry*, *19*(21), 5012–5040.

Devadas, V. V., Khoo, K. S., Chia, W. Y., Chew, K. W., Munawaroh, H. S. H., Lam, M. K., … & Show, P. L. (2021). Algae biopolymer towards sustainable circular economy. *Bioresource Technology*, *325*, 124702.

Fasciotti, M. (2017). Perspectives for the use of biotechnology in green chemistry applied to biopolymers, fuels and organic synthesis: From concepts to a critical point of view. *Sustainable Chemistry and Pharmacy*, *6*, 82–89.

Ferreira, F. V., Cividanes, L. S., Gouveia, R. F., & Lona, L. M. (2019). An overview on properties and applications of poly (butylene adipate-co-terephthalate)-PBAT based composites. *Polymer Engineering & Science*, *59*(s2), E7–E15.

Hernández, N., Williams, R. C., & Cochran, E. W. (2014). The battle for the "green" polymer. Different approaches for biopolymer synthesis: bioadvantaged vs. bioreplacement. *Organic& Biomolecular Chemistry*, *12*(18), 2834–2849.

Jha, A., & Kumar, A. (2019). Biobased technologies for the efficient extraction of biopolymers from waste biomass. *Bioprocess and Biosystems Engineering*, *42*, 1893–1901.

Komesu, A., de Oliveira, J. A. R., da Silva Martins, L. H., Maciel, M. R. W., & Maciel Filho, R. (2017). Lactic acid production to purification: A review. *BioResources*, *12*(2), 4364–4383.

Kong, L., Li, G., Wang, H., He, W., & Ling, F. (2008). Hydrothermal catalytic conversion of biomass for lactic acid production. *Journal of Chemical Technology & Biotechnology: International Research in Process, Environmental & Clean Technology*, *83*(3), 383–388.

Koutinas, A. A., Vlysidis, A., Pleissner, D., Kopsahelis, N., Garcia, I. L., Kookos, I. K., … & Lin, C. S. K. (2014). Valorization of industrial waste and by-product streams via fermentation for the production of chemicals and biopolymers. *Chemical Society Reviews*, *43*(8), 2587–2627.

Li, S., Deng, W., Wang, S., Wang, P., An, D., Li, Y., … & Wang, Y. (2018). Catalytic transformation of cellulose and its derivatives into functionalized organic acids. *ChemSusChem*, *11*(13), 1995–2028.

Liu, S., Lu, H., Hu, R., Shupe, A., Lin, L., & Liang, B. (2012). A sustainable woody biomass biorefinery. *Biotechnology Advances*, *30*(4), 785–810.

Maharana, T., Mohanty, B., & Negi, Y. S. (2009). Melt-solid polycondensation of lactic acid and its biodegradability. Progress *in Polymer Science*, *34*(1), 99–124.

Ochoa-Gómez, J. R., & Roncal, T. (2017). Production of sorbitol from Biomass. In: Fang, Z., Smith, R. L. & Qi, X. (Eds.), *Production of Platform Chemicals from Sustainable Resources* (pp. 265–309). Singapore: Springer.

Patel, M. A., Ou, M. S., Harbrucker, R., Aldrich, H. C., Buszko, M. L., Ingram, L. O., & Shanmugam, K. (2006). Isolation and characterization of acid-tolerant, thermophilic bacteria for effective fermentation of biomass-derived sugars to lactic acid. *Applied and Environmental Microbiology*, *72*(5), 3228–3235.

Razali, N., & Abdullah, A. Z. (2017). Production of lactic acid from glycerol via chemical conversion using solid catalyst: A review. *Applied Catalysis A*: *General*, *543*, 234–246.

Rose, M., & Palkovits, R. (2011). Cellulose-based sustainable polymers: State of the art and future trends. *Macromolecular Rapid Communications*, *32*(17), 1299–1311.

Sadasivuni, K. K., Saha, P., Adhikari, J., Deshmukh, K., Ahamed, M. B., & Cabibihan, J. J. (2020). Recent advances in mechanical properties of biopolymer composites: A review. *Polymer Composites*, *41*(1), 32–59.

Sarkar, S., Ponce, N. T., Banerjee, A., Bandopadhyay, R., Rajendran, S., & Lichtfouse, E. (2020). Green polymeric nanomaterials for the photocatalytic degradation of dyes: A review. *Environmental Chemistry Letters*, *18*, 1569-1580.

Schnepp, Z. (2013). Biopolymers as a flexible resource for nanochemistry. *Angewandte Chemie International Edition*, *52*(4), 1096–1108.

Sharma, S. K., & Mudhoo, A. (Eds.). (2011). *A Handbook of Applied Biopolymer Technology*: *Synthesis, Degradation and Applications* (No. 12). Cambridge: Royal Society of Chemistry.

Stoica, D., Alexe, P., Ivan, A. S., Moraru, D. I., Ungureanu, C. V., Stanciu, S., & Stoica, M. (2022). Biopolymers: Global carbon footprint and climate change. In: Nadda, A. K., Sharma, S., & Bhat, R. (Eds.), *Biopolymers*: *Recent Updates, Challenges and Opportunities* (pp. 35–54). Springer International Publishing.

Yang, X., Qiao, C., Li, Y., & Li, T. (2016). Dissolution and resourcfulization of biopolymers in ionic liquids. Reactive and Functional Polymers, 100, 181-190

Tan, J., Abdel-Rahman, M.A., Sonomoto, K. (2017). Biorefinery-Based Lactic Acid Fermentation: Microbial Production of Pure Monomer Product. In: Di Lorenzo, M., Androsch, R. (eds) *Synthesis, Structure and Properties of Poly(lactic acid)*. Advances in Polymer Science, vol 279. Springer, Cham. https://doi.org/10.1007/12_2016_11

van Beilen, J. B. (2008). Transgenic plant factories for the production of biopolymers and platform chemicals. *Biofuels*, *Bioproducts and Biorefining*, *2*(3), 215–228.

Wang, S., Zhang, L., Semple, K., Zhang, M., Zhang, W., & Dai, C. (2020). Development of biodegradable flame-retardant bamboo charcoal composites, Part I: Thermal and elemental analyses. *Polymers*, *12*(10), 2217.

Weng, Y. X., Wang, L., Zhang, M., Wang, X. L., & Wang, Y. Z. (2013). Biodegradation behavior of P (3HB, 4HB)/PLA blends in real soil environments. *Polymer Testing*, *32*(1), 60–70.

Yabushita, M., Kobayashi, H., & Fukuoka, A. (2014). Catalytic transformation of cellulose into platform chemicals. *Applied Catalysis B*: *Environmental*, *145*, 1–9.

Yadav, A., Mangaraj, S., Singh, R., Das, S.K., Kumar, N., & Arora, S. (2018). Biopolymers as packaging material in food and allied industry. *International Journal of Chemical Studies*, *6*(2), 2411–2418.

Yang, L., Su, J., Carl, S., Lynam, J. G., Yang, X., & Lin, H. (2015). Catalytic conversion of hemicellulosic biomass to lactic acid in pH neutral aqueous phase media. *Applied Catalysis B*: *Environmental*, *162*, 149–157.

Yates, M. R., & Barlow, C. Y. (2013). Life cycle assessments of biodegradable, commercial biopolymers-A critical review. *Resources*, *Conservation and Recycling*, *78*, 54–66.

Zhao, X., Zhang, L., & Liu, D. (2012). Biomass recalcitrance. Part I: The chemical compositions and physical structures affecting the enzymatic hydrolysis of lignocellulose. *Biofuels*, *Bioproducts and Biorefining*, *6*(4), 465–482.

8 Blending Techniques for Preparing Cost-minimized Bioplastics and Biocomposites

Orebotse Joseph Botlhoko, Raphaahle Caroline Mphahlele, Rakgoshi Lekalakala and Sudhakar Muniyasamy

8.1 INTRODUCTION

Bioplastics or biodegradable polymers such as poly(lactic acid) (PLA), poly(ε-caprolactone) (PCL), poly(glycolic acid) (PGA), poly(hydroxybutyrate-co-hydroxyvalerate) (PHBV), and poly(butylene succinate) (PBS) are increasingly receiving intensive attention for various packaging applications, due to plastic pollution, diminishing landfill space, animal health and safety issues as results of plastics disposal. For this reason, biodegradable polymers are promising packaging materials due to their biodegradability, biocompatibility and environmental friendliness point of view. On the other hand, biodegradable polymers will address the sustainability issues associated with petroleum-based polymers. However, biodegradable polymers have some drawbacks that limit their use as neat biodegradable polymers. For example, PLA is a brittle thermoplastic polymer with low degree of crystallinity, highly hygroscopic thermal instability and very slow biodegradation rate in relation to cellulose and other biobased polymers (Modi et al., 2013; Mo et al., 2023). While PCL is a ductile biodegradable polymer with a moderate degree of crystallinity, very low melting temperature (60°C), and poor modulus (Modi et al., 2013; Chrissafis et al., 2007). PGA has insufficient toughness, a high friction coefficient, difficulty in processing and relatively high production cost. PGA is characterized by high crystallinity; as a result, it is not soluble in most organic solvents; the exceptions are highly fluorinated organic solvents such as hexafluoroisopropanol. It has a high melting temperature (225°C), and it is highly sensitive to hydrolytic degradation; hence, it requires careful control of processing conditions (Gunatillake et al., 2003). PHBV is also a brittle thermoplastic polymer

DOI: 10.1201/9781003304142-8

with low thermal stability (Modi et al., 2013). Hence, its melt processing technique results in thermal degradation of the polymer (PHBV) near or above the melting temperature of about 171°C (Modi et al., 2011). Therefore, it should be processed through the solvent mixing method in order to avoid thermal degradation when processing through melt mixing. Finally, PBS has low molar mass, low melt viscosity and strength, and high crystallinity, limiting the action of degradation enzymes (Mo et al., 2023). Polybutylene adipate terephthalate (PBAT) has Low stiffness and barrier properties (Mo et al., 2023). To overcome the above-mentioned drawbacks, these biodegradable polymers must be blended with other polymers and filled with nanoparticles for better performance. The addition of certain lignocellulosic fillers may enhance the rate of biodegradation of the polymer material, while maintaining the environmental friendliness of the neat material. It is also important to note that most of the biodegradable polymers exhibit very poor barrier to oxygen transmission rate (OTR) and water vapour transmission rate (WVTR) when compared to that of petroleum-based polymers (Figure 8.1). Therefore, there is a need for improvement of biodegradability and barrier properties of biodegradable polymer by blending and incorporating nanoparticles to compensate for the above-mentioned drawbacks and determine their viability.

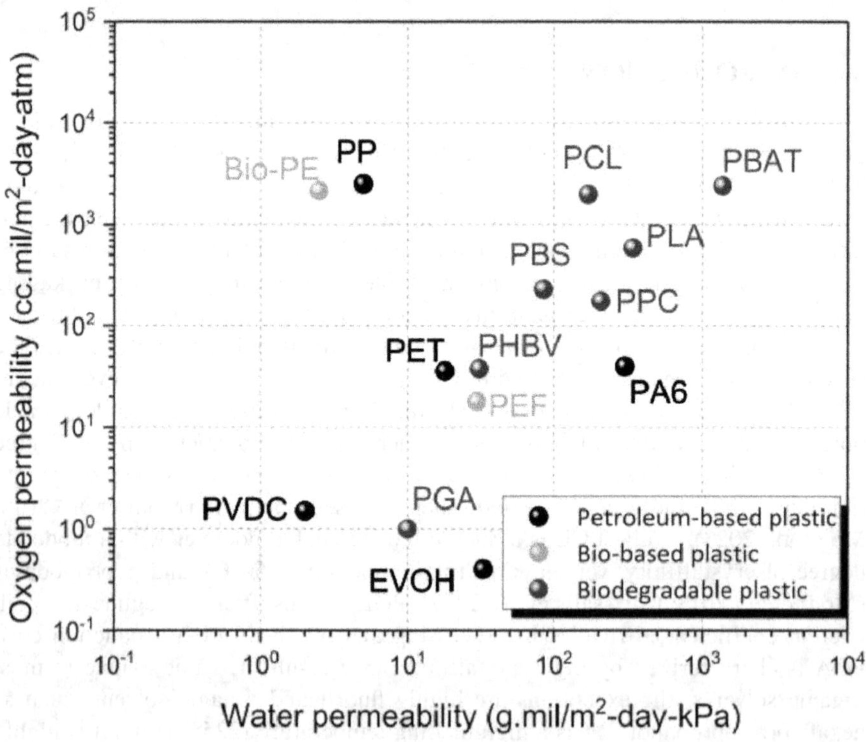

FIGURE 8.1 A comparison between oxygen/water transition rate of selected biodegradable, bio-based, and petroleum-based polymers (Wu et al., 2021). (Copyright 2021, Elsevier B.V.)

Among other factors, there are several aspects that must be tested in order to determine the viability of biobased materials. This includes establishing the mechanical properties of the material, thus making the material comparable to similar polymers and composites. This fundamental strategy could largely determine the range of applications of the novel material. Other important factors that must be determined are the thermal and rheological properties of the polymer material. In this rational, a proper understanding of these properties is essential for establishing appropriate processing techniques and modification particles/blending systems for a novel lightweight polymer composite material with improved functionalities. With such a variety of factors affecting the performance of polymer composites, fibres such as glass and carbon fibre have long been incorporated into composites to create strong, lightweight materials.

8.2 SYNTHETIC AND SEMI-SYNTHETIC POLYMERIC MATERIALS-BASED COMPOSITES

8.2.1 Polymer Matrix Materials

The main component of any lignocellulosic polymer composite is the polymer matrix. The matrix is the component that makes up the majority of the weight of the composite material. In biodegradable polymer composites, the polymer matrix is most often a bioplastic, an inherently biodegradable, or natural resource-based polymer. For each application, certain matrix polymers may have thermal, mechanical and barrier properties that make them favourable over others. The polymer must also be processed under conditions that are compatible with the desired filler materials. For example, it would be imperative to process a filler with low thermal stability using a polymer that can be processed at a low temperature.

PLA is a three-carbon-membered aliphatic polyester polymer with hydroxyl and carbonyl at the end. It is produced from the polymerization of lactic acid and lactide molecules, which can be derived from the fermentation of agricultural products, such as corn starch (Henton et al., 2005). As such, PLA is an entirely biobased and biodegradable polymer which is derived from renewable resources. Under physiological conditions, PLA degrades into lactic acid, a natural product of the body. As such, it is suitable for biomedical applications, and as a result, it has been successfully applied in sutures and bone fracture repair (Athanasiou et al., 1996).

Poly(hydroxybutyrate), or PHB, and its co-polymer PHBV, are biodegradable polyesters produced by specific bacterial strains. The polymers are formed as an energy storage material in certain bacterial species and can be produced using renewable resources (Byrom, 1987). Both polymers are biodegradable in physiological conditions and have non-toxic byproducts. As such, PHBV and PHB have been used in biomedical applications such as tissue scaffolding materials and in controlled drug release (Chen & Wu, 2005).

PCL is a six-carbon membered of a synthetic biodegradable polyester with hydroxyl terminated end groups produced by the polymerization of ε-caprolactone (Labet & Thielemans, 2009). This polymer has potential application in the biomedical field as a tissue engineering scaffold material due to its biodegradability in

TABLE 8.1
Characteristics of Biodegradable Matrix Polymers (Someya et al., 2004; Velde & Kiekens, 2002)

Matrix Polymer	Density (g/cm³)	Melting Temperature (°C)	Glass Transition Temperature (°C)
PLA	1.25	160	60
PHB	1.25	175	10
PHBV	1.25	165	5
PCL	1.11	55–60	−60
PBS	1.23	115	−35
PGA	1.6	225	40

TABLE 8.2
Mechanical Properties of Biodegradable Polymers (Someya et al., 2004; Velde & Kiekens, 2002)

Matrix Polymer	Tensile Strength (Mpa)	Young's Modulus (Gpa)	Elongation at Yield (%)
PLA	60	1.6	6
PHB	35	1.7	10
PHBV	25	1.2	20
PCL	16	0.4	550
PBS	33	0.7	320

physiological conditions (Williams et al., 2005). It is unique among other biodegradable polymers because of its relatively low melting temperature (Table 8.1).

PBS is a biodegradable polymer produced by the polymerization of succinic acid and 1,4-butanediol. Though it may be made from petroleum resources, it has also been shown that it can be synthesized entirely from biomass waste products (Tachibana et al., 2010). Because of its biodegradability and ability to be processed into sheets and films, PBS is applied as a biodegradable packaging/bagging material and in mulch films, mechanical properties are summatized in Table 8.2.

PGA is a biodegradable polyester, which is synthesized by the polymerization of glycolic acid and has received wide use as a suture in biomedical applications (Gilding & Reed, 1979) due to its capability to form fibres. PGA biodegrades in physiological conditions with non-toxic byproducts, making it especially suitable for use in surgery (Gilding & Reed, 1979). Moreover, PGA is characterized by relatively high melting temperature when compared to that of PHB, PLA, and PCL (Table 8.1).

8.2.2 Lignocellulosic Filler Materials

To create a lignocellulosic polymer composite, biological filler materials are required to be blended with a suitable polymer matrix. Polymer industry has much interest in filler materials ranging from natural fibres, such as flax, jute and hemp, to biofuel

TABLE 8.3
Properties of Common Lignocellulosic Fibres (Joshi et al., 2004; Ashori et al., 2006; Dhakal et al., 2007; David & Ragauskas, 2010; Mohanty et al., 2002)

Lignoellulosic Material	Lignin Content (%)	α-Cellulose Content (%)	Tensile Strength (MPa)	Young's Modulus (GPa)
Jute	11.8–12.9	59–71	400–800	10–30
Flax	2–5	62–72	780–1500	44
Hemp	4–6	70–76	690	30–60
Sisal	8–12	60–67	530–630	17–22
Kenaf	14–22	46–56	200–600	15–35
Cotton	2	90	400	12
Switchgrass	17–18	32–38	-	-

co-products such as soy meal (Diebel et al., 2012). These filters have several roles that can improve the properties of specific polymers for potential polymer industrial applications. First, the addition of a filler reduces the amount of polymer needed to produce a material; this is very important when using bioplastics, which are more costly than petroleum-based polymers. The addition of biomass to the polymer can increase its value, thus paving the way for new applications and exponential market growth of filler-polymer composite material. Fillers can also be used to modify and improve the mechanical qualities of polymer materials. Although it has not been researched extensively, biodegradation studies have shown that the incorporation of lignocellulosic filler materials can improve the biodegradability and compostability of polymer materials. Lignocellulosic fibres are a form of biomass that can possess considerable mechanical properties (Table 8.3), particularly their tensile strength. Fibres are low-cost, renewable filler materials and are the most important structural compound in plant cells, specifically in cell walls. However, fibres vary in their composition of lignin and cellulose. Importantly, fibres from woody plants tend to have higher lignin content, while cellulose, the main component of the cell wall, has a high concentration in green plants. Furthermore, fibres have the capability to improve the tensile properties of polymer materials and expand their application.

8.3 PROCESSING TECHNIQUES

To properly blend the polymer matrix and filler material to produce polymer composite, specific conditions must be achieved using specialized machines or chemicals and glassware. The schematic illustration of the steps involved in melt blending, solution blending, and in situ polymerization methods are presented in Figure 8.2.

8.3.1 Melt Blending

Melt blending method is the most industrially favourable polymer processing method due to its high-capacity polymer production (scalability), economically viable and

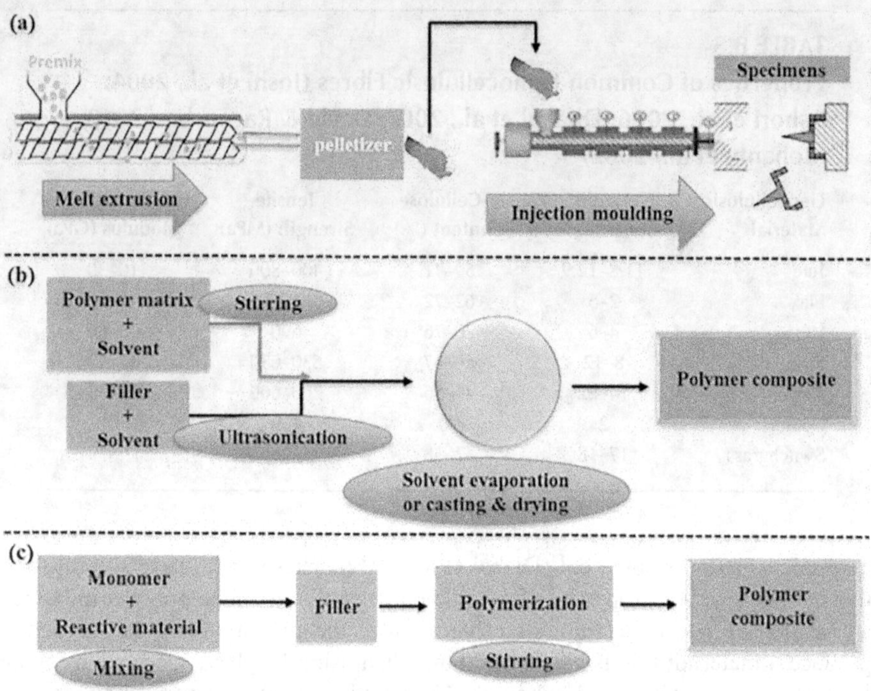

FIGURE 8.2 The schematic illustration of the melt blending, solution blending, and in situ polymerization methods (a–c).

environmentally friendly. In this method, polymers are generally dry mixed with filler materials in a container; since moisture initiates polymer degradation during melt processing, the material must be pre-dried before mixing (≤1% moisture content is ideal). The melt processing requires a certain temperature and pressure in order to properly melt the polymer matrix, a property that is different for every polymer (Table 8.1). As mentioned previously, the processing temperature must be above the polymer melting temperature (ideally ±20°C) but must not exceed the thermal degradation temperature of either the filler or the polymer.

A blending mechanism is required to ensure uniform distribution of the filler throughout the material. The blending mechanism uses mechanical mixing to properly disperse the filler through the melted polymer. To further ensure consistency, the filler material needs to be of a consistent size and shape, which reduces any imperfections in the polymer system. As they are added to any process, the polymer and filler materials must be carefully weighed to ensure that the mass ratio of polymer to filler remains uniform throughout the process. A common blending mechanism is the single- or dual-screw compounder. This device consists of one or two long threaded shafts housed within a tight-fitting cylinder. The cylinder temperature can be adjusted to provide an appropriate temperature to fully melt the matrix polymer. The cylinder has one or more inputs, through which the materials are fed into the mixer, and a single output for the blended material (Figure 8.2a). The rotation of the screws mechanically blends the matrix and filler materials together and forces the composite material through the output (die nozzle). Proper melting and strong shear forces facilitate the dispersion and distribution of the filler into the polymer matrix. Although, the melt blending method is the most industrially favourable, its degree of dispersion of filler nanoparticles is not better than the one that can be achieved by solution and in situ polymerization methods (Zhang et al., 2015). Depending on the properties of the polymer and the die nozzle, it can potentially be extruded as a strand or sheet. Simultaneously, the stands will be quenched in a water bath, pelletized, dried, and compounded using injection moulding or hot presser to form a specific shape for testing or product application (Shukla, 2019). While the sheets will be physically cut into different testing specimens. On the other hand, the pelletized material can be blown into a thin film for various applications. Importantly, prior to testing the specimens, samples are generally kept for some time to allow the polymer chain relation process to take place.

8.3.2 Solution Blending Method

In this method, a polymer is dissolved in an appropriate common solvent; simultaneously, the filler material is separately dispersed in the appropriate common solvent. Afterwards, the two solutions are mixed together and under vigorous agitation/stirring, ultrasonication mixing is used to facilitate filler dispersion and distribution within the polymer matrix solution until the complete mixing occurs. This critical step is followed by the homogenization of polymer composite material and solvent evaporation process or casting and drying of the as-prepared polymer composites (Figure 8.2b) (Zhang et al., 2015; Shukla, 2019; Zhan et al., 2017). Generally, the solution blending method permits better filler dispersion into the polymer matrix than the melt blending method. Nevertheless, the solution blending method is not extensively utilized by polymer commercial players due to low production output and its environmental issues (destructive solvents limit industrial application of solution blending) (Zhang et al., 2015).

8.3.3 In situ Polymerization Method

In situ polymerization method is normally used to prepare polymer composites with inherent insoluble and thermally unsteady matrices that cannot be developed by solution/melting blending methods (Shukla, 2019). Herein, the monomer and reactive material are mixed, followed by the addition of the filler and polymerization under a continuous stirring process (Figure 8.2c). The in situ polymerization method usually results in better filler dispersion due to the high degree of interactions between the filler and polymer matrix, which provides as-prepared polymer composites with exclusive properties (Zhan et al., 2017). However, the method has several limitations, such as the use of solvents and long processing time (Zhang et al., 2015).

8.4 EFFECT OF BLENDING AND PROPOSED BIODEGRADATION MECHANISMS

After processing, the final composite material is output. Ideally, this material is a polymer matrix that is interspersed with evenly distributed particles of filler material.

One of the most important characteristics of the composite is the interaction of the polymer with the surface of the filler or fibre material. Poor interfacial adhesion between the matrix and fibre will reduce the mechanical strength of the composite by creating imperfections and inconsistencies within the material structure. This poor interaction can be overcome with the use of surface treatments on the filler prior to processing, in which the surface of the fibre or filler is modified either chemically or physically to make it more compatible with the matrix polymer. After surface treatment, the matrix will adhere better to the filler particles, creating a more consistent structure, which can result in an increase in mechanical properties (Liu et al., 2004). Though research on the subject is currently limited, it is thought that the incorporation of lignocellulosic materials into polymeric materials, especially bioplastics, may increase the biodegradability of the material. As their name suggests, lignocellulosic materials are traditionally biomass consisting mainly of cellulose and lignin. Due to their composition, lignocellulosic materials are inherently biodegradable. Cellulose is a polysaccharide, and it biodegrades via the enzymatic hydrolysis of its polymer chain by microorganisms as they metabolize the polysaccharide. Unlike cellulose, lignin is not very susceptible to hydrolytic chain cleavage and, thus, is much more difficult to break down. The exact mechanism of lignin biodegradation is not yet fully understood; however, it has been noted that lignin can be broken down by certain species of fungi (Kapich et al., 1999). In this case, lignin degrades by oxidation of its molecular chain by free radicals.

The addition of these materials to polymeric composite materials may enhance the biodegradability of the polymer. Though research on the subject is limited, it has been shown that when added to a polymer matrix, certain lignocellulosic fibres will dramatically increase the degradation rate of the material (Teramoto et al., 2004). It is suggested that the imperfections that are introduced to the matrix by the filler material, especially one that is not compatibilized, contribute to the increase in biodegradability. This is a result of preferential degradation of the matrix occurring in regions adjacent to particles of the filler material.

8.5 TECHNIQUES PRINCIPLE/REQUIREMENTS AND PROPERTIES

8.5.1 Thermal Properties and Requirements

Biopolymer blends secure their place in packaging and multiply industries due to their balanced mechanical, thermal and biodegradability with respect to their neat component. However, the immiscibility phenomenon and associated costs of the biopolymers may hinder their optimum performance and affect daily practical applications. Thermal analysis is one of the most significant properties used to reveal the immiscibility character of the polymer blends (binary and ternary) system. This includes thermogravimetric analysis (TGA) and differential scanning calorimetry (DSC). The TGA and DSC parameters of different polymer blend systems are summarized in Table 8.4. It is interesting to note that the thermal stability and miscibility aspects of binary and ternary polymer-based systems depend on the compact packing of the polymeric chains or segments, and configuration changes upon blending could indicate if the single-phase or two-phase system has been archived (de Matos Costa et al., 2020). For TGA, more than one-step degradation curve will

TABLE 8.4
Summary of the Thermal Properties of the Polymer Blends and Composites

Formulation	Melting Temperature (°C)	Crystallization Temperature (°C)	X_C (%)	Thermal Degradation (°C)	Ref.
PLA/PBS	178.8	–	48.7	–	
PLA/PBS/ tocopherol	177.6	–	49.5	–	Aversa et al. (2020)
PLA/PHB	178.8	–	52.6	–	
PLA/20PBS	143.7 & 152.4	–	10.5	364 & 394	
PLA/20PBS/ 1S-CNC	140.9 & 151.6	–	7.2	345 & 397	Luzi et al. (2016)
PLA/20PBS/ 1S-CNC	139.9 & 151.1	–	6.7	344 & 397	
PBAT	128	102	–	351	
PBAT/ 12-sulphonated copolyesters	116.4	–	–	322	Huang et al. (2020)
PBS	114	–	–	–	
10PBS/40PLA/ 50TPS	109.4 & 153.1	–	–	–	Zhen et al. (2011)
40PBS/10PLA/ 50TPS	112.3 & 153.9	–	–	–	

be noted, while for DSC, more than one melting and crystallization will be noted. In some instances, glass transition temperature peaks can be noted. Nofar et al. (2020) reported on the thermal behaviour of the SC-PLA/PBAT/PBSA ternary blend system. In this system, the crystallization of PLA, PBSA and PBAT was hindered due to the ability of the PBSA phase being able to encapsulate the PBAT droplets; hence, no crystal melting peak was detected for PBAT. On the other hand, blending PBS with PBAT can severely modify PBS's crystallization profile. In the de Matos Costa et al. (2020) case, the blending PBS with PBAT inhibited the crystallization of PBS at higher contents of PBAT. Interestingly, the melt crystallization temperature and rate of crystallization were enhanced at high PBS content. This behaviour was attributed to the ability of PBS and PBAT physical interaction to influence the reciprocal thermal behaviour. This means that the control of crystalline region, within the biopolymer matrix plays an important role in controlling the final properties of the binary and ternary blends. However, the physical interaction does not have much influence on thermal degradation behaviour. For example, PBAT is more thermally stable than PBS. Binary blends of PBS/PBAT were characterized by a single degradation peak without a significant thermal stability improvement due to poor compatibility factors. Therefore, the single degradation peak phenomenon was a confirmation of the partial physical compatibility between the two polymers.

The thermal degradation of polymers happens by four mechanisms: arbitrary chain scission, end-chain scission, chain-stripping and different procedures, like cross-linking. Scission includes the cleavage of a carbon-carbon bond in the foundation of the polymer to create two radicals (Surej Rajan et al., 2018). The phenomenon of polymeric chains or segments and configuration changes then influence the temperature profiles as well as the performance of the binary or ternary biopolymer systems. For example, the PLLA/PCL blend system suffers from poor mechanical properties due to the macro-phase separation and the poor adhesion phenomenon. This behaviour was indicated by thermal analysis. Where the sight enthalpy changes were observed in the blend system due to the improved perfection of PLLA crystals and reorganization of the biopolymer chains within the crystalline PLLA phase (Erba et al., 2001). Botlhoko et al. (2018) characterized the thermal behaviour of PLA/PCL blends. The dispersed PCL domains, which were acting as nucleating agents, promoted the crystallization of PLA chains. Moreover, the results indicated that the unique thermal stability of the dispersed PCL domains improved the thermal stability of the PLA/PCL blend system due to the governing mechanism change during the degradation process and activation energy improvements. Their results showed that the 40 wt% PLC-filled blend resulted in higher activation energy, indicating higher thermal stability. Oza et al. (2014) also investigated the thermal behaviour of hemp/PLA composites. Surface modification of hemp fibre was done for better mechanical bonding between the fibre and polymer component. Their results showed that the acetic anhydride modification PLA-based composites resulted in higher activation energy, indicating higher thermal stability arising from higher bond energy. In another work, Mofokeng and Luyt (2015) studied the morphology and thermal stability of melt-mixed PLA/PCL/TiO_2 nanocomposites. The results showed that the thermal stability of the PLA/PCL/TiO_2 nanocomposites improved with the addition of TiO_2 nanoparticles, as these acted as a degradation catalyst and retarded the escape of volatile degradation products.

In a more specific study, Mittal et al. (2015) showed that the thermal stability of PLA/PCL blends exhibited two-step degradation, while PLA/PCL/TPS ternary blends exhibited three-step degradation behaviour. The authors indicated that the thermal stability of the blend system increased with the addition of PCL due to its high thermal stability in relation to that of neat PLA and TPS. Moreover, the lower thermally stable TPS containing high moisture content and a high degree of intermixing of the components limited the maximum thermal stability improvement of the system. On the other hand, samples with large amount of TPS retained most of the char (endpoint of decomposition) in relation to that of their lower amount of TPS present. This confirms that the thermal stability of binary and ternary biopolymer systems does not only depend on the presence of the other component by weight percentage but also on the miscibility factor and polar solubility parameter between the biopolymers. The authors elaborated that the amount of incorporated PCL does not only enhance the thermal stability but also enhance the time taken for the material to degrade at fixed temperature conditions. However, the thermal stability of PCL component decreased significantly due to the presence of PLA and TPS. On the other hand, the ternary blend systems showed a reduction in melting temperature profiles with the addition of TPS. Generally, the melting point decrease in the biopolymer/TPS system is due to the amorphous morphology of TPS and the decrease in lamellar thickness as well as the thermodynamic factors such as polymer-polymer interaction proportion.

Hence, the need for modifications or plasticization of starch prior to the development of the binary and ternary system is important. Unfortunately, the thermal properties of binary biopolymers containing starch components cannot be discussed without an understanding of the fundamentals and properties of the native starch. Native starch is characterized by low thermal stability, poor mechanical integrity and humidity absorption. Due to these limitations, starch is mostly plasticized by the addition of different plasticizers, often with glycerol, to form TPS, as mentioned in the processing section. The spontaneous destructuration of the semi-crystalline structure of native starch and the formation of hydrogen bonds between the plasticizer and the starch makes TPS exclusive biodegradable polymer. However, this does not make TPS more thermally stable therefore TPS is often blended with other biopolymers to enhance their properties with the main focus on reducing the cost and enhancing their biodegradability rate. A recent study by Yun et al. (2016) showed that blending PBS with TPS worsened the thermal properties of neat PBS due to the presence of the extension of chain length and plasticization effects resulting from the glycerol component. This phenomenon concurs with the decrease of melting temperature when increasing the TPS component within the PBS matrix due to the hydrolysis completion within the amorphous region. This behaviour is subjected to the increase of TPS with the capability to reduce crystalline region in the PBS blend system; hence, thermal decomposition is possible (Yun et al., 2016). As previously discussed, the thermal properties of a polymer composite material are also an essential part of its processing and characterization. In brief, understanding of thermal properties allows proper processing conditions to be determined for the specific material, as well as establishing a temperature range in which the material will perform optimally.

8.5.2 Barrier Properties and Requirements

Barrier properties play an important role in food packaging, which could extend the shelf life and assure the preservation of food. Briefly, gas and vapour barrier films hold great potential for protection applications in food packaging, encapsulation of drugs and electronic/photovoltaic devices (Jiajie Wang et al., 2018). A functional gas and vapour barrier is defined as a layer that prevents substances such as oxygen and water vapour from migrating from the outside environment into the packaged food environment (Helanto et al., 2019). A variety of barrier materials have been developed to protect the package or prevent oxygen and water vapour molecules from penetrating a surface, to ensure a safety and prevention of rapid food spoilage. For instance, plastics are used to protect and keep food fresh, as well as extending the storage life (shelf life). Because of the global plastic pollution and other environmental concerns of plastics, biopolymers/starch films have garnered global research attention. Attributed to their safety, fast biodegradation rate and cost effectiveness of biopolymer-based starch film (Nazrin et al., 2020).

The overall principle of gas/water vapour permeation through any polymer-based matrix involves four steps: which are absorption of the permeating species into the polymer surface, solubility into the polymer-based matrix, diffusion through the polymer wall along a concentration gradient, and desorption from the outer surface (Wu et al., 2021). For instance, the polymer matrix could be the conventional polymers or the emerging biopolymer/starch films. Figure 8.3 shows the general mechanism of

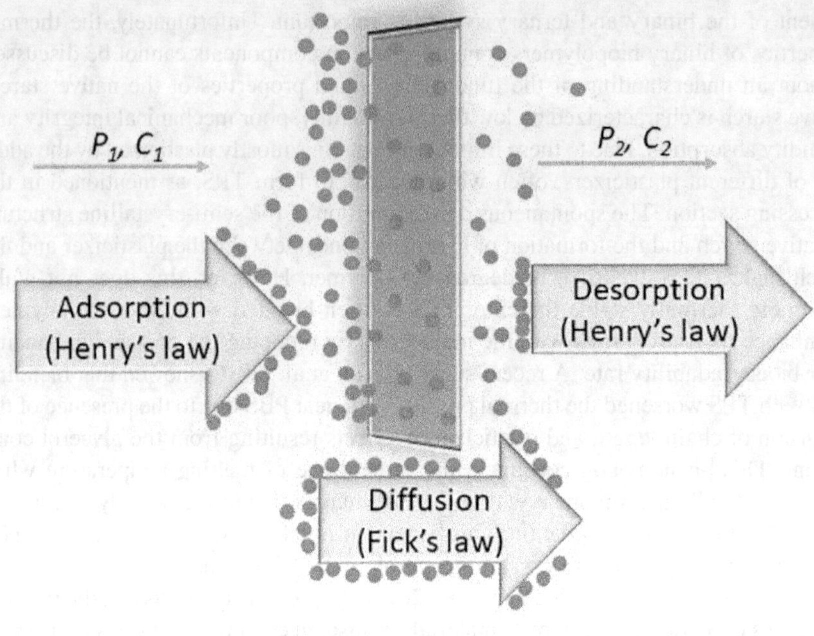

FIGURE 8.3 General mechanism of gas and vapour permeation through a packaging film.

gas and vapour permeation. The gas/water vapour permeation mechanism is based on the diffusion model described by Henry's and Fick's law involves the thickness, permeant pressure and different permeant concentration (Siracusa, 2012). Fick's law states the description of the permeation of gas/water vapour for one-dimensional diffusion (across) through a film (Siracusa, 2012; Vimala Bharathi et al., 2019):

$$J = -D \cdot \Delta C \tag{8.1}$$

Where J is the diffusion flux (mol (cm^2 s)), D is the diffusion coefficient (cm^2/s) and ΔC is the concentration gradient (mol/cm^3). However, if the diffusion is in the steady state condition, then Henry's law is applicable to describe the relationship and equilibrium phenomenon of the concentration of the gas and pressure at the interfaces:

$$C = kp \tag{8.2}$$

Where C is the solubility of a gas at a fixed temperature in a particular solvent, k is Henry's law constant and p is the partial pressure of the gas (Vimala Bharathi et al., 2019).

Barrier performance often measured according to the parameters described in Table 8.5. However, the accuracy and reproducibility of the measured results strongly depend on the consistency of the test conditions and nature of the sample (Jinwu Wang et al., 2018). The research or formulation of barrier film performance is

TABLE 8.5
Barrier Parameters and Corresponding Equations (Jinwu Wang et al., 2018)

Barrier Property and Equation	Unit
$WVTR = \left(\frac{\text{Weight passed through}}{\text{area} \cdot \text{time}}\right)$	g/m^2.day
$WVP = \left(\frac{\text{WVTR} \cdot \text{thickness}}{\text{Saturated pressure} \cdot \Delta\%\text{RH}}\right)$	g.µm/m^2.day.kpa
$OTR = \left(\frac{\text{Volume passed through}}{\text{Area.time}}\right)$	cm^3/m^2.day
$OP = \left(\frac{\text{OTR} \cdot \text{thickness}}{\text{Oxygen partial pressure difference}}\right)$	cm^3.µm/m^2.day.atm

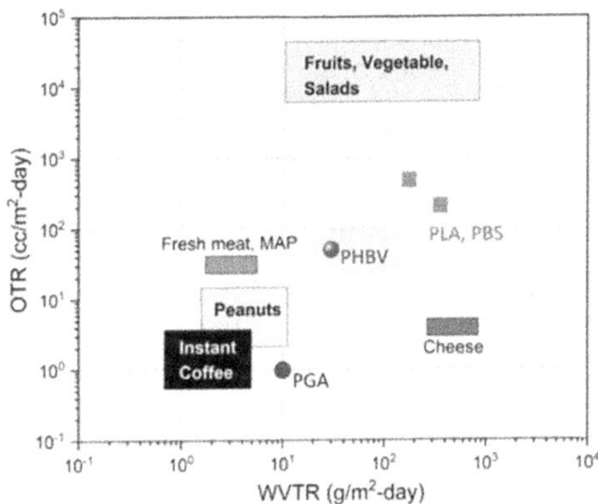

FIGURE 8.4 Requirements of barrier properties for different food packaging applications and a comparison between oxygen/water transition rate of selected biodegradable polymers (Wu et al., 2021). (Copyright 2021, Elsevier B.V.)

generally guided by the product requirements and the inherent barrier properties of the neat material (Figure 8.4). Based on Figure 8.4, it is clear that most of the biodegradable polymers not yet meet the requirements of the daily used food, such as fresh meat and instant coffee. By looking at Figures 8.1 and 8.4, the prescribed or blending biodegradable polymers will not satisfy the barrier-packaging problem. Therefore, blending biodegradable polymers with barrier nano-filler together with oxygen scavenger is the most interesting research topic. In this rational, the development of PGA, PLA, PBS, PBAT, PHBV binary and their ternary system such as PBS/PBAT, PBS/PLA, PLA/PBAT, PHB/PCL and PLA/PBAT-TPS through melt/solution blending has been extensively investigated. A recent study by Turco et al. (2019) showed that blending PLA with TPS worsened the barrier properties of neat PLA (both WVTR and OTR) due to the presence of the hydroxyl polar groups resulting from natural starch and plasticizers such as glycerol. The fundamental principle shows that materials with rich hydroxyl group will show higher interaction with water molecules. Hence, TPS absorbing water at high relative humidity signifies its hydrophilic nature (Dilkes-Hoffman et al., 2018). On the other hand, rich glycerol material as a plasticizing agent has the potential to enhance the intermolecular chain distance, improving macromolecular mobility; hence, a poor barrier is possible. Moreover, the amorphous phase of the PLA/TPS blend system provides easier water and oxygen molecules diffusion pathways (Turco et al., 2019). Importantly, the incorporation of the epoxidized oil into the PLA/TPS matrix improved the WVTR and OTR barrier properties effectively. In case of WVTR, the improvement was due to the ability of the epoxidized oil being able to reduce the interfacial adhesion between the PLA matrix and TPS minor phase (miscibility factor improvement). While the OTR barrier performance was enhanced due to the available hydrogen bonding which created a physical hindrance to the absorption and consequent oxygen molecules. The OTR improvement is also related to the good oxygen and carbon dioxide barrier properties of TPS. Nevertheless, TPS is hydrophilic, but its presence in the multilayer assembly, together with a coating strategy, coated TPS multilayer films can promote good water barrier properties (Dilkes-Hoffman et al., 2018; Chang et al., 2021).

In a more specific study, de Matos Costa et al. (2020) reported the water vapour and oxygen gas permeability values of neat PBS and PBAT to be (2.74×10^{-12} and 7.09×10^{-12} mol/m.s.Pa) and (0.98×10^{-16} and 4.41×10^{-16} mol/m.s.Pa), respectively. In the same study, the addition of 25%PBS improved both the WVTR and OTR properties of PBAT by 29% and 8.5%, while the addition of 25%PBAT only improved the OTR of PBS by 30%. The barrier improvement was attributed to the good interaction between the PBS and PBAT, and the subsequent higher crystallinity of PBS with respect to PBAT leads to a more tortuous path within the amorphous phase and reduces the diffusivity of the permeants. Apart from the tortuous path, oxygen properties also depend on the orientation of the polymer chains and the interaction between the minor and matrix polymer (Aversa et al., 2020). The oxygen and water vapour barrier properties of blends biodegradable polymer system are summarized in Table 8.6.

TABLE 8.6
Summary of the Barrier Properties Improvement

Sample ID	Oxygen Permeation (%)	Water Vapour Permeation (%)	Ref.
PBS75/PBAT25	30	−5.8	de Matos Costa et al. (2020)
PBS25/PBAT75	29	8.5	de Matos Costa et al. (2020)
MTPS/PBAT	74	86.8	Chang et al. (2021)
PLA/PBS/tocopherol	22	*	Aversa et al. (2020)
PLA/PHB	41	*	Aversa et al. (2020)
PLA/20PBS/1s-CNC	40	30	Luzi et al. (2016)
PLA/20PBS/3s-CNC	47	21.5	Luzi et al. (2016)
PBAT/12-sulphonated copolyesters	42	−84	Huang et al. (2020)

- It indicates the decrease in percentage; * no data.

8.5.3 Mechanical Properties and Requirements

More recently, natural fibres have been investigated as potential reinforcement materials in polymeric materials. Wollerdorfer and Bader (1998) investigated the effects of blending natural fibres, including cellulose, oil palm, ramie, flax and jute, with several different biodegradable polymers to develop tough composites. To accomplish this, a twin-screw extruder was used to produce fibre-based polymer blends, which were then pelletized followed by injection moulding into different specimens (e.g. Tensile testing specimen). Importantly, prior to testing the samples were kept for some time to allow the polymer chain relation process to take place. Author results showed that the incorporation of natural fibres had a significant influence on the Young's modulus of the composite (Wollerdorfer & Bader, 1998). PBS showed increased strength only when combined with ramie fibres, an improvement of 19%. With oil palm fibres, the strength was significantly decreased by over 50%; this was attributed to a high level of impurities in the fibres. However, the other blends did not change significantly from the pure PBS biopolymer. PBS composites did not show any significant change in tensile strength; the most significant change was a 16% improvement in the jute composite. Biocell plastic demonstrated a significant increase in all composites tested. The most improved composite was cellulose fibre. Biocell samples with varying levels of flax were also tested; the strength improved from 15% flax to 25% flax, but increasing the flax to 35% brought no significant increase. Composites with TPS showed the greatest improvement of any polymer; every composite had more than double the tensile strength of pure TPS, the greatest being the 20% flax composite, which increased by over 300%. This result was also produced with a 15% flax blend with 1% resin added. In Bioplast fibre composites, there was no significant change in tensile strength for any fibre. This was attributed to the incomplete disintegration of starch grains. Composites made with Mater-Bi plastics showed an increase in tensile strength similar to that of the TPS composites. The blend with ramie fibre improved tensile strength by 64%. Flax also improved the tensile strength, and it was shown that over 15% flax, changing the amount of flax did not affect the tensile strength.

Wollerdorf and Bader (1998) concluded that polysaccharide and starch blends are favourable for composites with natural fibres, while polyesters are not favourable. This is because the polysaccharides adhere much better to the fibres due to their similar chemical structures.

8.5.4 Biodegradability Properties and Requirements

Unlike their mechanical properties, the biodegradability of lignocellulosic polymeric composites is a subject that has not yet been thoroughly investigated. Some existing research focuses on how natural fibre affects the biodegradability of bioplastics. Teramoto et al. (2004) conducted a study on the incorporation of abaca fibre into PCL, PLA, PHBV and PBS polymers that affected their rate of biodegradation. For each polymer, the prepared samples were neat polymer, polymer with 10% untreated abaca, and polymer with 10% acetic anhydride treated abaca. Each sample was prepared using a twin rotary mixer and then formed into thin plate specimens using a hot press. To study the biodegradation rate, each plate was buried in soil and incubated at a specific humidity level. Throughout the 180th day of the experiment, samples were removed from the soil, weighed and photographed at 30-day intervals. Measuring the weight loss of a specimen over time showed its rate of biodegradation. Their experiment yielded very interesting results; in both PBS and PHBV specimens, the addition of abaca fibre significantly improved the rate of biodegradation of the material. Both untreated and acetic anhydride-treated abaca fibre had this result, although the effect was much more pronounced in the untreated abaca composites. PCL and PLA polymer composites showed a slight improvement when untreated abaca fibre was added, but this was attributed to the degradation of the abaca alone and did not indicate the abaca promoted the polymer matrix degradation. Interestingly, the neat PLA and PLA/acetic anhydride-treated abaca samples did not show any weight loss within the 180-day period of the experiment. They concluded that the addition of abaca fibres significantly promoted biodegradation in PBS and PHBV polymers. Also, they stated that the faster degradation rate of untreated abaca to acetic anhydride-treated abaca could be due to the cracks in the surface of the untreated abaca composites. The cracks were not present in the treated abaca composites because treated abaca adhered better to the polymer matrix.

8.6 FUTURE PERSPECTIVES

In the future, the melt processing technique will be integrated with a robotic system using 4IR technology to improve processing speed and automate the melt processing system. On the other hand, the extruder manufacturers will produce a unique extruder with multi-nozzle pin holes for active production, thus improving product output, time reduction and cost-saving strategy. While for solvent-based processing methods, less concentrated solvents will be used in order to adapt to the green economy agenda. The modification of filler before blending with different biodegradable polymers will not only improve the biocompatibility and performance of the as-prepared material but also greatly expand their application fields. Hence, bio-based polymer and filler-derived nanocomposites with unique structures, properties and functions will replace the commodity polymer materials due to green economy agenda.

8.7 CONCLUSION

Polymer material needs to be chemically or physically modified to meet the requirements of a direct product line in terms of performance, cost and sustainability and to achieve a wider range of applications. However, chemical modification is largely discouraged due to the non-environmental friendliness and scalability issues. Hence, physical modification or melt processing is industrially favourable method, due to its environmentally friendliness and potential for scaleup production of filler-polymer composite material. Excellent interfacial adhesion between the matrix and filler has the potential to improve the mechanical strength of the polymer composite due to good dispersion and distribution within the material structure.

REFERENCES

Ashori, A., Harun, J., Raverty, W., & Yusoff, M. N. M. (2006). Chemical and morphological characteristics of Malaysian cultivated kenaf (Hibiscus cannabinus) fiber. *Polymer – Plastics Technology and Engineering*, *45*(1), 131–134. https://doi.org/10.1080/03602550500373782

Athanasiou, K. A., Niederauer, G. G., & Agrawal, C. M. (1996). Sterilization, toxicity, biocompatibility and clinical applications of polylactic acid/polyglycolic acid copolymers. *Biomaterials*, *17*(2), 93–102. https://doi.org/10.1016/0142-9612(96)85754-1

Aversa, C., Barletta, M., Puopolo, M., & Vesco, S. (2020). Cast extrusion of low gas permeability bioplastic sheets in PLA/PBS and PLA/PHB binary blends. *Polymer – Plastics Technology and Materials*, *59*, 231–240. https://doi.org/10.1080/25740881.2019.1625396

Botlhoko, O. J., Ramontja, J., & Ray, S. S. (2018). A new insight into morphological, thermal, and mechanical properties of melt-processed polylactide/poly(ε-caprolactone) blends. *Polymer Degradation and Stability*, *154*, 84–95. https://doi.org/10.1016/j.polymdegradstab.2018.05.025

Byrom, D. (1987). Polymer synthesis by microorganisms: technology and economics. *Trends in Biotechnology*, *5*(9), 246–250. https://doi.org/10.1016/0167-7799(87)90100-4

Chang, C. C., Trinh, B. M., & Mekonnen, T. H. (2021). Robust multiphase and multilayer starch/polymer (TPS/PBAT) film with simultaneous oxygen/moisture barrier properties. *Journal of Colloid and Interface Science*, *593*, 290–303. https://doi.org/10.1016/j.jcis.2021.03.010

Chen, G. Q., & Wu, Q. (2005). The application of polyhydroxyalkanoates as tissue engineering materials. *Biomaterials*, *26*(33), 6565–6578. https://doi.org/10.1016/j.biomaterials.2005.04.036

Chrissafis, K., Antoniadis, G., Paraskevopoulos, K. M., Vassiliou, A., & Bikiaris, D. N. (2007). Comparative study of the effect of different nanoparticles on the mechanical properties and thermal degradation mechanism of in situ prepared poly(ε-caprolactone) nanocomposites. *Composites Science and Technology*, *67*, 2165–2174. https://doi.org/10.1016/j.compscitech.2006.10.027

David, K., & Ragauskas, A. J. (2010). Switchgrass as an energy crop for biofuel production: A review of its ligno-cellulosic chemical properties. *Energy and Environmental Science*, *3*(9), 1182–1190. https://doi.org/10.1039/b926617h

de Matos Costa, A. R., Crocitti, A., Carvalho, de L. H., Carroccio, S. C., Cerruti, P., & Santagata, G. (2020). Properties of biodegradable films based on poly(butylene succinate) (PBS) and poly(butylene adipate-co-terephthalate) (PBAT) blends. *Polymers*, *12*, 2317.

Dhakal, H. N., Zhang, Z. Y., & Richardson, M. O. W. (2007). Effect of water absorption on the mechanical properties of hemp fibre reinforced unsaturated polyester composites. *Composites Science and Technology*, *67*(7), 1674–1683. https://doi.org/10.1016/j.compscitech.2006.06.019

Diebel, W., Reddy, M. M., Misra, M., & Mohanty, A. (2012). Material property characterization of co-products from biofuel industries: Potential uses in value-added biocomposites. *Biomass and Bioenergy*, *37*, 88–96. https://doi.org/10.1016/j.biombioe.2011.12.028

Dilkes-Hoffman, L. S., Pratt, S., Lant, P. A., Levett, I., & Laycock, B. (2018). Polyhydroxyalkanoate coatings restrict moisture uptake and associated loss of barrier properties of thermoplastic starch films. *Journal of Applied Polymer Science*, *135*, 46379. https://doi.org/10.1002/app.46379

Erba, R. D., Groeninckx, G., Maglio, G., Malinconico, M., & Migliozzi, A. (2001). Immiscible polymer blends of semicrystalline biocompatible components: Thermal properties and phase morphology analysis of PLLA/ PCL blends. *Polymers*, *42*, 7831–7840.

Gilding, D. K., & Reed, A. M. (1979). Biodegradable polymers for use in surgery-polyglycolic/poly(actic acid) homo- and copolymers: 1. *Polymer*, *20*(12), 1459–1464. https://doi.org/10.1016/0032-3861(79)90009-0

Gunatillake, P. A., Adhikari, R., & Gadegaard, N. (2003). Biodegradable synthetic polymers for tissue engineering. *European Cells and Materials*, *5*, 1–16. https://doi.org/10.22203/eCM.v005a01

Helanto, K., Matikainen, L., Talja, R., & Rojas, O. J. (2019). Bio-based polymers for sustainable packaging and biobarriers: A critical review. *BioResources*, *14*, 4902–4951. https://doi.org/10.15376/biores.14.2.Helanto

Henton, D. E., Gruber, P., Lunt, J., & Randall, J. (2005). Polylactic acid technology. *Natural Fibers, Biopolymers, and Biocomposites*, *48674*(23), 527–577. https://doi.org/10.1002/1521-4095(200012)12:23<1841::aid-adma1841>3.3.co;2–5

Huang, F., Wu, L., & Li, B. G. (2020). Sulfonated biodegradable PBAT copolyesters with improved gas barrier properties and excellent water dispersibility: From synthesis to structure-property. *Polymer Degradation and Stability*, *182*, 21–26. https://doi.org/10.1016/j.polymdegradstab.2020.109391

Joshi, S. V., Drzal, L. T., Mohanty, A. K., & Arora, S. (2004). Are natural fiber composites environmentally superior to glass fiber reinforced composites? *Composites Part A: Applied Science and Manufacturing*, *35*(3), 371–376. https://doi.org/10.1016/j.compositesa.2003.09.016

Kapich, A. N., Jensen, K. A., & Hammel, K. E. (1999). Peroxyl radicals are potential agents of lignin biodegradation. *FEBS Letters*, *461*(1–2), 115–119. https://doi.org/10.1016/S0014-5793(99)01432-5

Labet, M., & Thielemans, W. (2009). Synthesis of polycaprolactone: A review. *Chemical Society Reviews*, *38*(12), 3484–3504. https://doi.org/10.1039/b820162p

Liu, W., Mohanty, A. K., Askeland, P., Drzal, L. T., & Misra, M. (2004). Influence of fiber surface treatment on properties of Indian grass fiber reinforced soy protein based biocomposites. *Polymer*, *45*(22), 7589–7596. https://doi.org/10.1016/j.polymer.2004.09.009

Luzi, F., Fortunati, E., Jiménez, A., Puglia, D., Pezzolla, D., Gigliotti, G., Kenny, J. M., Chiralt, A., & Torre, L. (2016). Production and characterization of PLA_PBS biodegradable blends reinforced with cellulose nanocrystals extracted from hemp fibres. *Industrial Crops and Products*, *93*, 276–289. https://doi.org/10.1016/j.indcrop.2016.01.045

Mittal, V., Akhtar, T., & Matsko, N. (2015). Mechanical, thermal, rheological and morphological properties of binary and ternary blends of PLA, TPS and PCL. *Macromolecular Materials and Engineering*, *300*, 423–435. https://doi.org/10.1002/mame.201400332

Mo, A., Zhang, Y., Gao, W., Jiang, J., & He, D. (2023). Environmental fate and impacts of biodegradable plastics in agricultural soil ecosystems. *Applied Soil Ecology*, *181*, 104667. https://doi.org/10.1016/j.apsoil.2022.104667

Modi, S., Koelling, K., & Vodovotz, Y. (2011). Assessment of PHB with varying hydroxyvalerate content for potential packaging applications. *European Polymer Journal*, *47*, 179–186. https://doi.org/10.1016/j.eurpolymj.2010.11.010

Modi, S., Koelling, K., & Vodovotz, Y. (2013). Assessing the mechanical, phase inversion, and rheological properties of poly-[(R)-3-hydroxybutyrate-co-(R)-3-hydroxyvalerate] (PHBV) blended with poly-(l-lactic acid) (PLA). *European Polymer Journal*, *49*, 3681–3690. https://doi.org/10.1016/j.eurpolymj.2013.07.036

Mofokeng, J. P., & Luyt, A. S. (2015). Morphology and thermal degradation studies of melt-mixed poly(lactic <producer>acid) (PLA)/poly(ε-caprolactone) (PCL</producer>) biodegradable polymer blend nanocomposites with TiO_2 as filler. *Polymer Testing*, *45*, 93–100. https://doi.org/10.1016/j.polymertesting.2015.05.007

Mohanty, A. K., Misra, M., & Drzal, L. T. (2002). Sustainable bio-composites from renewable resources: Opportunities and challenges in the green materials world. *Journal of Polymers and the Environment*, *10*(1), 19–26. https://doi.org/10.1023/A:1021013921916

Nazrin, A., Sapuan, S. M., Zuhri, M. Y. M., Ilyas, R. A., Syafiq, R., & Sherwani, S. F. K. (2020). Nanocellulose reinforced thermoplastic starch (TPS), polylactic acid (PLA), and polybutylene succinate (PBS) for food packaging applications. *Frontiers in Chemistry*, *8*, 1–12. https://doi.org/10.3389/fchem.2020.00213

Nofar, M., Salehiyan, R., Ciftci, U., Jalali, A., & Durmuş, A. (2020). Ductility improvements of PLA-based binary and ternary blends with controlled morphology using PBAT, PBSA, and nanoclay. *Composites Part B: Engineering*, *182*, 107661. https://doi.org/10.1016/j.compositesb.2019.107661

Oza, S., Ning, H., Ferguson, I., & Lu, N. (2014). Effect of surface treatment on thermal stability of the hemp-PLA composites: Correlation of activation energy with thermal degradation. *Composites Part B: Engineering*, *67*, 227–232. https://doi.org/10.1016/j.compositesb.2014.06.033

Shukla, V. (2019). Review of electromagnetic interference shielding materials fabricated by iron ingredients. *Nanoscale Advances*, *1*(5), 1640–1671. https://doi.org/10.1039/c9na00108e

Siracusa, V. (2012). Food packaging permeability behaviour: A report. *International Journal of Polymer Science*, *2012*(ID 302029), 1–11. https://doi.org/10.1155/2012/302029

Someya, Y., Nakazato, T., Teramoto, N., & Shibata, M. (2004). Thermal and mechanical properties of poly (propylene carbonate) nanocomposites with various organo-modified montmorillonites. *Journal of Applied Polymer Science*, *91*(3), 1463–1475. https://doi.org/10.3724/sp.j.1105.2008.01123

Surej Rajan, C., Varghese, L. A., & Joseph, S. (2018). A review on mechanical and thermal properties of binary and ternary thermoplastic immiscible polymer blends. *Asian Journal of Applied Science and Technology (AJAST)*, *2*, 1021–1028.

Tachibana, Y., Masuda, T., Funabashi, M., & Kunioka, M. (2010). Chemical synthesis of fully biomass-based poly(butylene succinate) from inedible-biomass-based furfural and evaluation of its biomass carbon ratio. *Biomacromolecules*, *11*(10), 2760–2765. https://doi.org/10.1021/bm100820y

Teramoto, N., Urata, K., Ozawa, K., & Shibata, M. (2004). Biodegradation of aliphatic polyester composites reinforced by abaca fiber. *Polymer Degradation and Stability*, *86*(3), 401–409. https://doi.org/10.1016/j.polymdegradstab.2004.04.026

Turco, R., Ortega-Toro, R., Tesser, R., Mallardo, S., Collazo-Bigliardi, S., Boix, A. C., Malinconico, M., Rippa, M., Di Serio, M., & Santagata, G. (2019). Poly (lactic acid)/thermoplastic starch films: Effect of cardoon seed epoxidized oil on their chemicophysical, mechanical, and barrier properties. *Coatings*, *9*. https://doi.org/10.3390/coatings9090574

Velde, K. Van De, & Kiekens, P. (2002). Biopolymers: Overview of several properties and consequences on their applications. *Polymer Testing*, *21*(4), 433–442.

Vimala Bharathi, S. K., Rohini, B., Moses, J. A., & Anandharamakrishnan, C. (2019). Nanocomposite for Food Packaging: Principles and Application. In *Food Nanotechnology*. pp. 275–307. CRC Press. https://doi.org/10.1201/9781315153872-13

Wang, Jiajie, Pan, T., Zhang, J., Xu, X., Yin, Q., Han, J., & Wei, M. (2018). Hybrid films with excellent oxygen and water vapor barrier properties as efficient anticorrosive coatings. *RSC Advances*, *8*, 21651–21657. https://doi.org/10.1039/c8ra03819h

Wang, Jinwu, Gardner, D. J., Stark, N. M., Bousfield, D. W., Tajvidi, M., & Cai, Z. (2018). Moisture and oxygen barrier properties of cellulose nanomaterial-based films. *ACS Sustainable Chemistry and Engineering*, *6*, 49–70. https://doi.org/10.1021/acssuschemeng.7b03523

Williams, J. M., Adewunmi, A., Schek, R. M., Flanagan, C. L., Krebsbach, P. H., Feinberg, S. E., Hollister, S. J., & Das, S. (2005). Bone tissue engineering using polycaprolactone scaffolds fabricated via selective laser sintering. *Biomaterials*, *26*(23), 4817–4827. https://doi.org/10.1016/j.biomaterials.2004.11.057

Wollerdorfer, M., & Bader, H. (1998). Influence of natural fibres on the mechanical properties of biodegradable polymers. *Industrial Crops and Products*, *8*(2), 105–112. https://doi.org/10.1016/S0926-6690(97)10015-2

Wu, F., Misra, M., & Mohanty, A. K. (2021). Challenges and new opportunities on barrier performance of biodegradable polymers for sustainable packaging. *Progress in Polymer Science*, *117*, 101395. https://doi.org/10.1016/j.progpolymsci.2021.101395

Yun, I. S., Hwang, S. W., Shim, J. K., & Seo, K. H. (2016). A study on the thermal and mechanical properties of poly (butylene succinate)/thermoplastic starch binary blends. *International Journal of Precision Engineering and Manufacturing - Green Technology*, *3*, 289–296. https://doi.org/10.1007/s40684-016-0037-z

Zhan, C., Yu, G., Lu, Y., Wang, L., Wujcik, E., & Wei, S. (2017). Conductive polymer nanocomposites: A critical review of modern advanced devices. *Journal of Materials Chemistry C*, *5*(7), 1569–1585. https://doi.org/10.1039/c6tc04269d

Zhang, M., Li, Y., Su, Z., & Wei, G. (2015). Recent advances in the synthesis and applications of graphene-polymer nanocomposites. *Polymer Chemistry*, *6*(34), 6107–6124. https://doi.org/10.1039/c5py00777a

Zhen, Z., Ying, S., Hongye, F., Jie, R., & Tianbin, R. (2011). Preparation, characterization and properties of binary and ternary blends with thermoplastic starch, poly(lactic acid) and poly(butylene succinate). *Polymers from Renewable Resources*, *2*, 49–62. https://doi.org/10.1177/204124791100200201

9 End-of-life Options of Biobased Plastic Materials

Nomvuyo Nomadolo and Sudhakar Muniyasamy

9.1 WORLD PLASTICS PRODUCTION, CONSUMPTION AND ITS RELATED PLASTIC WASTE DISPOSAL ISSUES

Since ancient times, humans have been known to have benefited from the use of polymers, but it was not until the 20th century that the development of plastics expanded, yielding 15 new classes of polymers being synthesised (Andrady & Neal, 2009). This led to the exponential growth of plastic in the 21st century, and it is reported that the production of plastics increased from 1.5 million metric tons (Mt) in the 1950s to 359 million tons in 2018. The success of plastic as a material has been substantial due to several factors: they have proved to be versatile in many different applications, ease of large-scale production, low cost, strong yet light in weight, offers great water resistance capacity, resilient, robust and can be moulded to suit several applications (Ali et al., 2021; Verma & Vinoda, 2016). As a result, plastics are a crucial part of the economy and affect almost every industry in the world. Figure 9.1 illustrates the distribution of plastic per application.

Plastic has superior properties compared to other commonly used packaging materials such as paper, glass or even wood. The lightweight nature of plastic, among other properties, permits less amount used to package a product, leading to reduced transportation costs and low amount of waste material generated. Reports show that

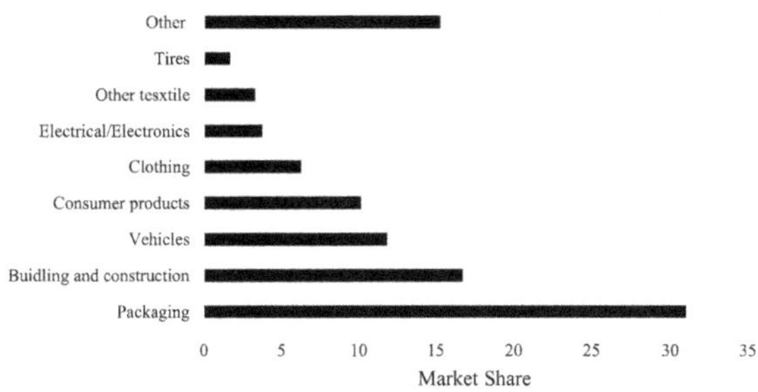

FIGURE 9.1 Distribution of plastic consumption worldwide in 2019, by application, information collected from Statista (Andrady & Neal, 2009; Ali et al., 2021; Verma & Vinoda, 2016).

DOI: 10.1201/9781003304142-9

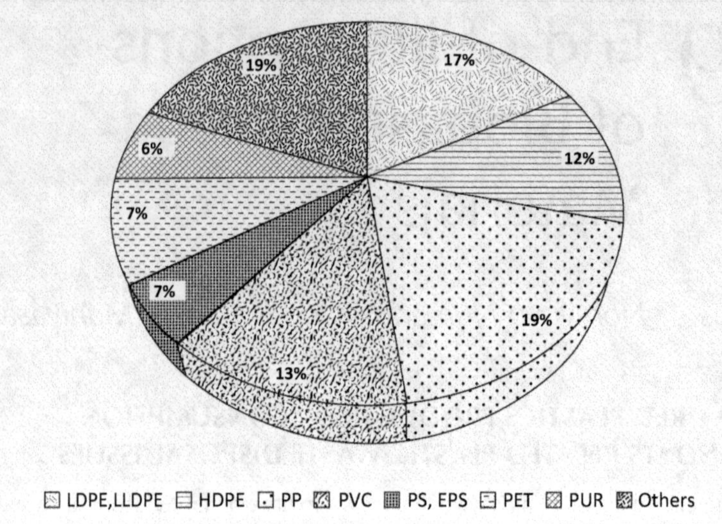

FIGURE 9.2 World plastic materials demand by resin types 2006 (Maddah, 2016).

replacing glass bottles with plastic bottles for beverage packaging in airlines has led to saving over USD 1 million in fuel costs, just because of the difference in weight (Halden, 2010). It is notable that the biggest share of plastic produced is directed towards packaging items, which is approximately 50%–55% of the overall plastic production. Packaging forms an integral part of all applications since it serves to protect and secure, allows safe handling and enhances the usability of products. The common plastic polymer resins used for both packaging and other single-use products include polypropylene, polyethylene (PE), polystyrene (PS), polyvinyl chloride (PVC), phenolic resin and polyurethane (PU). The polymers are relatively cheap and can be moulded into a variety of rigid products (like yoghurt cups, food trays, margarine tubs and others) and flexible barrier film (used for food wraps, packaging of snacks, candy and others). The consumption of plastic packaging materials and single-used products has been one of the key drivers in the economy of the world and, as such, is in demand (Figure 9.2).

Indubitable, plastics deliver many economic benefits; however, the production and use rates follow a linear model of ‘take, make, use, and dispose of, and this model is a main course of waste, environmental degradation, climate change and natural resource depletion, which ultimately have adverse human health effects’. The number of plastics in circulation is calculated to increase from 236 to 417 million tons per year by 2030 (Schyns & Shaver, 2021).

9.2 PLASTIC POLLUTION

According to literature, around 4% of oil and gas produced worldwide is used as feedstock for plastic, while another 4% is used to provide energy for plastic manufacturing, and it is projected that the total oil consumption will increase from

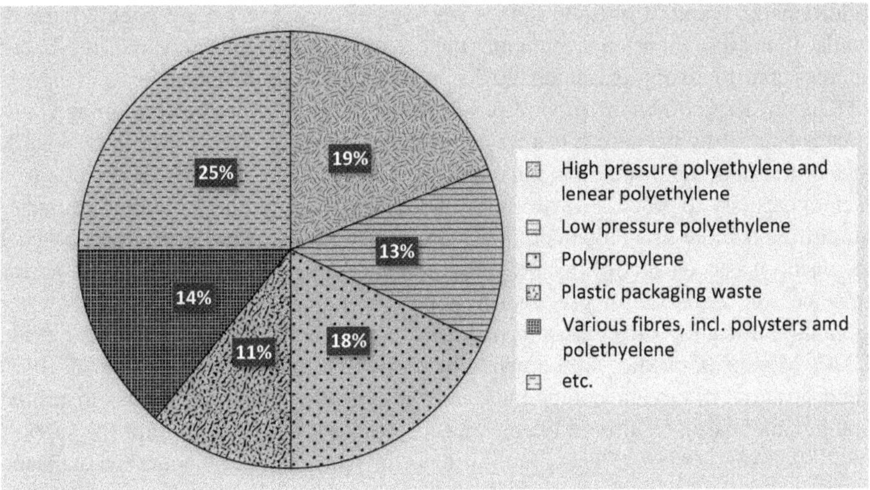

FIGURE 9.3 Structure of plastic waste in the world by types of materials, 2015 (Geyer et al., 2017).

the current 8% to 20% by 2050 according to the consumption patterns observed presently (Kaplan, 1998). Consequently, plastic is one of the major contributors to greenhouse gas emissions, from the extraction to processing of fossil fuels. Additionally, oil and gas are non-renewable resources, and this is concerning particularly because the bulk of plastic produced every year is directed towards the production of packaging items and/or other short-lived products that are habitually discarded within a year of manufacture. The sustainability of short-term plastic production is heavily dependent on a non-renewable resource. The first observation alone is mainly concerning as it contributes to the depletion of oil and gas, and the second observation is that most produced plastics are for short-term applications, which then leads to substantial quantities of plastic being discarded indiscriminately. The plastic consumption rate is due to the rise in living standards, which is motivated by both population growth and rapid economic growth. Figure 9.3 shows plastic waste generated according to the polymer resin used.

The sturdiness of plastic results in the accrual of plastic waste on land and oceans, thereby causing an adverse effect on land, oceans, animals, and humans on a global scale. While many countries around the world are attempting to manage the current volumes of plastic waste, the enormity of the problem is very complex and requires a plastic waste management system to be holistically evaluated. Plastic are typically land-filled together with municipal solid waste (MSW), this being a common practice across the world. Literature reports that each year, a staggering amount of about 318 million tonnes of plastic resin is produced. Inevitably, 218 million tonnes of this plastic constitutes the total annual solid plastic waste (PSW), compounding the MSW the most (Schyns & Shaver, 2021). Globally, only 15% of plastic on average is recycled, whilst approximately 40% of plastic is mismanaged every year and ultimately discharged to the environment by stormwater washouts, littering, and so forth.

Additionally, about 50 million tonnes per year of plastic waste are openly burned, whilst 10 million tonnes per year enter the ocean's aquatic environment. Land-based sources contribute approximately 90% of plastics entering the oceans.

The most favourable method of plastic waste disposal in many developing countries is landfilling because it is a waste management option that is low-cost, regardless of the huge land required for it and the bad land aesthetic. However, due to a lack of capacity to process waste and properly equipped landfills, in many countries landfilling is more of an 'open dumping site' for solid waste, and as such, the impact of plastic waste on landfilling presents several issues, such as leachates of toxins into the soil, emission of greenhouse gases, migration of microplastic into water streams, infiltration in food chain leading to a decline in human health (Ru et al., 2020). Moreover, plastics are known to be carriers of toxins like heavy metals, high concentrations of organic compounds, and pathogens, posing a serious risk to animal and human health, resulting in depleted ecosystem services required for the support of all life. Another side effect of landfilling is the open incineration method of plastic waste disposal. The open burning or incineration of plastics has negative effects: it releases CO_2 and soot (black carbon) – two of the most potent gasses that initiate climate change. Furthermore, it is reported that exposure to open burning of plastic containing chlorinated and brominated additives poses serious health issues such as liver problems and impairment of immune, reproductive and endocrine systems.

About 80% of plastics in the ocean are from land-based sources. Plastics from landfills enter the ocean through poor management and improper disposal of sewages and plastic wastes, coastal landfill operations, and litter carried through streams and rivers. The remaining 20% are litters generated from ships/boats through recreational activities, discharge of litter, fishing nets, etc. (Mattsson et al., 2015). Our oceans serve as a sink for plastic particles, and it is estimated that 5.25 trillion micro and nanoplastics contaminate the global sea surface (Mattsson et al., 2017). In the marine environment, plastics are broken down into microplastics, which threaten marine biodiversity. Furthermore, microplastics can end up in the food chain, with potentially damaging effects on human health, because they may also accumulate high concentrations of persistent organic pollutants, POPs and other toxic chemicals and potentially serve as a pathway for their transfer to aquatic organisms, and consequently human beings. According to the literature, microplastic is emerging as a source of soil pollution (Halden, 2010). The impacts of microplastics in soils, sediments, and freshwater could have a long-term damaging effect on terrestrial ecosystems globally, causing adverse effects on organisms, such as soil-dwelling invertebrates and fungi, needed for important ecosystem services and functions (Geyer et al., 2017). It is reported that up to 895 microplastic particles per kilogram have been found in organic fertilizers used in agricultural soils, and a minimum of 730,000 tonnes of microplastics are transferred every year to agricultural lands in Europe and North America from urban sewage sludges used as farm manure, with potentially direct effects on soil ecosystems, crops, and livestock or through the presence of toxic chemicals. Moreover, microplastics are an emerging freshwater contaminant that may degrade water quality and consequently affect water availability and harm freshwater fauna. The contamination of tap and bottled water by microplastics is already widespread, and the World Health Organization is assessing the possible effects on human health.

9.3 PLASTIC WASTE MANAGEMENT

Plastic waste management refers to managing the plastic waste generated and processing it to make it reusable. Since petroleum-based plastic is predominately non-biodegradable, the practical alternative is to recycle and/or reuse, such as the case of thermoplastic, where the products can be reshaped into new desirable products. Thermosets, on the other hand, pose a challenge since most of the plastic waste generated is made up of thermosetting plastic. These plastics cannot be remoulded into new products. However, thermosets are usually used in long-term applications like construction equipment panels, electrical housings, insulators, cell tower tops, heat shields, and circuit breakers; therefore, thermosets are not major contributors to plastic waste. It is the once-off used thermoplastics that are a huge problem. Managing plastic waste is a challenging task as plastic constitutes a huge fraction of MSW, this has instigated much debate on how plastic waste should be managed. Their abundant presence in the waste stream poses critical challenges, especially without effective waste management.

Many countries accumulate waste plastic pyramids requiring proper disposal; however, due to the cost of waste disposal, many rely on indiscriminate plastic dumping. Mismanaged waste has been defined as material that is either littered or inadequately disposed of. Inadequately disposed waste is not formally managed and includes disposal in dumps or open, uncontrolled landfills, where it is not fully contained. Mismanaged waste could eventually enter the ocean via inland waterways, wastewater outflows, and transport by wind or tides. Based on the literature review, there are four dominant technologies applied in waste management: mechanical recycling, feedstock recycling, incineration, and landfilling. A study by *Jambeck et al.* indicated that out of 192 countries, South Africa was ranked the 11th country with the highest level of mismanaged plastic waste entering the ocean (Geyer et al., 2017), with China, Indonesia, and the Philippines taking the first, second and third position, respectively. The Department of Environmental Affairs (Baloy et al., 2012) reported that landfilling remains the dominant technology in South Africa, where 90% of waste generated is directed towards landfills, much of this waste being plastic. Additionally, most of the landfills are unlicensed, which implies that there are inflated risks in terms of environmental damage (land, soil, and air), the health and safety of humans, animals and plants (Godfrey & Oelofse, 2017). Although South Africa is a leading country in the African circular plastic industry in terms of plastic bags and recycling, it was reported that in 2019, plastic production in South Africa amounted to 1,841,700 tons, only 27% of plastic waste was collected and 337,700 tons of this plastic waste was recycled, that is just 18% of recycled plastic waste. This demonstrates the plastic waste management challenges faced by the country and as a continent.

An unprecedented population growth is experienced by the African continent, with predictions that there will be 2.5 billion people on the continent by 2050 (Ayeleru et al., 2020; Nyathi & Togo, 2020). This implies that every minute, there will be 80 babies being born in Africa, making Africa the biggest contributor to the future growth of the population. The increasing trend in per capita consumption, urbanisation, and population growth is concerning when combined with the lack of infrastructure and legislation to manage waste generation. Additionally, Africa is faced with a diverse of challenges, which include a shortage of skilled waste management (WM) experts and facilities, lack of a conducive environment for business and governance, and poor PW and municipal solid waste management (MSWM) in general. The poor economic growth in the region ultimately increases the level of poverty and is compounded by an increasing population, which adversely affects MSWM since the residents and their local governments are unable to afford WM strategies, and this makes funding the services a bit difficult.

One of the most developed countries, like Europe, has plastic WM challenges as well. Although not as serious as in the African context, it has been reported that of the 25.8 million tonnes of plastic waste generated in Europe every year, under 30% is collected for recycling; 31% ends up in landfills, and 39% is incinerated (Schyns & Shaver, 2021). Illustrating beyond doubt, there is a need for improved plastic WM systems globally. Considering global value chains and trade in plastics, aligning research to assist with the material design and the regulation of chemicals will be key to improving the circularity of plastics.

9.4 RECYCLING

9.4.1 Recycling of Plastic

Recycling of plastic is fundamental in the quest to obtain the green ecology target, since recycling saves energy, conserves natural resources, and decreases various types of pollution (Vollmer et al., 2020). According to the ISO and the American Society of Testing and Materials, plastic recycling can be classified into four categories: primary recycling, secondary recycling, tertiary recycling, and quaternary recycling. Primary recycling entails the extrusion of pre-consumer polymer or pure polymer streams (in-house waste) to produce products of the same material. This method is considered the best method of recycling because it uses less energy and fewer resources while fossil fuels are retained. The method is typically referred to as close-loop recycling because the quality and properties of the recycled material from closed-loop processes are very close to the original material. Thus, it can be used as feedstock for high-added-value product manufacturing. Consequently, primary recycling is considered the most economically viable. The input waste is usually a single type of plastic with slight contamination. The primary recycling process consists of waste transformation utilising extrusion, where the plastic is melted and re-granulated. So this recycling method is also called re-extrusion.

Secondary recycling involves size reduction, cleaning, sorting, and regranulation to a suitable feedstock size since material feedstock usually comes from other sources like plastic solid waste (PSW). This technique is favourably referred to as mechanical recycling. Thus, the new recycled material can be converted into new plastic products, replacing virgin polymers or a portion of virgin polymers. Secondary recycling often leads to deterioration of the properties of the recyclable material, resulting in the degradation of polymers, heterogeneity of plastic wastes, and the presence of low molecular weight compounds (degradation products, additives, and contaminants). Consequently, there is a need to develop a suitable technology for upgrading the properties of the waste materials, to make them suitable for new applications. Oftentimes, the challenges are overcome by the addition of new components during the melting process, such as virgin polymers, fillers, fibres, compatibilizers, or various additives (peroxides, coupling agents, reactive polymers, stabilisers and/or antioxidants) to improve their properties and provide them characteristics that facilitate their use as new materials in the plastics industry. While primary recycling is the most environmentally friendly and economical plastic WM, this process, however, requires only clean, non-degraded, and homogeneous plastic, thereby limiting its application. Consequently, secondary recycling is preferred over other plastic recycling processes.

Tertiary recycling is also referred to as chemical recycling since it involves processes that chemically convert polymer chaining into smaller molecules that are used as feedstock to produce fuels. Chemical recycling includes e-polymerisation or heating or pyrolysis, that is, pyrolysis and gasification. Finally, Quaternary recycling consists of the incineration of plastic waste and the recovery of energy through heat and electricity production; thus, this process is also known as energy recovery.

9.4.2 Mechanical Recycling

Based on an extensive literature review where several methods of plastic WM options were evaluated, it was concluded that mechanical recycling (primary and secondary recycling) is generally the best option, and it is an important component of the circular economy (Lamberti et al., 2020; Jang et al., 2020). Mechanical recycling represents one of the most successful processes as it offers many advantages like simplicity, requires low investment, its technological parameters are easily controlled, and the process is robust (Arikan & Ozsoy, 2015; Resch-Fauster et al. 2017; Maris et al., 2018). The process of recovering plastic waste by means of mechanical processing into secondary raw material without significantly depreciating the properties of the material is referred to as mechanical recycling. The process is widely applied due to its economic and technical feasibility.

Mechanical recycling process can be configured to be either a closed or an open-loop system, depending on the input waste and output products of the process. In an open-loop mechanical recycling process, the input waste is generally made up of mixed plastic waste, which are primarily compatible with each other and the output products are secondary products applied differently from their pre-recycled products. Often, open-loop recycling is also referred to as downcycling, as it results in the degradation of material properties due to heat, chemical reactions, or physical crushing occurring during the recycling process. Additionally, remnants of incompatible polymers, inks, plastic additives, and so forth also contribute towards the decline of polymer property profiles during reprocessing, producing materials which are suitable only for less demanding applications like pipelines, agricultural products, etc. Open-loop recycling process is very instrumental in mitigating the plastic waste pollution and for extending the serving life span of plastic material for economical purposes. Closed-loop mechanical recycling on the other hand plays a key role in supply chain sustainability. Since the closed-loop system is designed to ensure that all materials in the manufacturing process is fully used. The input waste is of the same material and the output products are also of the same purpose. Oftentimes, high-end products such as computers are recycled in closed-loop systems.

Mechanical recycling does not come without shortcomings; for example, plastic waste streams are made up of a concoction of mixed and contaminated plastic waste with different grades of plastics making it difficult to sort and consequently recycle. This poses a challenge due to the variations in properties such as melt flow index and the presence of certain additives (such as halogenated compounds and SVHCs), which may limit their use and market application. Plastic waste, either during its utilisation time or during its time in the waste stream, might absorb persistent organic pollutants (POPs), hindering the effectiveness of recycling plastic waste into functional plastic. Additionally, polymer degradation occurs during the recycling process. For mechanical recycling, this might include thermal-mechanical degradation, and therefore, for certain material types (e.g. PET), the materials can only be recycled for a limited number of cycles, resulting in the material quality loss after a number of cycles of recycling, resulting in the lower-grade polymers. Finally, the other challenge is the lack of standardisation across recycling operations and organizations, most especially in developing countries. This does not only lead to waste of expenditure and money, but it further contributes to plastic WM issues.

9.5 BIODEGRADATION AND COMPOSTABILITY

An emerging WM option for plastic waste is biodegradation and/or composting of plastic. Packaging plastic usually ends up in organic streams or natural environment; thus, biodegradation as an end-of-life cycle option is particularly of great benefit. Polymers with fast biodegradation rates and efficient mechanical and gas-barrier properties offer a great alternative to petroleum-based plastic used currently and would be a promising solution for plastic waste generated from packaging. Biodegradable plastics are seen by many as a promising solution to the plastic waste problem because they increase soil fertility and reduce the accumulation of bulky plastic materials in the environment, which invariably will minimize injuries to wild animals, they decrease the cost of WM, and they offer sustainability of resources (Haider et al., 2019; Jiang & Zhang, 2012).

Biodegradation is the degradation of macromolecular chains by the action of microorganisms. Biodegradation mainly occurs via a three-step process: fragmentation, hydrolysis, and assimilation. The process can take place in different environments, such as soil, aquatic media and, of course, in humans, so long as the conditions are conducive. The fragmentation step entails the breaking down of a high molar mass chain into small pieces or microplastics of inconsistent molar masses by weathering, UV-irradiation, microorganisms and so on. The hydrolysis step then follows suit, where the ester bond of the polymer breaks down into carboxylic and alcohol groups, forming short-chain polymers or oligomers that are soluble as they possess polar functional chain ends. The hydrolysis step leads to a further decrease in molar mass, mechanical strength and changes in the crystallinity. In the third and final step, the polar chain ends of oligomers are mineralised by microorganisms, ultimately producing carbon dioxide (CO_2) and/or methane (CH_4), water, and biomass (Figure 9.4).

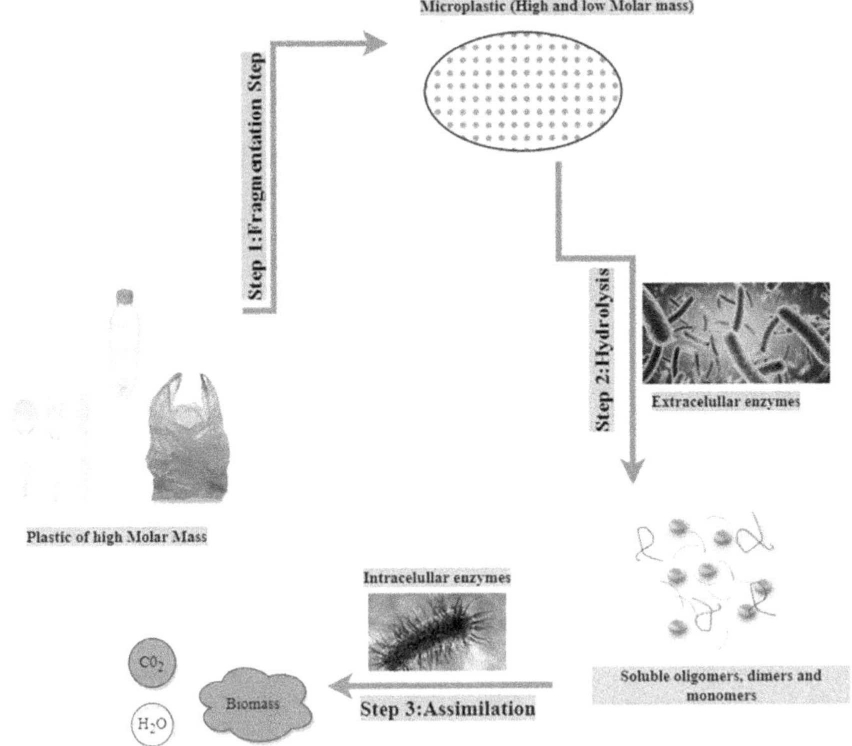

FIGURE 9.4 Schematic representation of biodegradation mechanisms.

The polymeric materials that can decompose into carbon dioxide (CO_2), methane (CH_4), water, inorganic compounds, or biomass by either the action of microbial enzymes or by non-enzymatic processes, such as chemical hydrolysis are therefore called biodegradable polymers (Bahl et al., 2021). This phenomenon leads to the two classifications of biodegradable polymers, which are (i) polymers that are inherently biodegradable, where direct enzymatic degradation reaction is enabled by the chemical nature of the polymer (e.g., starch, cellulose, and chitin) and (ii) Polymers whose degradation process is first stimulated by either photo-oxidation or thermal oxidation upon exposure to UV or heat, respectively (Sivan, 2011). The process of biodegradation is mainly controlled by two overarching factors that can affect the degradation of biodegradable polymers: the polymer's characteristic properties and the environment in which biodegradation is taking place. Figure 9.5 breaks down these factors further (Ojeda et al., 2009).

Firstly, there are two types of environmental conditions in which biodegradation can take place, namely, aerobic and anaerobic. These are categorized by the presence or absence of oxygen, respectively. According to ASTM methods, O_2 levels are less than 6% in aerobic biodegradation, and microorganisms use oxygen during the metabolism process to break down the polymer chains (ASTM, 2012). Subsequently, carbon dioxide, other gasses, and water are emitted into the environment. Biodegradation in a compost pile is predominantly aerobic. ASTM defines composting as a process where

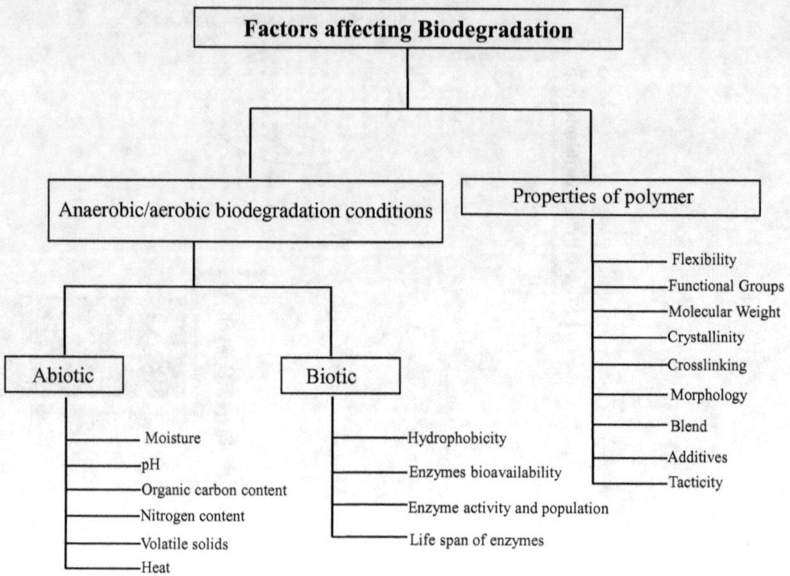

FIGURE 9.5 Factors affecting biodegradation process.

biodegradable materials, such as manure and leaves, are decomposed (Standard, 2015). A compostable polymer is defined as a plastic capable of undergoing biological decomposition in a compost site such that the plastic is not visually distinguishable and breaks down to carbon dioxide, water, inorganic compounds, and biomass at a rate consistent with a known compostable material such as cellulose. The biodegradation and composting processing of polymers are distinguished from each other by the rates of disintegration and microbial assimilation steps; as such, compostable polymers are innately biodegradable, but not all biodegradable polymers are compostable. In an anaerobic environment, there is no oxygen present, and, in this case, methane gas (CH_4) and water are generated and released instead of CO_2; examples of anaerobic environments include sewages, landfills where CH_4 is collected (Kijchavengkul et al., 2009; Seng et al., 2013). Anaerobic environments are also referred to as non-aerated environments; typically, anaerobic digesters are referred to as WM options that can serve as energy recovery from plastic waste.

Secondly, biodegradation is also influenced by polymer properties such as the type of a polymer, crystallinity, molecular weight, morphology, tacticity, and so forth, all of which have a bearing on the type of biodegradation behaviour that will take place. For example, polymers derived from biomass feedstock like PHA, PLA, and lignan-based polyutherane harvested from straw, grape pornance, and plants, respectively, tend to biodegrade faster than polymers derived from fossil fuels like PE, PET, and so forth. This is because, in naturally occurring types of polymers, microbes can use their existing enzymes to break the polymer chain at specific locations and use it as a source of energy. Additionally, biodegradable polyesters, such as PLA have ester bonds that are susceptible to chemical degradation involving water via a hydrolysis process, resulting in a faster biodegradation process. Finally, the success or failure of biodegradation of polymers is determined by abiotic (physicochemical conditions) and biotic (i.e. enzymatic nature) degradation setting. Concretely, biotic and abiotic factors act synergistically to decompose organic matter. For example, macronutrients are important for the cellular metabolisms of microorganisms, and thus, the macronutrient (including nitrogen, potassium, phosphorus, oxygen, or hydrogen) controls microbial growth. Another abiotic factor that is synergistic with abiotic factors is temperature; the bioavailability, hydrocarbon distribution, solubility, and diffusion rates increase with increasing temperature and therefore increase the biodegradation rate. While the pH variable determines the type of microorganism that will be responsible for polymer assimilation, ultimately breaking it down to CO_2, water, etc. Interestingly, at low pH values, the group of microorganisms that are metabolically active are called acidophilic microorganisms while at pH above neutrality, microorganisms with optimal activity are called alkaliphilic microorganisms.

9.6 BIOBASED POLYMERS, BIODEGRADABLE AND COMPOSTABLE POLYMERS

To curb the depletion of natural gas and oil used to make petroleum-based plastic and to minimize the burden of plastic waste on landfills, the production of plastics from biobased material with a high degree of biodegradation and compostability has emerged as a beacon of light in plastic waste crisis. Sustainability has become the keyword in the design and manufacturing of new products in every field and application. Sustainable development provides growth for both ecological integrity and social equity to meet basic human needs through viable economic development over time. Biobased polymers offer enormous potential for the sustainable development of plastic products because they favour life cycle assessments, can meet consumer demands, increase support of materials supplements, offers optimized land use, improve energy efficiency and minimize environmental pollution (Kaplan, 1998; Endres & Siebert-Raths, 2011).

Biopolymers are defined by two main criteria: the source of the raw materials and the biodegradability of the material. These two criteria lead to the classification of biopolymers into three groups: (1) biopolymers made from renewable resources (biobased) and are biodegradable, (2) biopolymers made from renewable resources (biobased) but are not biodegradable, and finally, (3) biopolymers made from fossil fuels and being biodegradable (Al Hosni et al., 2019; Mohanty et al., 2022; Thakur et al., 2018). Therefore, biopolymers can be defined as polymers derived from renewable and non-renewable biodegradable polymers. Consequently, biopolymers are organised into three groups. The first group of biopolymers entails naturally derived biomass polymers, where there is the direct use of biomass as a polymeric material. This may include chemically modified ones such as cellulose, cellulose acetate, poly(hydroxybutyrate (P3HB), starches, chitin, poly(lactic acid) (PLA), soy protein, so forth; The second group is of bio-engineered polymers where microorganisms and plants are used to synthesise polymers, such as poly(hydroxy alkanoates) (PHAs), poly (glutamic acid), poly(butylene succinate) (PBS), and PLA blends, this group can also be referred to as partially biobased; Finally, the third group is comprised of fossil fuel-based polymers such as polyvinyl alcohol (PVOH), PBS, bio-polyolefins, bio-poly(ethylene terephthalic acid) (bio-PET) (Luengo et al., 2003; Wei & McDonald, 2016).

Advantageously, both the first- and second-class biopolymers offer direct use without any purification steps required which allows for the more efficient production of polymers at a relatively lower cost as they are derived from natural sources. This feature is critical in situations that require high biodegradability with moderate mechanical properties. However, the chemical structure of the first- and second-class biopolymers designs have limited flexibility that restricts their dependability. Leading to the monomers used in third class polymers, where both chemical and biological processes are employed to harvest molecules from natural resources. Allowing monomers to have versatile chemical structures, suitable to be introduced into the existing production system of petroleum-derived polymers, evidently, the third class of biopolymers has the most potential.

Polyesters have become a research area of interest, particularly aliphatic ones, as they have demonstrated comparable properties to conventional petroleum-based polymers. Examples of polyesters include PBAT, PLA, PBS, PHB, and so on. Polyesters are a class of polymers identified by the ester bonds between components of monomers. The polyesters that favour biodegradation reaction are linear aliphatic polyesters since they are mostly hydrophobic polymers due to the open semicrystalline structure that encourages assimilation by microbes (Wei & McDonald, 2016; Manavitehrani et al., 2016). Furthermore, linear polyesters can be physically, chemically, and mechanically tuned to suit a number of applications. In fact, the ease of processing these polyesters into desired structures by the use of current technologies used by conventional plastic is an added advantage that warrants their integration into the polymer industry. Currently, there are several successful companies that produce biopolymers at an industrial scale indicating the brewing transformation that is occurring around the world.

The two major applications of biopolymers are the biomedical fields and the ecological field (Manavitehrani et al., 2016). Typically, biopolymers are used in four main sections (i) medical disposable products like catheters, syringes, blood bags, etc.; (ii) materials supporting surgical operation, including adhesives, sealants, and sutures; (iii) prostheses for tissue replacements like dental implants, intraocular lenses, breast implant, and finally, (iv) permanent or temporary artificial organs assists such as artificial heart, artificial kidney, vascular graft. The applications of biopolymers in the ecological field cover a variety of industries such as agricultural and forestry, fisheries, packaging, and toiletry. Table 9.1 details the applications of biopolymers in different ecological industries.

The concentrated application areas are those where plastic is used once and then thrown away like in packaging or in toiletries and daily necessities like refuge bags, cups, margarine containers and so forth. These single-use plastic contributes the most to plastic pollution, burdening the landfills and the incineration processes. For biopolymers to be successfully applied in the ecological or any other industry, their product needs to be just as effective as conventional non-biodegradable/petroleum-based plastic products in terms of the mechanical strength and other material performances required by the end user. It also possesses the ability to be degraded to low molecular weight compounds such as H_2O and CO_2 and non-toxic by-products by microorganisms living in the earth's environments after their use. Therefore, a quest towards finding thermomechanical, chemical and mechanically stable biopolymers is what has motivated scientists all over the world to conquer the goliath that affects mankind and planet earth, that is, plastic waste.

TABLE 9.1
Applications of Biopolymers

Application	Field	Examples
Composting	Food package	Wrapping plastic, containers, bottles, retail bags, forks, cups
		Cutlery and disposable plates
	Toiletry	Diapers, feminine hygiene products
	Daily necessity	Refuge bags, cups
		Food storing containers like margarine containers, milk bottles, refuse bags
Industrial applications	Agriculture, forestry	Mulch films,
		Delivery systems for
		Fertilizers and pesticides
		Vegetation nets
	Fisheries	Fishing lines, fishing gear
		Fish hooks, and nets

9.7 BIOPOLYMER BLENDS AND COMPOSITES

In 1998, the European Commission predicted that global production of biopolymers, with packaging application being the major market, will increase from 260 to 1500 kilotons between 2007 and 2011. A decade later, a staggering 2.1 million tons of biopolymers were produced globally, and the packaging sector remains the biggest market. Impressively, 18.7% of PLA is manufactured, making PLA the number one produced biopolymer. Starch-based polymer blends follow with 16.4%, then 13.5% PBAT and its blends are produced for various applications while PBS accounts for 4.1% of the 2.1 million tons of biopolymer (Figure 9.6) (Mohanty et al., 2000; Mtibe et al., 2021; Fredi & Dorigato, 2021).

This illustrates the monumental growth of biopolymers and the strong environmental tide that is pushing for minimizing the use of fuel-based polymers. However, the production of biopolymers is only 0.63% of the global production of petroleum-based polymers, which was reported to be 335 million tons in 2017. The anticipated exponential growth of biopolymers is hugely impeded by inferior mechanical, thermal, and barrier properties as well as the relatively high production cost of some biopolymers in comparison to conventional polymers. For example, PBAT has mechanical properties that are comparable to those of low-density polyethylene (LDPE) such as high-stress resistance, high elasticity, good impact resistance, and good processing properties, which makes it suitable for processing with the same infrastructure as polyethylene. In addition, PBAT has tunable water barrier properties, and it is completely biodegradable in the environment (França et al., 2019; Khan et al., 2020; Makadia & Siegel, 2011). These properties render PBAT a highly versatile polymer that can be employed in many applications. Nevertheless, the high production cost,

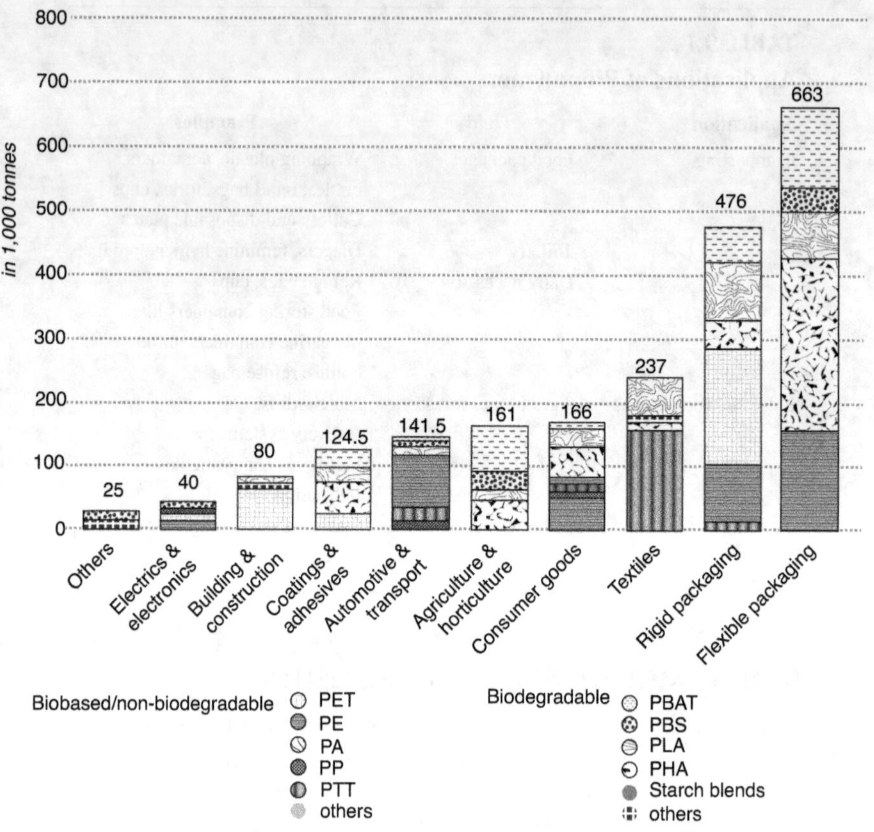

FIGURE 9.6 Global production of biopolymers by market share [44].

as well as moderate thermomechanical properties, are challenges that hinder the complete application of PBAT. PLA is another type of biopolymer that is extremely flexible, and it has found application in different sectors which include biomedical applications where it is used to produce scaffolds, bone fixing material, sutures, and PLA drugs encapsulation with the controlled release mechanism. The extensive use of PLA is due to thermal and mechanical properties comparable to poly(ethylene terephthalate) (PET), with the addition of biodegradation coupled with biocompatibility advantage. Despite these PLA benefits, the absolute application of PLA is restrained due to its inherent shortcomings, such as low crystallinity and ductility, as well as moderate barrier and mechanical performance. These are the types of challenges faced by biopolymers. An immense effort has been made to enhance the thermomechanical and mechanical properties while also investigating different pathways for reducing the production cost of biopolymers to improve the commercial potential of biopolymers. Biocomposite and biopolymer blends are the research advances that are enthralled with mitigating the limitations of biopolymers (Husárová et al., 2014; Baghaei et al., 2013; Schyns & Shaver, 2021; Zhang & Thomas, 2011).

A biocomposite is a polymer that has been reinforced with natural agents that are within the micro and nano size range to improve the properties and commercial importance of biopolymers (Thiyagu et al., 2022). This approach results in an environmentally friendly composite with superior and robust properties. Replacing conventional plastics with biodegradable materials comes with the great challenge of designing a material that exhibits structural and performance stability during storage and use. Additionally, the material must be susceptible to environmental and microbial degradation upon disposal. Evaluations of appropriate natural agents have led to extensive explorations of natural fibres, with a special focus on the agricultural waste repurposed to make fibre-reinforced polymer composites of commercial value and market appeal. Types of natural fibres used include hemp, jute, sisal, henequen, etc. (Mohanty et al., 2001; Coombes et al., 2002), and these fibres have been steadily explored for polymer reinforcement, resulting in the application of biocomposite in building materials, circuit boards, domestic sector, and automotive application since the early 2000s and there's a potential of biocomposite to stimulate an increase in demand. The natural fibres consist of lignin, cellulose, and hemicellulose produced from biomass; thus, fibres are also referred to as lignocellulosic. They are in abundance as every plant matter is composed of approximately 40%–60% lignocellulosic matter. The role of lignin in a plant is to decrease the sorption of water and consequently enhance thermal stability, while cellulose is a good mechanical property reinforcer. Examples of PBAT biopolymer composite include PBAT-Hemp, PBAT/Lignin, PBAT/Cellulose, and PHBV/PBAT-Hemp. The biocomposite has enhanced modulus and significantly decreased the cost of the end products while maintaining the eco-friendly benefit (Zhang et al., 2015; da Silva et al., 2019; Gupta et al., 2021; Singh et al., 2014).

Another well-known technique employed to improve the biopolymer challenges is blending, where two or more polymers are mixed to create a new polymer with specific properties without requiring new polymer synthesis. This method is usually cheaper and less time-consuming compared to creating new polymeric materials with new properties or developing new monomers and/or new polymerisation routes. Another advantage of polymer blends is that the properties of the material can be improved by combining the polymer components and changing the blend composition. For example, in a mulch application, highly promising materials of PLA/PBAT blend with supreme toughness of PBAT coupled with the mechanical strength of PLA produces a material that is comparable to PE film in terms of tensile strength, Young modulus, and elongation at break (Zhang et al., 2019; Gao et al., 2021; Touchaleaume et al., 2021; Dammak et al., 2020). The newly attained good mechanical strength from blending the two polyesters enables the successful application of PLA/PBAT mulch film for different crops like potatoes, cotton, vineyards, and so forth. In addition, the biodegradation by-products are reported to have soil nutrition value, and this grants the biopolymer blend an added value that is superior to PE mulch films. Blending PBAT with starch is a route taken to reduce the cost of the biopolymer while also improving or maintaining an acceptable balance of the mechanical, chemical, and biodegradation nature of the materials. Literature reports improved mechanical and rheological properties of the PBAT/TPS blend that are acceptable for packaging films. Other promising blends of large commercial interest and comparable cost include PLA/poly-caprolactone (PCL) blend, PHA/PLA/starch,

PBAT/PBS, PHA/PBS, and PLA/PBS blends (Rocha et al., 2018; Kwiecien et al., 2018; Al-Itry et al., 2014; Boonprasertpoh et al., 2017; Hu et al., 2022; Mdlalose et al., 2020 Jiang et al., 2020; Abdelrazek et al., 2016).

9.8 CONCLUDING REMARKS AND FUTURE PERSPECTIVE

Replacing petroleum-based polymers with biopolymers will undoubtedly lead to the reduction of carbon footprint and conservation of renewable resources. However, the end-of-life (EoL) evaluations are of utmost importance since they inform the sustainability of bioplastic in terms of EoL management, which can then render bioplastics an environmentally friendly alternative. Accordingly, factors used to determine the best EoL options are the type of polymer, the market size of the polymer, the processing infrastructure, and the collection system. Moreover, the European regulations on WM state that the management of waste should follow a precise hierarchy indicating a priority order in the policy and legislation for WM and prevention. The priority order is as follows: prevention > reuse > recycling > other recovery methods, for example, energy recovery; and > disposal (Figures 2.1 and 9.7).

The primary priority for both non-biodegradable and biodegradable polymers is recycling EoL option. Recycling of biopolymers research is slowly gaining traction as this alternative can valorise discarded materials and thus make way for a circular lifecycle concept. Mechanical recycling outperforms the other material recovery methods for the reasons mentioned in Section 7.4.2; therefore, it is imperative to focus research on the mechanical recycling of biopolymers. The secondary priority, only applicable to biopolymers, is assigned to organic recycling, that is, aerobic industrial composting and anaerobic digestion, where the biogenic carbon evolved from these processes is closing the carbon loop. If these two EoL options are not feasible, then biopolymers are routed to energy recovery for the production of renewable energy, but it is not favoured since it results in material loss.

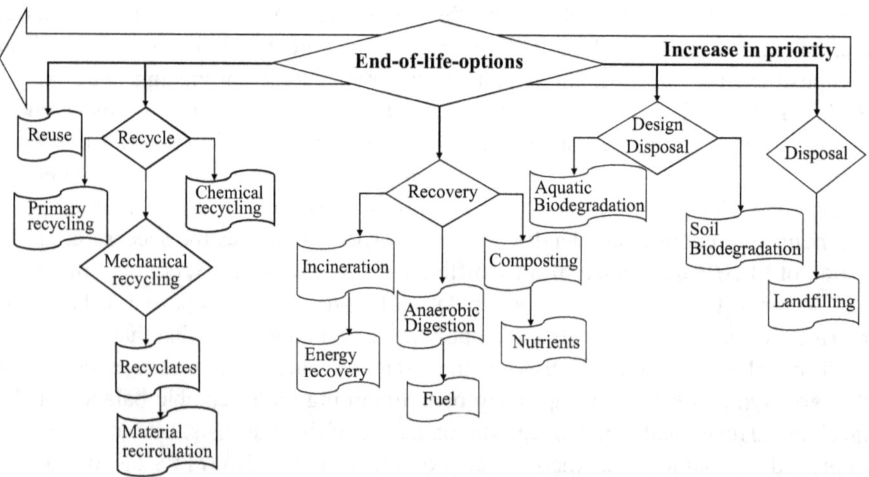

FIGURE 9.7 Schematic representation of the EoL options of biopolymers.

REFERENCES

Abdelrazek, E. M., Hezma, A. M., El-Khodary, A., & Elzayat, A. M. (2016). Spectroscopic studies and thermal properties of PCL/PMMA biopolymer blend. *Egyptian Journal of Basic and Applied Sciences*, *3*(1), 10–15.

Al Hosni, A. S., Pittman, J. K., & Robson, G. D. (2019). Microbial degradation of four biodegradable polymers in soil and compost demonstrating polycaprolactone as an ideal compostable plastic. *Waste Management*, *97*, 105–114.

Al-Itry, R., Lamnawar, K., & Maazouz, A. (2014). Reactive extrusion of PLA, PBAT with a multi-functional epoxide: Physico-chemical and rheological properties. *European Polymer Journal*, *58*, 90–102.

Ali, S. S., Elsamahy, T., Al-Tohamy, R., Zhu, D., Mahmoud, Y. A. G., Koutra, E., & Sun, J. (2021). Plastic wastes biodegradation: mechanisms, challenges and future prospects. *Science of the Total Environment*, *780*, 146590.

Andrady, A. L., & Neal, M. A. (2009). Applications and societal benefits of plastics. *Philosophical Transactions of the Royal Society B: Biological Sciences*, *364*(1526), 1977–1984.

Arikan, E. B., & Ozsoy, H. D. (2015). A review: Investigation of bioplastics. *Journal of Architecture and Civil Engineering*, *9*(1), 188–192.

ASTM, D. (2012). *1557-12. Test Method for Laboratory Compaction Characteristics of Soil Using Modified Effort (56,000 Ft-lbf/ft³ (2,700 Kn-m/m³))*. ASTM International: West Conshohocken, PA.

Ayeleru, O. O., Dlova, S., Akinribide, O. J., Ntuli, F., Kupolati, W. K., Marina, P. F., ... & Olubambi, P. A. (2020). Challenges of plastic waste generation and management in sub-Saharan Africa: A review. *Waste Management*, *110*, 24–42.

Baghaei, B., Skrifvars, M., & Berglin, L. (2013). Manufacture and characterisation of thermoplastic composites made from PLA/hemp co-wrapped hybrid yarn prepregs. *Composites Part A: Applied Science and Manufacturing*, *50*, 93–101.

Bahl, S., Dolma, J., Singh, J. J., & Sehgal, S. (2021). Biodegradation of plastics: A state of the art review. *Materials Today: Proceedings*, *39*, 31–34.

Baloy, O., Sihaswana, D., Maringa, M., Sibande, J., Oelofse, S., & Schubert, S. (2012). *National Waste Information Baseline Report*. Department of Environmental Affairs: Pretoria.

Boonprasertpoh, A., Pentrakoon, D., & Junkasem, J. (2017). Investigating rheological, morphological and mechanical properties of PBS/PBAT blends. *Journal of Metals, Materials and Minerals*, *27*(1), 1–11.

Coombes, A. G. A., Verderio, E., Shaw, B., Li, X., Griffin, M., & Downes, S. (2002). Biocomposites of non-crosslinked natural and synthetic polymers. *Biomaterials*, *23*(10), 2113–2118.

da Silva, J. B. A., Santana, J. S., de Almeida Lucas, A., Passador, F. R., de Sousa Costa, L. A., Pereira, F. V., & Druzian, J. I. (2019). PBAT/TPS-nanowhiskers blends preparation and application as food packaging. *Journal of Applied Polymer Science*, *136*(26), 47699.

Dammak, M., Fourati, Y., Tarrés, Q., Delgado-Aguilar, M., Mutjé, P., & Boufi, S. (2020). Blends of PBAT with plasticized starch for packaging applications: Mechanical properties, rheological behaviour and biodegradability. *Industrial Crops and Products*, *144*, 112061.

Endres, H. J., & Siebert-Raths, A. (2011). Engineering biopolymers. *Engineering Biopolymers*, *71148*, 3–15.

França, D. C., Almeida, T. G., Abels, G., Canedo, E. L., Carvalho, L. H., Wellen, R. M., ... & Koschek, K. (2019). Tailoring PBAT/PLA/Babassu films for suitability of agriculture mulch application. *Journal of Natural Fibers*, *16*(7), 933–943.

Fredi, G., & Dorigato, A. (2021). Recycling of bioplastic waste: A review. *Advanced Industrial and Engineering Polymer Research*, *4*(3), 159–177.

Gao, X., Xie, D., & Yang, C. (2021). Effects of a PLA/PBAT biodegradable film mulch as a replacement of polyethylene film and their residues on crop and soil environment. *Agricultural Water Management*, *255*, 107053.

Geyer, R., Jambeck, J. R., & Law, K. L. (2017). Production, use, and fate of all plastics ever made. *Science Advances*, *3*(7), e1700782.

Godfrey, L., & Oelofse, S. (2017). Historical review of waste management and recycling in South Africa. *Resources*, *6*(4), 57.

Gupta, A., Chudasama, B., Chang, B. P., & Mekonnen, T. (2021). Robust and sustainable PBAT-Hemp residue biocomposites: Reactive extrusion compatibilization and fabrication. *Composites Science and Technology*, *215*, 109014.

Haider, T. P., Völker, C., Kramm, J., Landfester, K., & Wurm, F. R. (2019). Plastics of the future? The impact of biodegradable polymers on the environment and on society. *Angewandte Chemie International Edition*, *58*(1), 50–62.

Halden, R. U. (2010). Plastics and health risks. *Annual Review of Public Health*, *31*, 179–194.

Hu, D., Xue, K., Liu, Z., Xu, Z., & Zhao, L. (2022). The essential role of PBS on PBAT foaming under supercritical CO2 toward green engineering. *Journal of CO_2 Utilization*, *60*, 101965.

Husárová, L., Pekařová, S., Stloukal, P., Kucharzcyk, P., Verney, V., Commereuc, S., ... & Koutny, M. (2014). Identification of important abiotic and biotic factors in the biodegradation of poly (l-lactic acid). *International Journal of Biological Macromolecules*, *71*, 155–162.

Jang, Y. C., Lee, G., Kwon, Y., Lim, J. H., & Jeong, J. H. (2020). Recycling and management practices of plastic packaging waste towards a circular economy in South Korea. *Resources, Conservation and Recycling*, *158*, 104798.

Jiang, G., Wang, F., Zhang, S., & Huang, H. (2020). Structure and improved properties of PPC/PBAT blends via controlling phase morphology based on melt viscosity. *Journal of Applied Polymer Science*, *137*(31), 48924.

Jiang, L., & Zhang, J. (2012). Biodegradable polymers and polymer blends. In S. Ebnesajjad (ed.), *Handbook of Biopolymers and Biodegradable Plastics*: *Properties, Processing and Applications* (pp. 109--128). William Andrew.

Kaplan, D. L. (1998). *Introduction to Biopolymers from Renewable Resources* (pp. 1–29). Springer: Berlin

Khan, H., Kaur, S., Baldwin, T. C., Radecka, I., Jiang, G., Bretz, I., ... & Kowalczuk, M. (2020). Effective control against broadleaf weed species provided by biodegradable PBAT/PLA mulch film embedded with the herbicide 2-methyl-4-chlorophenoxyacetic acid (MCPA). *ACS Sustainable Chemistry & Engineering*, *8*(13), 5360--5370.

Kijchavengkul, T., Kale, G., & Auras, R. (2009). Degradation of biodegradable polymers in real and simulated composting conditions. In M. C. Celina, N. C. Billingham, & J. S. Winggings (eds.), *ACS Polymer Degradation and Performance*, *Symposium Series 1004*, 31–39. American Chemical Society (ACS)

Kwiecien, I., Adamus, G., Jiang, G., Radecka, I., Baldwin, T. C., Khan, H. R., ... & Kowalczuk, M. (2018). Biodegradable PBAT/PLA blend with bioactive MCPA-PHBV conjugate suppresses weed growth. *Biomacromolecules*, *19*(2), 511–520.

Lamberti, F. M., Román-Ramírez, L. A., & Wood, J. (2020). Recycling of bioplastics: Routes and benefits. *Journal of Polymers and the Environment*, *28*, 2551–2571.

Luengo, J. M., Garcıa, B., Sandoval, A., Naharro, G., & Olivera, E. R. (2003). Bioplastics from microorganisms. *Current Opinion in Microbiology*, *6*(3), 251–260.

Maddah, H. A. (2016). Polypropylene as a promising plastic: A review. *American Journal of Polymer Science*, *6*(1), 1–11.

Makadia, H. K., & Siegel, S. J. (2011). Poly lactic-co-glycolic acid (PLGA) as biodegradable controlled drug delivery carrier. *Polymers*, *3*(3), 1377–1397.

Manavitehrani, I., Fathi, A., Badr, H., Daly, S., Negahi Shirazi, A., & Dehghani, F. (2016). Biomedical applications of biodegradable polyesters. *Polymers*, *8*(1), 20.

Maris, J., Bourdon, S., Brossard, J. M., Cauret, L., Fontaine, L., & Montembault, V. (2018). Mechanical recycling: Compatibilization of mixed thermoplastic wastes. *Polymer Degradation and Stability*, *147*, 245–266.

Mattsson, K., Ekvall, M. T., Hansson, L. A., Linse, S., Malmendal, A., & Cedervall, T. (2015). Altered behavior, physiology, and metabolism in fish exposed to polystyrene nanoparticles. *Environmental Science & Technology*, *49*(1), 553–561.

Mattsson, K., Johnson, E. V., Malmendal, A., Linse, S., Hansson, L. A., & Cedervall, T. (2017). Brain damage and behavioural disorders in fish induced by plastic nanoparticles delivered through the food chain. *Scientific Reports*, *7*(1), 11452.

Mdlalose, L., Chauke, V., Nomadolo, N., Msomi, P., Setshedi, K., Chimuka, L., & Chetty, A. (2020). Metal oxide nanocomposites for adsorption and photoelectrochemical degradation of pharmaceutical pollutants in aqueous solution. In O. Ama, & S. Ray (eds.),*Nanostructured Metal-Oxide Electrode Materials for Water Purification: Fabrication, Electrochemistry and Applications* (pp. 167–189). Springer: Cham.

Mohanty, A. K., Misra, M. A., & Hinrichsen, G. I. (2000). Biofibres, biodegradable polymers and biocomposites: An overview. *Macromolecular Materials and Engineering*, *276*(1), 1–24.

Mohanty, A. K., Misra, M., & Drzal, L. T. (2001). Surface modifications of natural fibers and performance of the resulting biocomposites: An overview. *Composite Interfaces*, *8*(5), 313–343.

Mohanty, A. K., Wu, F., Mincheva, R., Hakkarainen, M., Raquez, J. M., Mielewski, D. F., ... & Misra, M. (2022). Sustainable polymers. *Nature Reviews Methods Primers*, *2*(1), 46.

Mtibe, A., Motloung, M. P., Bandyopadhyay, J., & Ray, S. S. (2021). Synthetic biopolymers and their composites: Advantages and limitations-An overview. *Macromolecular Rapid Communications*, *42*(15), 2100130.

Nyathi, B., & Togo, C. A. (2020). Overview of legal and policy framework approaches for plastic bag waste management in African countries. *Journal of Environmental and Public Health*, 2020, 1–8.

Ojeda, T. F., Dalmolin, E., Forte, M. M., Jacques, R. J., Bento, F. M., & Camargo, F. A. (2009). Abiotic and biotic degradation of oxo-biodegradable polyethylenes. *Polymer Degradation and Stability*, *94*(6), 965–970.

Resch-Fauster, K., Klein, A., Blees, E., & Feuchter, M. (2017). Mechanical recyclability of technical biopolymers: Potential and limits. *Polymer Testing*, *64*, 287–295.

Rocha, D. B., Souza de Carvalho, J., de Oliveira, S. A., & dos Santos Rosa, D. (2018). A new approach for flexible PBAT/PLA/$CaCO_3$ films into agriculture. *Journal of Applied Polymer Science*, *135*(35), 46660.

Ru, J., Huo, Y., & Yang, Y. (2020). Microbial degradation and valorization of plastic wastes. *Frontiers in Microbiology*, *11*, 442.

Schyns, Z. O., & Shaver, M. P. (2021). Mechanical recycling of packaging plastics: A review. *Macromolecular Rapid Communications*, *42*(3), 2000415.

Seng, B., Hirayama, K., Katayama-Hirayama, K., Ochiai, S., & Kaneko, H. (2013). Scenario analysis of the benefit of municipal organic-waste composting over landfill, Cambodia. *Journal of Environmental Management*, *114*, 216–224.

Singh, S., Deepak, D., Aggarwal, L., & Gupta, V. K. (2014). Tensile and flexural behavior of hemp fiber reinforced virgin-recycled HDPE matrix composites. *Procedia Materials Science*, *6*, 1696–1702.

Sivan, A. (2011). New perspectives in plastic biodegradation. *Current Opinion in Biotechnology*, *22*(3), 422–426.

Standard, A. S. T. M. (2015). *D5338-15, Standard Test Method for Determining Aerobic Biodegradation of Plastic Materials under Controlled Composting Conditions, Incorporating Thermophilic Temperatures, ASTM B.* ASTM International: West Conshohocken, PA.

Thakur, S., Chaudhary, J., Sharma, B., Verma, A., Tamulevicius, S., & Thakur, V. K. (2018). Sustainability of bioplastics: Opportunities and challenges. *Current Opinion in Green and Sustainable Chemistry*, *13*, 68–75.

Thiyagu, T. T., Gokilakrishnan, G., Uvaraja, V. C., Maridurai, T., & Prakash, V. A. (2022). Effect of SiO_2/TiO_2 and ZnO nanoparticle on cardanol oil compatibilized PLA/PBAT biocomposite packaging film. *Silicon*, *14*(7), 3795–38081–14.

Touchaleaume, F., Martin-Closas, L., Angellier-Coussy, H., Chevillard, A., Cesar, G., Gontard, N., & Gastaldi, E. (2016). Performance and environmental impact of biodegradable polymers as agricultural mulching films. *Chemosphere*, *144*, 433–439.

Verma, R., & Vinoda, K. S. (2016). Toxic pollutants from plastic waste – A review. *Procedia. Environmental Sciences*, *35*, 701–708.

Vollmer, I., Jenks, M. J., Roelands, M. C., White, R. J., van Harmelen, T., de Wild, P., ... & Weckhuysen, B. M. (2020). Beyond mechanical recycling: Giving new life to plastic waste. *Angewandte Chemie International Edition*, *59*(36), 15402–15423.

Wei, L., & McDonald, A. G. (2016). Accelerated weathering studies on the bioplastic, poly (3-hydroxybutyrate-co-3-hydroxyvalerate). *Polymer Degradation and Stability*, *126*, 93–100.

Zhang, J., Choi, Y. S., Yoo, C. G., Kim, T. H., Brown, R. C., & Shanks, B. H. (2015). Cellulose-hemicellulose and cellulose-lignin interactions during fast pyrolysis. *ACS Sustainable Chemistry & Engineering*, *3*(2), 293–301.

Zhang, M., & Thomas, N. L. (2011). Blending polylactic acid with polyhydroxybutyrate: The effect on thermal, mechanical, and biodegradation properties. *Advances in Polymer Technology*, *30*(2), 67–79.

Zhang, M., Jia, H., Weng, Y., & Li, C. (2019). Biodegradable PLA/PBAT mulch on microbial community structure in different soils. *International Biodeterioration & Biodegradation*, *145*, 104817.

10 Economics of Bioplastics and Biobased Products

S. Vishnu Suba, Pandi Sangavi, Kathiresan Nachammai and Kulanthaivel Langeswaran

10.1 INTRODUCTION

Plastics are a polymer composite with polymers as a significant component. Polyethylene, poly styrofoam, polypropylene, polyvinyl chloride, and polytetrafluoroethylene (Teflon) are synthetic polymers. According to their reactivity to heat, all plastics may be categorized into thermosetting and thermoplastics (Ibrahim et al., 2021). Overuse of plastics has had a huge impact on the environment throughout time; it is estimated that 34 million tonnes of plastic are produced each year, with only 7% recycled and the remaining 93% discarded in oceans and landfills (Sushmitha et al., 2016). As a result, new biodegradable and ecologically friendly materials have been created to replace traditional plastics. Bioplastics are made from waste, biomass, and renewable sources like jackfruit, sugar cane, organic waste, rapeseed oil, cellulose from plants, oil palm empty fruit bunch, vegetable oil,and so on (Sidek et al., 2019). Bioplastics include qualities such as odor barrier and ease of molding that make them ideal alternatives to traditional plastics. Bioplastics also have unique biodegradable, ecologically friendly, energy-efficient, and compostable (Shamsuddin et al., 2017). Bioplastics have also been made from animal biomass (whey and chitosan) and protein- or oil-rich plant biomass. PLA (polylactic acid), PHA, PHB, and starch blends are examples of bioplastics, as are microbial polymers such as polynucleotides, polypeptides, and polysaccharides (Ibrahim et al., 2021). Biodegradation of compostable plastic occurs in industrial composting facilities that adhere to tight guidelines. The majority of bioplastics are made from renewable resources. Bioplastics are hence biobased materials. Petroleum-based plastics are known as non-bioplastics (Niaounakis, 2013). Bioplastics are often promoted as a long-term, ecologically benign replacement for regular plastics. However, since it must avoid upsetting potential food sources, making bioplastics becomes the most challenging job. This scenario might be improved by using non-food resources for this purpose. They are known as second-generation bioplastics. Furthermore, many PLA bioplastics are only biodegradable under specified temperature and humidity settings since the properties of these materials are only fixed under these conditions (Figure 10.1) (Zulkafli, 2014).

DOI: 10.1201/9781003304142-10

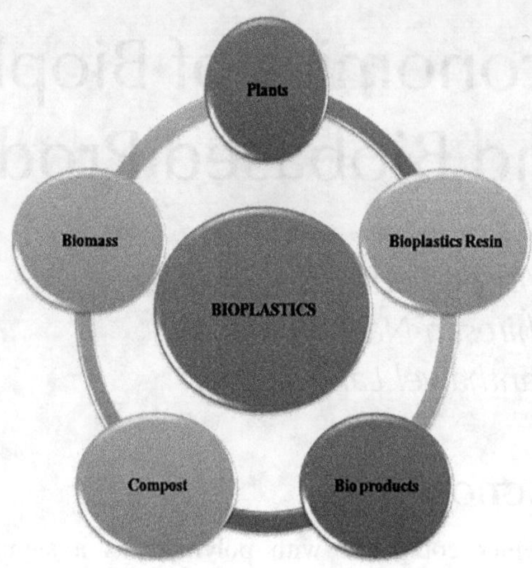

FIGURE 10.1 Illustration of bioplastics.

10.2 LIFE CYCLE ASSESSMENT OF BIOPLASTICS

Multiple environmental concerns about the extensive usage of conventional polymers prompted the development of biobased and biodegradable plastic products. It is critical to compare the environmental performance of new plastics to the performance of the traditional polymers they replace using methodologies such as life cycle assessment (LCA). The raw materials acquisition phase, the production phase, the usage phase, and the end-of-life (EoL) phase are all considered in an LCA. Because of the relative shortness of a single-use product's use time and the cheap cost of individual monitoring, it is difficult to estimate its usage phase and end of life (Morris & Hicks, 2021). Because most biobased plastics are designed to replace petrochemical plastics, a precise comparison of the environmental efficiency of these diverse polymers using LCA is essential. To compare biobased plastics to petrochemical plastics, each plastic's "full" life cycle must be represented, which may be difficult due to long production-use-reuse/recycling value chains (Bishop et al., 2021). Previous LCA studies examined the environmental performance of various bioplastic EoL alternatives (Pawelzik et al., 2013), critical components of LCA methodology for biobased materials (Spierling et al., 2018), and aspects of comparative environmental efficiency between bioplastics and petrochemical plastics (Figure 10.2).

10.3 DEMANDS OF BIOPLASTICS

China is the world's most significant bioplastic producer, outperforming industrialized nations such as Germany, the United States, Japan, and others. Growing awareness of environmental conservation, biobased or natural resources for producing products, and the formation of diverse rules across nations for optimal use of natural

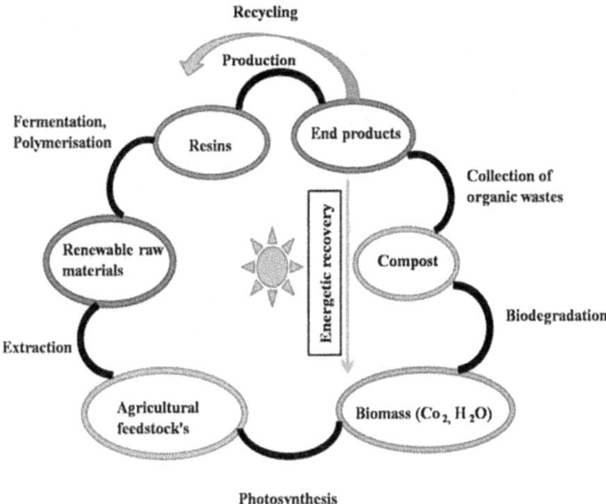

FIGURE 10.2 Life cycle assessment of bioplastics.

resources and waste management have increased demand for bioplastics over the previous decade. Consumer desire for biodegradable materials and rising environmental concerns will propel the biodegradable packaging industry forward. Food and beverage packaging is predicted to generate the fastest growth in the biodegradable packaging industry. Biobased biodegradable polymers are used in various industries, including fibers, medicine, packaging, and agriculture. The packaging industry has a huge need for biobased biodegradable plastics, and it is expected that this trend will continue in the next five years. Agriculture and medicine are two more significant areas that will drive demand for biobased biodegradables. PHA, cellulose, polyester, and PLA are among the product types that make up the worldwide biobased biodegradable plastics market. Polyester and PLA account for more than 60% of the worldwide market for biobased biodegradable plastics. The fact that these biobased biodegradable polymers are durable and cost-effective is one of the main reasons for their popularity. According to Research and Markets, the worldwide bioplastic market reached 1.6 million Mt in 2015 and is expected to reach almost 6.1 million Mt in 2020, representing a 30% compound annual growth rate (CAGR) during the five years from 2015 to 2020. The research discusses using renewable resources to develop monomers that can replace petroleum-based monomers, such as sugarcane feedstock for polyester and polyethylene production. Ethanol, a significant product in Brazil, is produced by a tiny chemical reaction. Plastics created from renewable resources such as biomass or food crops are the topic of this paper. Bioplastics made from animal resources may perhaps be developed in the future.

10.4 POSSIBLE CHALLENGES OF BIOPLASTICS

Bioplastics seem to be an enticing, ecologically friendly choice since enzymes present in certain bacteria swiftly degrade them. These new biomaterials were made using aliphatic polyesters, different polypeptides, and several polysaccharides derived from plant or animal feedstocks. Bioplastics may be made by beginning with biobased monomers (such as lactic acid) generated by fermentation or traditional chemistry and then polymerizing them in a second phase (Sabbah& Porta, 2017). Even though biodegradable plastics are thought to be environmentally friendly, they may affect the ecosystem in specific ways. At landfill sites, greenhouse gases such as methane and carbon dioxide are released in significant quantities as they degrade. Engineering polymers may address this by decomposing slowly or collecting the leaked methane and using it as fuel elsewhere. Plants such as maize, potatoes, and others make starch-based bioplastics. This puts enormous strain on agricultural products, which must meet the demands of an ever-increasing population. Crops must be planted to create plastics, which may result in deforestation (Mamidala et al., 2021). Using natural materials with environmental benefits is appealing to both businesses and consumers. Bioplastics are crucial in agriculture, consumer electronics, a packaging business, gastronomy, and automotive. However, they account for a small percentage of overall plastic production (currently about 1% of the about 300 million tonnes of plastic produced annually). Biodegradable plastics are often thought to solve the waste issue; however, biodegradability is only another quality of the material. They should be used when a low-cost technique of disposing of an object once it has fulfilled its function is required.

10.4.1 EoL Options for Bioplastics

Bioplastics are a diverse group of materials with a broad range of qualities. FOR EXAMPLE, biobased PE and biobased PET may be mechanically recycled in existing recycling systems. Plastics that are biodegradable and compostable may be recycled naturally (industrial composting and anaerobic digestion). In recovery streams, all bioplastics may be handled (incineration and the production of renewable energy due to the biobased origin). Bioplastic trash is recovered in the same way conventional plastic waste is recovered, depending on the kind of product, the bioplastic material used, and the quantities and recycling and recovery technologies available. Studies examine at least two EoL options. Each research looks at 3–to 4 distinct therapy alternatives on average. Rossi (Rossi et al., 2015) offers the most range of EoL alternatives, with eight distinct possibilities. Chemical recycling, industrial composting, Mechanical recycling, direct fuel substitution in plants, landfill without energy recovery, incineration with energy recovery, incineration without energy recovery, landfill with energy recovery, solvent-based recycling, landfill with energy recovery, anaerobic digestion, landfill with energy recovery, landfill with energy recovery (Figure 10.3).

10.5 COST OF MANUFACTURING BIOPLASTICS

Bioplastic is now more expensive than traditional plastic, but experts think that the price will soon equal the price of a barrel of crude oil. Assume that biopolymer manufacturers can produce at the exact cost of traditional plastic while also extracting the required biomass. We may anticipate a steady increase in bioplastic supply in the future decades in such a situation. Bioplastics have the potential to be as inexpensive as plastic, resulting in increased adoption. Biobased plastics have higher manufacturing

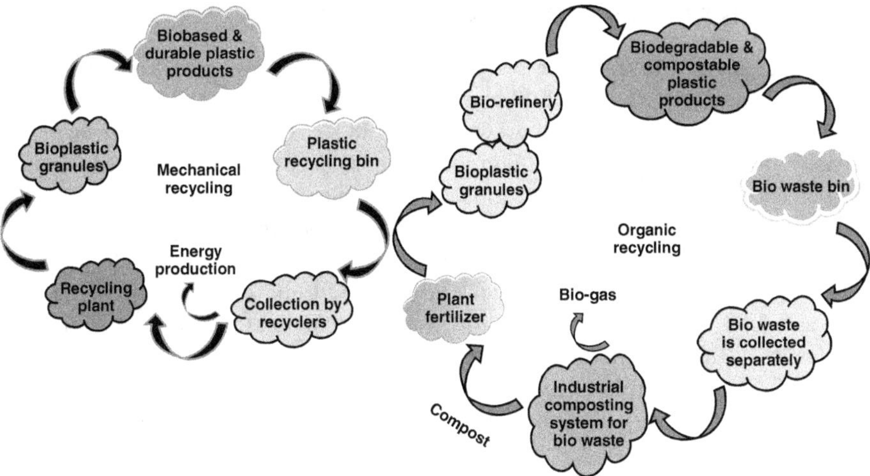

FIGURE 10.3 Closing the loop – end-of-life options for bioplastics.

costs than their fossil-based equivalents, particularly in the sector of new polymers with biodegradability potential. Because of the complexity of the manufacturing processes, the causes for this cannot be linked to a single input or technology right once. Representative estimates are needed to understand better the market-side barriers to developing the biobased industry. Unfortunately, there is presently very little data available in this area. There is presently no accessible information on the evolution of typical manufacturing costs of various polymers at the industry level, for example (Wellenreuther et al., 2021). Apart from the increased price, one reason for biobased plastics' low commercial adoption is environmental concerns over resource exploitation (Brizga et al., 2020). Currently, industrial-scale biobased plastics production depends almost solely on first-generation feedstocks like maize grain and sugarcane. Feedstock agriculture has a detrimental influence on the overall environmental balance due to the loss of carbon sinks due to indirect land-use change and the damages caused by nutrient leakage into surrounding ecosystems (Ögmundarson et al., 2020). Fortunately, alternative production methods using agricultural by-products or waste materials have been accessible for quite some time from a technology standpoint. However, the cost advantage of existing manufacturing processes is still seen as a hurdle from an economic standpoint (Brodin et al., 2017). The production costs are highly influenced by raw material prices and bioplastics manufacturing technology advancements. Unit costs are also influenced by manufacturing capacity and related economies of scale. Bioplastics' market share is now in its infancy, accounting for barely 1% of the overall global plastics market. Much work needs to be done to normalize the use of bioplastics. Manufacturers are also resistant since they are constantly under pressure to produce the best quality and value at the lowest possible cost. However, as more manufacturers switch to bioplastics, we should significantly improve these costs. As previously stated, the cost of bioplastics will remain consistent with the cost of traditional polymers generated from fossil fuels.

10.6 MARKETING STRATEGIES OF BIOPLASTICS

The global market revenue for biobased and biodegradable bioplastics was estimated at USD 21.1 billion in 2017 and is expected to increase to USD 68.5 billion by 2024, representing an annual growth rate of 18.8%. Global bioplastic production capacity increased from 1.49 million tonnes in 2012, with non-biodegradable bioplastics accounting for 61.7% and biodegradable bioplastics accounting for 38.3%, to 2.11 million tonnes in 2019, with non-biodegradable and biodegradable bioplastics accounting for 44.5% and 55.5%, respectively (García-Depraect et al., 2021). On the other hand, bioplastic feedstock is expected to have little competition with food and feed, which now account for just 0.02% of total arable land (European Bioplastics, 2019). Global bioplastics production capacity is expected to grow from roughly 2.42 million tonnes in 2021 to over 7.59 million tonnes in 2026, according to the latest industry estimates obtained by European Bioplastics in conjunction with the nova-Institute. As a result, bioplastics will surpass 2% of worldwide plastic output for the first time. In 2021, the packaging application dominated the market with a revenue share of more than 62.0%. Food and beverage packaging, home care products, films and sheets, personal care packaging, and other uses all employ bioplastics. Some of the most common bioplastics used for packaging are starch blends, PLA, PBAT, PET, PE, and PBS. Consumer demand for environmentally friendly packaging and the global waste crisis drives manufacturers to incorporate bioplastics in packaging (Figure 10.4).

Bioplastics are primarily employed in food packaging in the packaging sector. Bioplastics are frequently used to produce bottles, jars, containers, and the packaging of fresh foods. PLA plastic bottles are long-lasting, disposable, and have qualities

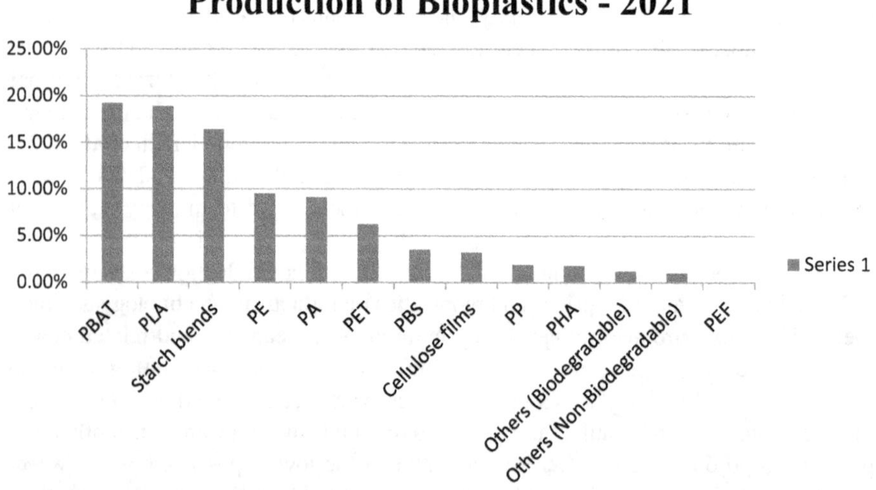

FIGURE 10.4 Production of bioplastics – 2021.

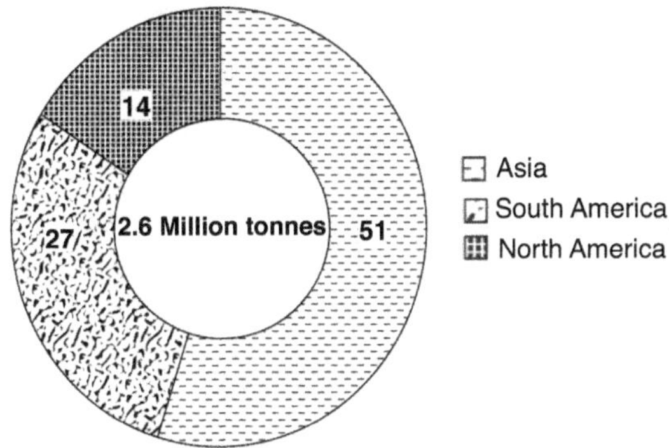

FIGURE 10.5 Production of bioplastics globally.

like transparency and gloss. To maintain product quality and extend their regional presence, most of the market's major companies have integrated their raw material manufacturing and distribution activities. This gives businesses a competitive edge in cost savings, boosting profit margins. Companies are researching and developing new industrial plastics to stay competitive and meet evolving end-user needs. If bioplastics were subsidized and politically promoted similarly to biofuels, global growth might be 10%–20% (Chinthapalli et al., 2019). Globally, the CAGR for bioplastic packaging is 18%. Drop-in polymers processed on conventional equipment (such as bio PU) and cost-competitive ones with existing large-scale facilities will see the most demand and capacity expansion (such as PLA blends and cellulose). In the future years, the businesses making these materials have stated scale-up goals of 60 kt and 75 kt, respectively (Figure 10.5).

10.7 ECONOMIC IMPACTS OF BIOPLASTICS

One of the primary environmental issues that governments and organizations must address today is the development of plastic trash and consequent uncontrolled plastic contamination. In 2019, worldwide plastic output reached about 370 million Mt (Plastics Europe, 2020). The great majority of plastic goods that enter the worldwide market are durable materials; polypropylene and polyethylene, in particular, are the most widely used polyolefins, with the packaging being their primary use. Compared to conventional plastics, bioplastic goods are generally sold on their recyclable features, but their biodegradability and compostability promise economic and environmental advantages. Bioplastics' additional benefits would be irrelevant if they were too costly to manufacture.

Metabolix'sMirel bioplastic costs nearly twice as much as a comparable heteroplastic. NatureWorks LLC's Ingeo, a bioplastic, is slightly more costly than petroleum-based polymers. Bioplastics are always more costly than petroleum-based polymers to replace, but costs have declined as research advances and oil prices grow more volatile. A possible advantage of bioplastics is price stability. Bioplastic prices should continue to reduce as the business matures and more efficient manufacturing techniques emerge (Anjali Cordeiro, 2008). A move to bioplastics is

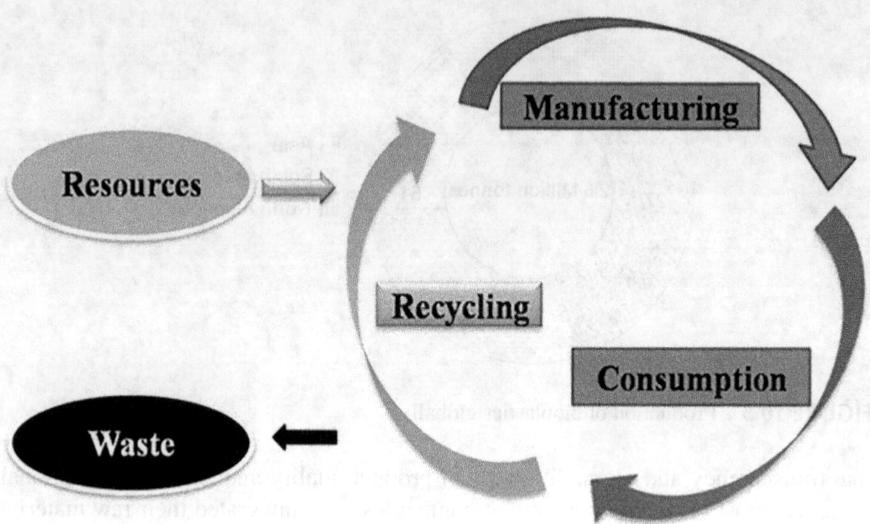

FIGURE 10.6 Bioplastics' role in the New Plastic Economy.

expected to have regional economic consequences. Bioplastics manufacturing units are likely to be erected closer to the country's key agricultural regions to reduce transportation expenses. This would result in a shift of industry from the United States' coastal regions, which are close to oil refineries, to the Midwest. Jobs would be unaffected since it is anticipated that as many jobs would be created as would be destroyed. Existing plastics manufacturing units might be converted to use bioplastics. Oil production is one example of such a geographical shift impact. As crude oil production switched to the Middle East in the 1950s, American refining facilities went from the Gulf coast to the ports of the northeast Atlantic coast to save money. Transportation costs were lowered because of its proximity to major seaports and significant petroleum consumption hubs in the United States. Not all refineries were relocated, and not all jobs were destroyed (Figure 10.6) (Abrahamson, 1952).

10.8 CIRCULAR ECONOMY OF BIOPLASTICS

A circular economy is "restorative and regenerative by design, aiming to retain goods, components, and materials at their greatest usefulness and value at all times, differentiating between technological and biological cycles," according to the Ellen MacArthur Foundation. The circular economy offers a possible alternative to the traditional linear economy manufacturing model, using, discarding, taking, making, and discarding. In the circular economy, resources are kept for as long as possible to ensure product, component, and material efficacy and value. Protecting and expanding natural capital, maximizing resource outputs, and boosting system performance are the three components of a circular economy. These three elements are the foundations of a booming circular economy. To protect and increase natural capital, society

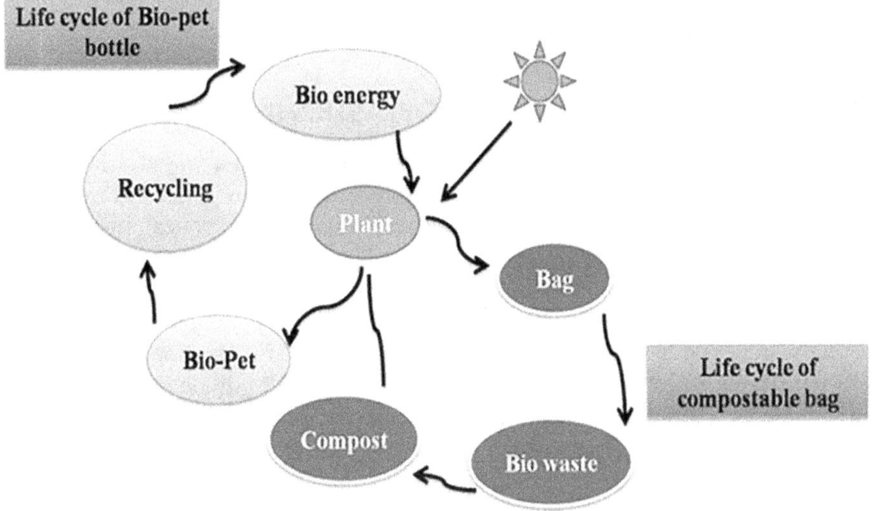

FIGURE 10.7 Circular economy of bioplastics.

must regulate the use of limited resources while also balancing and promoting the rise of renewable resources. The European Commission's Circular Economy Action Plan, released in 2020, describes the significant areas in which the economic model is evolving (European Commission, 2020). Reusability and recyclability should be considered while designing products, which means they should be more durable, repairable, and recyclable. Packaging will be reduced to the bare minimum, confined to specialized uses, and made recyclable (Di Bartolo et al., 2021). Manufacturing of single-use items will be restricted, and the disposal of unsold items will be forbidden. More than 100 million euros have been set aside in the Horizon 2020 Research Programme to finance R&D on alternative feedstocks. (European Commission, 2018). In the past five years, the European Committee for Standardization (CEN) has developed several standardized standards, including processes for asserting biodegradability and compostability and assessing the biobased content of plastics. Still, it is accepted that additional regulations are needed and that biodegradable plastic use may have both beneficial and destructive consequences (Figure 10.7) (Sander et al., 2020).

10.8.1 Principles

Plastics would be created from renewable or recycled materials in an ideal circular economy. Most plastic materials, on the other hand, have a linear lifespan. Seventy-nine percent of all plastic manufactured has ended up in landfills or the environment, 12% being burned or recycled (9%). Although recycling has grown since the 1980s, non-fiber plastic recycling has remained stable at 18%, and textile fibers are nearly never recycled (Geyer et al., 2017). Biodegradation is an EoL option for readily hydrolyzable polymers such as aliphatic esters like polylactic acid (PLA). Still, it should only be used in regulated industrial settings to assure complete digestion without uncontrollable side effects like contamination leakage or micro-plastic production (OECD, 2018).

10.9 STATISTICAL REPORT OF BIOPLASTICS IN INDIA

Bioplastics are critical in moving the plastics industry from a wasteful linear to a circular economy. Bioplastics market size is predicted to grow as their usage grows in terms of environmental and economic conditions and practical implications. Scientists from IIT-Guwahati have manufactured biodegradable plastic using domestic technology for the first time in India. The breakthrough comes as a massive shot for solid waste management in a nation where growing pollution levels remain a severe worry. The bioplastics market in India is expected to develop at a CAGR of 23.91%, from US$208.475 million in 2019 to US$754.648 million in 2025. Raising environmental consciousness is one of the major factors likely to boost the Indian bioplastics market to new heights during the forecast period.

Furthermore, key industry companies are being enticed to join this market to respond to rising consumer awareness while bolstering their market position, which is another factor expected to drive market expansion. The increased emphasis on sustainability that many firms have turned their attention to lessen their carbon footprint is another driving element expected to assist market expansion. Furthermore, the Indian government's stance on plastic, as exemplified by then-environment minister Harsh Vardhan's announcement on World Environment Day in 2018 that single-use plastics would be phased out by 2020, later revised to 2022, is expected to open up new avenues for the Indian bioplastics market, boosting its growth. As of 2019, 18 states throughout the country have prohibited single-use plastics. Total Corbion PLA, a 50/50 joint venture between Total and Corbion, stated in September 2019 that it would enter the Indian bioplastics market in technical cooperation with Konkan Speciality Poly Products Pvt Ltd, a polymers and chemicals company based in Mangalore, India. The bioplastics market in India is still in its infancy. Only a few companies in India are working in the bioplastics sector. The state-level body of the National Green Tribunal has given the government until August 31, 2019, to enforce the plastic ban. Truegreen, Plastobags, Ecolife, and Evergreen already produce bioplastics in India. The Indian bioplastics market has benefited from several technological developments, which have witnessed great growth. It is up to each individual to decrease their use of conventional plastics in their daily life as much as feasible (India Bioplastics Market, 2021).

10.10 IMPACTS OF BIOPLASTICS ON THE GLOBAL LEVEL

Bioplastics change the chemical composition of soil in the terrestrial environment, creating interactions between soil components and bioplastic particles, as well asother contaminants that pollute the soil compartment. At different weathering stages, bioplastic particles are made up of numerous polymers and occur in various forms and sizes (Koelmans et al., 2019). Taking bioplastics into consideration, the following materials were produced in the largest quantities in 2020: starch blends (18.7%), PBS (4.1%), PE (10.5%), PTT (9.2%), PLA (18.7%), PA (11.9%), PET (7.8%), and PBAT (13.5%) (European Bioplastics, 2020). The worldwide market for

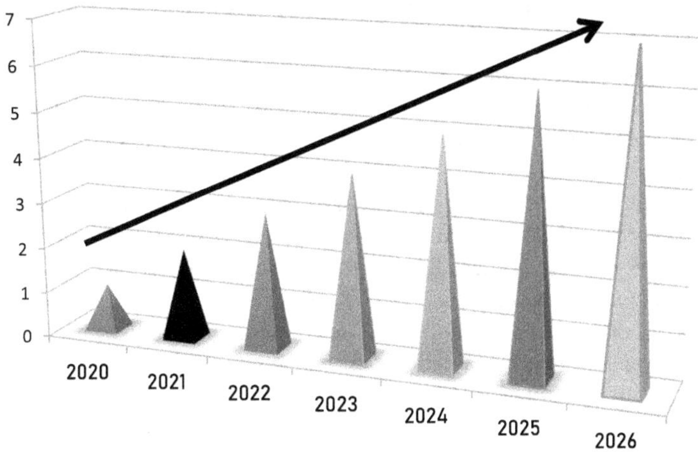

FIGURE 10.8 Global bioplastic market.

bioplastics is expected to grow every day, from USD 9.2 billion now to USD 20 billion by 2026 (Intrado Globe Newswire, 2021). Bioplastics' manufacturing capacity was above 2.1 million tons in 2020, and it is expected to increase to 2.9 million tons by 2025 (European Bioplastics, 2021). The packaging sector is the largest user of bioplastics today, accounting for about 1 million tons, or 47% of all bioplastics produced. Only 0.7 million hectares, or 0.015% of the world's arable land, are used to cultivate feedstock for bioplastic precursors. According to research by European Bioplastics, global biodegradable plastics output has been comparatively higher than non-biodegradable plastics since 2019 and is expected to continue in the next years. PLA, PHA, starch blends, and other biodegradable polymers currently account for more than 64% (over 1.5 million tonnes) of global bioplastic manufacturing capacity. Biodegradable plastics are predicted thanks to the significant development of polymers such as PBAT (polybutylene adipate terephthalate) and PBS (polybutylene succinate) and the steady growth of polylactic acids to climb to over 5.3 million in 2026 (PLAs). Biobased, non-biodegradable plastics account for around 36% of global bioplastics production capacity (about 865 thousand tonnes). Biobased PE (polyethylene) and biobased PET (polyethylene terephthalate), as well as biobased PA, are examples of drop-in solutions (polyamides). Their relative proportion is predicted to shrink even further in 2026, to over 30% (Figure 10.8).

10.10.1 Impact of COVID-19 on the Worldwide Bioplastics and Biopolymers Market

Packaging, consumer products, automotive and transportation, textiles, agriculture, and horticulture are among businesses that employ bioplastics and biopolymers. However, industries all across the globe have been harmed by the continuing epidemic. Human resources shortages, logistical constraints, material scarcity, and other constraints hindered the industry's expansion in the preceding year. Due to the onset of the COVID-19 pandemic and its steady spread, sectors that provide essential services, such as securely transferring food and supplies to customers, are being more impacted. Because of the worldwide effect, demand for several forms of packaging that are considered non-essential has dropped dramatically.

On the other hand, demand for necessary packagings, such as e-commerce shipping, has skyrocketed. These developments have created new problems for the packaging industry. Because of these factors, the impact of the Covid-19 epidemic on the packaging sector will be uneven. The COVID-19 epidemic has disrupted the usual operations of consumer products firms. The pandemic has influenced the successful marketing model, which has aided the correct functioning of the company model. According to the Institute for Supply Management Research, 76% of organizations had to cut sales objectives by 23% due to supply chain interruptions. Some industries, such as domestic cleaning and frozen foods, have witnessed a rise in customer demand, while others have seen a dip in sales and a significant reduction in food traffic in retail outlets. With the shift in customer demand toward e-commerce, businesses needed to adapt. As a result, the whole consumer products business has been driven into the digital era.

10.11 CONCLUSION

Bioplastics are quickly expanding due to their evident benefits in various applications. These benefits will rise as the oil supply tightens. They have a reduced carbon impact than their oil-based counterparts. Bioplastics have exceptional biodegradability, which may assist the globe in dealing with the growing issue of trash, especially in rivers and oceans. Plant-based bioplastics can be recycled much like their conventional counterparts, contributing to the development of a more sustainable global economy. Bioplastics R&D are producing high-quality products for several sectors. To adequately encourage the usage of bioplastic, it is essential to regulate waste disposal. Recycling seems to be the best choice for disposing of biobased goods to reduce environmental effects while preserving renewable resources. The environmental impact of bioplastic manufacturing and disposal is reduced when recycled. It is worth mentioning that the knowledge presented in this research may be valuable for environmental reliability owing to inappropriate bioplastic management and uses. PHA materials are the primary resource for replacing traditional plastic in most technical applications. PHA production costs are now too expensive. However, more technology and sourcing studies may lower manufacturing costs, allowing greater diversity and heterogeneity in bioplastic applications.

ACKNOWLEDGMENT

Dr. K. Langeswaran acknowledges MHRD-RUSA 2.0 [F.24/51/2014-U, Policy (TN Multi-Gen), Department of Education Government of India] for financial support.

Data Availability – Not Applicable

REFERENCES

Abrahamson, S. R. (1952). The shifting geographic center of petroleum production and its effect on pricing systems. *Economic Geography*, 28(4), 295–301.

Bishop, G., Styles, D., & Lens, P. N. (2021). Environmental performance comparison of bioplastics and petrochemical plastics: A review of life cycle assessment (LCA) methodological decisions. *Resources, Conservation and Recycling*, 168, 105451.

Brizga, J., Hubacek, K., & Feng, K. (2020). The unintended side effects of bioplastics: Carbon, land, and water footprints. *One Earth*, 3(1), 45–53.

Brodin, M., Vallejos, M., Opedal, M. T., Area, M. C., & Chinga-Carrasco, G. (2017). Lignocellulosics as sustainable resources for the production of bioplastics – A review. *Journal of Cleaner Production*, 162, 646–664.

Chinthapalli, R., Skoczinski, P., Carus, M., Baltus, W., de Guzman, D., Käb, H., & Ravenstijn, J. (2019). Biobased building blocks and polymers-global capacities, production and trends, 2018–2023. *Industrial Biotechnology*, 15(4), 237–241.

Cordeiro, A. (2008). Bioplastics begin to get real. *The Wall Street Journal*, 10–22.

Di Bartolo, A., Infurna, G., & Dintcheva, N. T. (2021). A review of bioplastics and their adoption in the circular economy. *Polymers*, 13(8), 1229.

European Bioplastics. (2019). Nova-Institute, Bioplastics Facts, and Figures. https://www.european-bioplastics.org/

European Bioplastics. (2020). Bioplastics Market Development Updates. https://www.european-bioplastics.org/

European Bioplastics. (2021). Bioplastics Market Data. https://www.european-bioplastics.org/

European Commission. (2018). *A European Strategy for Plastics in a Circular Economy*; European Commission: Belgium.

European Commission. (2020). *A New Circular Economy Action Plan for a Cleaner and More Competitive Europe COM 98 Final*; European Commission: Brussels, 2020.

García-Depraect, O., Bordel, S., Lebrero, R., Santos-Beneit, F., Börner, R. A., Börner, T., & Muñoz, R. (2021). Inspired by nature: Microbial production, degradation, and valorization of biodegradable bioplastics for life-cycle-engineered products. *Biotechnology Advances*, 53, 107772.

Geyer, R., Jambeck, J. R., & Law, K. L. (2017). Production, use, and the fate of all plastics ever made. *Science Advances*, 3(7), e1700782.

Ibrahim, N. I., Shahar, F. S., Sultan, M. T. H., Shah, A. U. M., Safri, S. N. A., & Mat Yazik, M. H. (2021). Overview of bioplastic introduction and its applications in product packaging. *Coatings*, 11(11), 1423.

India Bioplastics Market. (2021). India Bioplastics Market – Industry Analysis and Forecast (2019–2026) by Product and Application. https://www.maximizemarketresearch.com/market-report/india-bio-plastics-market/21457.

Intrado Globe Newswire. (2021). *Bioplastics Market Size Expected to Reach USD 19.93 Billion by 2026*; Fortune Business Insights(tm): Pune, Maharashtra.

Koelmans, A. A., Nor, N. H. M., Hermsen, E., Kooi, M., Mintenig, S. M., & De France, J. (2019). Microplastics in freshwaters and drinking water: Critical review and data quality assessment. *Water Research*, 155, 410–422.

Mamidala, A., Kamma, R., & Pydimalla, M. (2021). *Journal of Emerging Technologies and Innovative Research*, 8(5).

Morris, M. I. R., & Hicks, A. L. (2021). A human-centered review of life cycle assessments of bioplastics. *The International Journal of Life Cycle Assessment*, 27, 1–16.

Niaounakis, M. (2013). Definition and assessment of (bio) degradation. In *Biopolymer Reuse, Recycling and Disposal*; Elsevier: Amsterdam; pp. 77–94.

Ögmundarson, Ó., Sukumara, S., Laurent, A., & Fantke, P. (2020). Environmental hotspots of lactic acid production systems. *GCB Bioenergy*, 12(1), 19–38.

Organisation for Economic Co-operation and Development (OECD). (2018). *Improving Plastics Management: Trends, Policy Responses, and the Role of International Co-operation and Trade. OECD Environmental Policy Paper No. 12* (OECD, 2018).

Pawelzik, P., Carus, M., Hotchkiss, J., Narayan, R., Selke, S., Wellisch, M., & Patel, M. K. (2013). Critical aspects of bio-based materials' life cycle assessment (LCA)-Reviewing methodologies and deriving recommendations. *Resources, Conservation and Recycling*, 73, 211–228.

Plastics Europe. (2020). *Plastics – The Facts*. pp. 1–64. Available online: (https://plasticseurope.org/wp-content/uploads/2021/09/Plastics_the_facts-WEB-2020_versionJun21_final.pdf accessed on November 1, 2020).

Rossi, V., Cleeve-Edwards, N., Lundquist, L., Schenker, U., Dubois, C., Humbert, S., & Jolliet, O. (2015). Life cycle assessment of end-of-life options for two biodegradable packaging materials: Sound application of the European waste hierarchy. *Journal of Cleaner Production*, 86, 132–145.

Sabbah, M., & Porta, R. (2017). Plastic pollution and the challenge of bioplastics. *Journal of Applied Biotechnology & Bioengineering*, 2(3), 00033.

Sander, M., Filatova, T., & Weber, M. (2020). *Biodegradability of Plastics in the Open Environment*; ETH Zurich: Zürich, Switzerland.

Shamsuddin, I. M., Jafar, J. A., Shawai, A. S. A., Yusuf, S., Lateefah, M., & Aminu, I. (2017). Bioplastics as better alternative to petroplastics and their role in national sustainability: A review. *Advances in Bioscience and Bioengineering*, 5(4), 63–70.

Sidek, I. S., Draman, S. F. S., Abdullah, S. R. S., & Anuar, N. (2019). Current development on bioplastics and its prospects: A preliminary review. *INWASCON Technology Magazine*, 1, 3–8.

Spierling, S., Knüpffer, E., Behnsen, H., Mudersbach, M., Krieg, H., Springer, S., & Endres, H. J. (2018). Bio-based plastics – A review of environmental, social and economic impact assessments. *Journal of Cleaner Production*, 185, 476–491.

Sushmitha, B. S., Vanitha, K. P., & Rangaswamy, B. E. (2016). Bioplastics – A review. *International Journal of Modern Trends in Engineering and Research*, 3(4), 411–413.

Wellenreuther, C., Wolf, A., & Zander, N. (2021). *Cost Structure of Bio-based Plastics: A Monte-Carlo-Analysis for PLA* (No. 197). HWWI Research Paper Hamburg, Germany.

Zulkafli, N. N. (2014). *Production of Bioplastic from Agricultural Waste* (Doctoral dissertation, UMP, Malaysia).

11 Circular Economy on Bioplastics and Biobased Polymers

Abhispa Bora, T. Angelin Swetha and A. Arun

11.1 INTRODUCTION

The annual increase in world population poses challenges to global food security and environmental issues, both of which have an impact on the feasible development targets (Talan & Tyagi, 2020). The use of non-renewable resources to generate energy depletes natural resources and contributes to the release of greenhouse gases, which harms the environment. As a result, these global challenges necessitate an immediate solution, with the circular bioeconomy playing a major part in a low-carbon economy that will undoubtedly aid in the resolution of these issues (Leong et al., 2021).

Plastics manufacturing surged from 245 million metric tonnes in 2008 to 359 million metric tonnes in 2018 and is anticipated to triple by 2050, accounting for over a fifth of world oil usage (Chia et al., 2020). Despite the enormous manufacture of plastics ever since 1950s, no viable strategy for dealing with the disposal challenges caused by plastic waste has been implemented. Plastic recycling rates are poor when compared to the amount of plastic produced, with the majority of it ending up in landfills. Plastics are the most difficult to decompose when compared to aluminium, papers, fruits, and leathers. This is due to the persistence of plastics in nature for centuries before degrading (X. Chen & Yan, 2020).

Inadequate active solid-waste management to identify plastic waste results in serious environmental, human, and animal health problems. As a result, there is a lot of interest in the development of bioplastics, which are biodegradable and derived from plant, animal, and microbial sources (Kalia et al., 2000).

Biobased resources for plastic production are an alternative to fossil-based resources. The benefits of using biobased plastics include the conservation of fossil resources and the elimination of carbon dioxide emissions. Biobased plastics, in comparison to fossil-based plastics, have the possibility for a closed-loop in a truly circular economy because the biogenic carbon taken up by a plant is released back into the atmosphere after use (Spierling et al., 2018).

Bioplastics, also known as "biopolymers" have been the subject of significant research and debate on a global scale for quite some time. The scarcity of fossil fuels drives the development of biobased products, while the potential to reduce pollution and ease organic waste collection stimulates the development of biodegradable and compostable plastics (Di Bartolo et al., 2021). Bioplastics are already being used

DOI: 10.1201/9781003304142-11

in the market as packaging, compost bag, and carriers; they are also used in agriculture and horticulture industries, as well as the automotive and electronic industries. Bioplastics are critical for increasing sustainability, which can be defined as a balance between a firm's economic, environmental, and social aspects and can be applied to a wide range of disciplines (Yadav et al., 2020).

To increase the efficiency of resources and management of wastes, it is essential to elevate and follow a circular economy. In place of fossil fuels, alternatives such as biomass, municipal waste streams, and industrial waste streams can provide a sustainable carbon supply (Yadav et al., 2020). It will not only benefit the environment by lowering the waste disposal expenditure and wastes, but it will also help cut overall production expenses by using wastes as substrates (Jiang et al., 2012). The usage of non-renewable materials and the production of wastes are reduced in a circular economy, while the reprocessing and regeneration of the materials are enhanced (Rosenboom et al., 2022).

This chapter mainly concentrates on the detrimental effects of using conventional plastics and the role of bioplastics and biobased polymers in a circular economy. The different biobased polymers are also discussed, along with the end-of-life treatment options for bioplastics.

11.2 DETRIMENTAL EFFECTS OF USING CONVENTIONAL PLASTICS

In today's world, one of the most serious environmental issues that humans face is uncontrollable plastic pollution caused by the production of numerous plastic wastes. The varieties of plastic products that are available in the market are made up of imperishable materials, mainly polyethylene and polypropylene. These two materials are the dominant polyolefins in the market (Plastics Europe, n.d.). In 2017, about 8300 metric tons of plastics were generated globally. Plastics that are chemically synthesised do not degrade and thus end up in landfills (Geyer et al., 2017). According to the reports of the United Nations Environment Programme (UNEP), out of all the plastics generated, about 9% of the plastics are recycled while 12% of the plastics are ignited (UNEP, 2016). The remaining plastics are dumped into the environment and thereby polluting the land as well as the marine environment.

The most prevalent cause of garbage generation in the world is packaging, which generated 146 million tonnes of waste in 2015. Out of 146 million tonnes, 141 million tonnes of waste were not recycled (96.6%). Additionally, the operating life of any industrial plastic component is the shortest for packaging (Geyer et al., 2017). Single-use plastics may only have a brief shelf life of a few minutes. Around 80% of the plastic trash that ends up in the ocean originates on land, usually from kerbsides and poorly maintained landfills that are destroyed by wind and sea tides (Gallo et al., 2018; Jambeck et al., 2015). Every year, almost 2 million tonnes of plastic garbage is washed into rivers in both developing and developed countries, due to a lack of effective collection and waste treatment infrastructure (Jambeck et al., 2015; Lebreton et al., 2017; Schmidt et al., 2017).

Plastic waste is highly persistent in the natural environment, mainly in seawater it is expected to take hundreds to thousands of years to break down (Gallo et al., 2018;

UNEP, 2016). Plastic waste in marine has a significant and negative impact on the ecology (Gregory, 2009). Plastic trash, with its extended half-life and hydrophobic nature, provides ideal circumstances for the expansion of different microbial species, thereby generating a "plastisphere" ecosystem. The plastic wastes are converted into minute residual forms called microplastics through UV-light degradation, microbial action, heat, and mechanical stress (Andrady, 2011). These microplastics are available in enormous amounts to wildlife, birds as well as numerous aquatic organisms (Andrady, 2011). Subsequently, these microplastics enter the food chain and, thus, pose a high threat to human health (Rochman et al., 2015; Smith et al., 2018).

Because of the microplastics particulate nature, these particles have the potential to absorb and carry pollutants such as hydrophobic organic chemicals like polycyclic aromatic hydrocarbons (PAHs), polychlorinated biphenyls (PCBs) and microplastics transport pathogens from one location to another (Ziccardi et al., 2016). Most marine microplastics (98%) arise from soil sources, notably from washing textile clothing (especially from Asia) and stripping of car tyres (predominantly from North America) (Boucher & Friot, n.d.; Henry et al., 2019; Jan Kole et al., 2017). Although there are now too little microplastic fragments in freshwater to cause harm, but increasing levels can have negative consequences (Ziccardi et al., 2016). When exposed to increased quantities of microplastic particles, freshwater organisms such as worms, amphipods, oysters, and crabs showed decreased development, inflammation, and cognitive function (Crump et al., 2020; Straub et al., 2017; Von Moos et al., 2012). The sources of information regarding the trophic transmission of microplastics include measurements of the amount of microplastics in field-collected species and controlled feeding experiments that aimed to replicate the transfer of microplastics through a synthetic food chain (Carbery et al., 2018). In some sea-floor ecosystems, gravitational sinking and bottom currents result in localised and concentrated deposits of microplastics (Kane et al., 2020).

To solve plastic pollution, we can increase the recycling and reusing processes of already manufactured plastics. Moreover, we can replace several classes of plastic items, particularly single-use products, with recyclable alternatives. A shift in our society's mentality and habits is also important factor to curb plastic pollution. At the same time, fossil resources are limited, and their consumption results in greenhouse gas emissions. Plastics made from renewable resources have been proposed as a way to reduce carbon emissions by absorbing carbon dioxide during their growth and to reduce the economy's reliance on fossil fuels. As part of technological advancement in the bioeconomy, the use of biodegradable plastics in specialised domains such as soil cover films, carrier bags, and single-use packaging is also suggested.

11.3 CIRCULAR ECONOMY

The circular economy, in broad terms, is an economic and production model that aims to maximise resource reuse and recycling, thereby extending product life cycles while minimising waste. The model was conceived as retaliation to the traditional economy, the linear economy, in which resources are used to generate products which are used and discarded as waste. The circular economy is a viable alternative to the classic linear economy model of manufacture, use, and discard. In the circular economy, resources are kept in use for as long as possible in order to maintain product, component, and material effectiveness and value at all times. Businesses can get the most value out of the things their customers use if they do it this way. The product and materials can then be recovered and regenerated after their maximum value has been attained. The circular economy is designed to be redeemable and regenerative, mirroring the biological world. Natural materials disintegrate into simple building elements, which can then be repurposed for new purposes.

A true circular economy is determined by renewable sources such as raw supplies and renewable energies and not by fossil resources (European Bioplastics, n.d.). In context of plastics, the aim of circular economy should be to utilise resources that are non-polluting in nature. The plastic items should be recycled more, and the post-use treatment of plastics should be based on reuse, recycling, and other environmentally acceptable disposable methods.

The European Commission's Circular Economy Action Plan, which was released in 2020, defines the main directions in which the economic model is evolving. Few main points of the document are briefly summarised here—Reusability and recyclability should be considered while designing products, which means they should be more durable, repairable, and recyclable. Packaging will be reduced, limited to certain applications, and made recyclable. Single-use item manufacture will be limited, and the disposal of unsold objects will be prohibited. Finally, importance will be given to the biobased sector because it enables greater circularity in plastic industry. However, it should be noted that sourcing, labelling, and use of biobased, biodegradable, and compostable plastics are emerging challenges for which the European Commission will develop a policy framework in the coming years.

The three components of the circular economy are preserving and enhancing natural capital, optimising resource outputs, and fostering system effectiveness (Di Bartolo et al., 2021). These three elements are the foundations of the circular economy's success. To protect and increase natural capital, society must regulate the use of finite resources while also balancing and promoting the rise of renewable resources. Resources must be carefully chosen and then processed utilising cost-effective technologies and procedures. These methods promote the flow of nutrients throughout the system, resulting in better regeneration conditions.

Within the circular economy, optimising resource yields entails the cycling of high-value products, components, and materials for both technical and biological cycles. It also entails going beyond Design for Manufacturing methods to include refurbishment and recycling in the design process. When manufacturers adopt these design principles, their products have a longer lifespan and are more suited for reuse. Businesses, governments, and consumers must manage land use, as well as water, air pollution, and climate change, in order to enhance system effectiveness—the final tenant of the circular economy. Managing these externalities helps to protect renewable feedstocks while also limiting the use of scarce resources.

11.4 BIOPLASTICS AND BIOBASED PLASTICS

The word "bioplastic" is regularly used interchangeably with the term biodegradable. Some bioplastics are biodegradable; however, now no longer they all are. Bioplastics are described as polymers that meet either one of the criteria: they may be biodegradable or they may be biobased (Tokiwa et al., 2009). The term "biobased" refers to a polymer which is made completely or in part from biomass, which includes any type of organic renewable material of biological source as well as organic waste. The term "biodegradable" refers to a material that can be broken down by microbes into natural substances such as biomass, water, and carbon dioxide. A biodegradable plastic, in a more particular sense, is a plastic substance that meets certain official biodegradability requirements, where a certain percentage of decomposition must be scientifically observed within a set length of time and under specific conditions. A biodegradable plastic, meanwhile, undergoes biodegradation in industrial composting facilities and must adhere to strict guidelines. Thus, bioplastics can be divided into three categories: biobased and biodegradable, solely biobased, and only biodegradable. Polylactic acid (PLA) (Garlotta, 2001; Madhavan Nampoothiri et al., 2010), biobased polybutylene succinate (bio-PBS) (Xu & Guo, 2010), and polyhydroxyalkanoates (PHAs) (Chanprateep, 2010) are few examples of bioplastics having both biobased and biodegradable nature. Apart from this, plastics made from starch, chitosan, cellulose, and lignin are other examples of biobased and biodegradable bioplastics. Biobased polyamides (bio-PP), polyethylene terephthalate (bio-PET), and polyethylene (bio-PE), are examples of bioplastics that are biobased but not biodegradable (Siracusa & Blanco, 2020). Bioplastics such as polycaprolactone (PCL) (Labet & Thielemans, 2009), polybutylene adipate terephthalate (PBAT) (Ferreira et al., 2019), PBS, and polyvinyl alcohol (PVA) (Aslam et al., 2018) are procured from fossil resources but are biodegradable in nature (Figure 11.1).

Polyethylene and other conventional polymers are fossil-based and long-lasting. It is also important to differentiate biobased plastics into two types – drop-in types and chemical unique types. Drop-in kinds are biobased plastics, such as bio-polyethylene terephthalate (bio-PET) or Bio-PE, that have the same chemical structure as their traditional counterparts and change solely in terms of feedstock (Di Bartolo et al., 2021). As a result, the processing of plastic products as well as the recycling procedures are

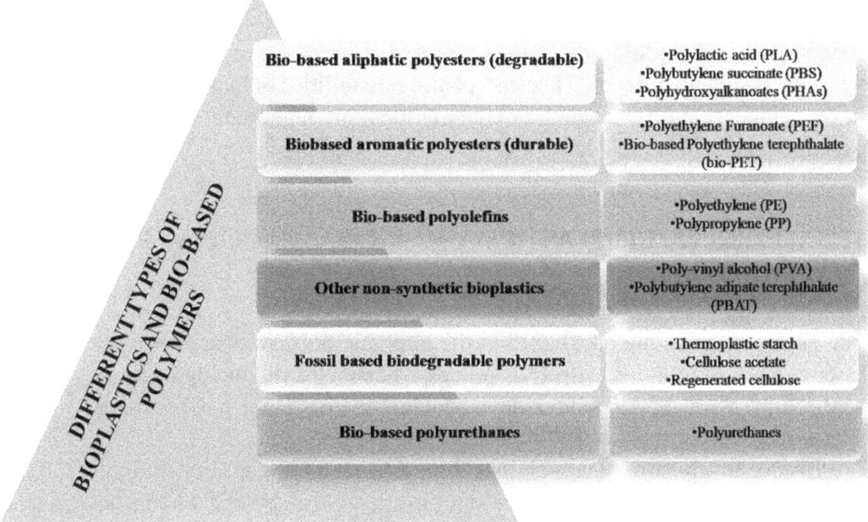

FIGURE 11.1 Different types of bioplastics and bio-based polymers.

the same for drop-ins, and the conventional infrastructure can be used in the same way. Chemical novel kinds such as PHA and PLA have no direct conventional counterparts and have their own set of features. As a result, they are typically unable to be integrated into current traditional recycling systems (Di Bartolo et al., 2021).

To date, the scientific community has developed four scientific methods for producing these environmentally friendly, sustainable, and biobased plastics. The methods are (i) partial alteration of naturally occurring polymers (starch, cellulose, pullulan); (ii) fabrication of bioplastics by growing and adapting microbial colonies isolatedfrom natural environments or developed through genetic engineering (PHA, PHB) (Mohanrasu et al., 2018; Nikodinovic-Runic et al., 2013); (iii) Monomers are made by de novo or fermentation methods from raw materials using traditional chemical procedures, with polymerisation occurring later (PLA, polyethylene); (iv) manufacturing of bioplastics by using polymers which are partially biodegradable such as PBS, PBAT, etc. (Rosenboom et al., 2022).

When compared to the size of the conventional plastics sector, the volume of bioplastics produced today is rather small. According to European Bioplastics, global bioplastic output was roughly 2 Mt in 2018, whereas global plastic production was around 360 Mt. Simultaneously, the global market for bioplastics is expected to increase rapidly over the next five years, increasing in volume by roughly 40%. Various types of bioplastics are already on the market and are manufactured by firms in Europe, the United States, and Asia. Corbion N.V. (Netherlands), Tianjin Guoyun (China), BASF (Germany), CJ Cheil Jedang (Korea), NatureWorks LLC (USA), and Novamont (Italy) are the most prominent manufacturers of bioplastics. Cellophane™ made from regenerated cellulose by Futamura Chemical Company (UK), and Nylon-11, made from castor oil by various producers, are two historically successful examples (Di Bartolo et al., 2021).

Building and construction, as well as flexible and rigid packaging, are among the application fields of bioplastics. Biobased plastics, such as bio-PET, account for the majority of the overall market share. Short-life uses, such as flexible and rigid packaging, currently account for around 70% of all applications. Long-term uses such as construction and building make up a minor portion of the total. Long-term applications are said to be hindered. The end-of-life possibilities of biobased plastics, as well as the application of the circular economy idea, are significant due to the focus on packaging and short life cycles.

11.5 TYPES OF BIOPLASTICS AND BIOBASED POLYMERS

11.5.1 Biobased Aliphatic Polyesters (Degradable)

The ester groups in the backbones of the aliphatic polyesters are easy to be cleaved by enzymatic activity or hydrolysis process. This helps in quickly degrading the aliphatic polyesters. PLA, polybutylene succinate (PBS), and polyhydroxyalkanoates (PHAs) are the most prevalent biobased aliphatic polyesters.

PLA is an aliphatic homopolymer with a production capacity of about 250,000 tonnes per year, making it the most cost-effective synthetic bioplastic (Chafran et al., 2019; Hottle et al., 2017; Morão & de Bie, 2019). PLA is commonly manufactured by lactic acid polycondensation, which is generated from sugar fermentation, or by ring-opening polymerisation of lactide, which is a cyclic dimer of lactic acid (Dusselier et al., 2015; Penczek et al., 2003). PLA is optically clear and has been used as a substitute for polyolefin films and polystyrene foams, as well as in single-use goods. PLA, on the other hand, is brittle and difficult to crystallise due to its small repeat unit and methyl side group. Before being treated, PLA is usually changed and mixed (for example, with additional biodegradable polymers or nucleating agents) (X. Chen & Yan, 2020).

Compared to other aliphatic copolyesters, PBS has a more adaptable molecular structure than PLA due to its longer hydrocarbon repeat units. As a result, PBS has similar material properties to polyolefins, such as reduced glass transition temperature and increased elongation at break (>500%) (Rosenboom et al., 2022). Although the feedstocks used to make PBS are often non-renewable, but the monomers of PBS, that is, butanediol and succinic acid, can be derived from sources which are environmentally friendly and renewable in nature. The generation of succinic acid from lignocellulosic sugars is being researched, and hydrocracking of starches and sugars yields butanediol (Dechent et al., 2020; Xu & Guo, 2010).

In the coming years, it is anticipated that the commercial market for PHAs, a novel class of biodegradable aliphatic polyesters, will increase to more than 100,000 tonnes annually (Tullo, 2021). PHAs may be produced by a variety of bacteria, including *Pseudomonas* and *Ralstonia*strains, as well as algae, rather than chemical synthesis (Mendhulkar & Shetye, 2017; Mohanrasu et al., 2018, 2020, 2021). PHAs are stored intracellularly by these microbes at up to 80% of their cell capacity (Künkel et al., 2016). Various carbon-rich feedstocks, such as affordable food scraps and liquefied plastic wastes, can be employed for culture, demonstrating the biological PHA production process' use in permitting circularity (Medeiros Garcia Alcântara et al., 2020; Nikodinovic-Runic et al., 2013). Most PHAs decay faster than PLA, making them appealing for applications that need biodegradation.

11.5.2 Biobased Aromatic Polyesters (Durable)

Polyethylene furanoate (PEF), a high-performance plastic similar to PET, is expected to enter the market in the coming years. The slightly altered semi-aromatic structure of PEF results in a higher glass transition temperature, gas diffusion barrier, and tensile strength, which could be beneficial for long- term storage packing (Loos et al., 2020). PEF, on the other hand, is more thermally sensitive and thus requires more care during processing (Burgess et al., 2014; Rosenboom et al., 2018). Under particular industrial composting settings (within nine months), PEF biodegrades faster than PET but is otherwise regarded as an equally durable polymer with less biodegradation in the environment (Loos et al., 2020). PEF is made by polycondensing the bio-derived monomers monoethylene glycol and 2,5-furandicarboxylic acid in the same way that PET is made (Rosenboom et al., 2018).

Alternately, cyclic PEF oligomers can be converted into PEF using ring-opening polymerisation, which can speed up reaction times and enhance molecular weight control (Carlos Morales-Huerta et al., 2016; Fleckenstein et al., 2018; Rosenboom et al., 2018). To optimise the glass transition temperature, mechanical strength, and degradability of PEF (Righetti et al., 2020; Terzopoulou et al., 2017), various molecular modifications and copolymerisations are being investigated. However, producing the monomer 2,5-furandicarboxylic acid at a low cost remains a challenge (Sajid et al., 2018).

PET with a bio-derived drop-in variation is known as bio-PET. Because of its similar qualities, it may be used directly in the textile (two-thirds) and beverage (one-third) markets (Pudack et al., 2020), as well as in PET recycling processes. PET can be manufactured by esterification of terephthalic acid with ethylene glycol. The terephthalic acid can be generated by microbes from biomass via. intermediates like para-xylene and 2,5-furandicarboxylic acid (Sajid et al., 2018).

11.5.3 Biobased Polyolefins

Polyolefins, such as PE and PP, account for more than half of all worldwide plastics production and more than 90% of all packaging materials (Hees et al., 2019). Because of their superior chemical stability and tailorable mechanical qualities, they are widely used. Chemically, biobased PE is similar to PE, thereby making it useable and recoverable on the current technologies as well as future recycling methods like thermolysis (Hees et al., 2019). Ethylene can be produced by number of methods such as conversion of methanol to olefins, ethanol dehydration from sugarcane and biomass steam cracking (Harmsen et al., 2014; Wang et al., 2020).

11.5.4 Fossil-Based Biodegradable Polymers

PVA is a widely used water-soluble polymer(1.2 Mt per year). Ethylene, used to make PVA, is commonly extracted from fossil energy but can also be derived from bioethanol. PVA is a single vinyl polymer that degrades quickly by the process of hydroxyl groups being converted to diketones, which are subsequently hydrolysed and cleaved (Ben Halima, 2016; Kawai & Hu, 2009; Matsumura et al., 1993).

PBAT is a biodegradable aromatic, aliphatic copolyester supplied by BASF (BadischeAnilin & Soda Fabric AG) Company (Germany) as Ecoflex and by many Asian vendors under various brand names. It is utilised in agricultural mulch films, which can take up to nine months to breakdown in the soil (Künkel et al., 2016; Zumstein et al., 2018).

Several biomedical applications have been explored for degradable fossil-derived polymers. Polycaprolactone is a biodegradable and biocompatible polymer commonly utilised in sutures and implantable pharmaceutical delivery systems (Kawai & Hu, 2009; Matsumura et al., 1993). In humans, polycaprolactone hydrolyses non-enzymatically in years and is biodegraded in seawater by fungi and bacteria in weeks (Labet & Thielemans, 2009; Woodruff & Hutmacher, 2010).

The simplest aliphatic ester is polyglycolic acid. The excellent gas barrier and fast industrial and marine breakdown rates make it a promising candidate for plastic packaging. Polyglycolic acid's manufacturing volumes are insignificant from a commodities standpoint, despite its large economic market dominance in the biomedical sector (Jem & Tan, 2020; Lamberti et al., 2020).

11.5.5 Other Non-synthetic Bioplastics

Polymers can be extracted directly from biomass, which is a very simple and cost-effective approach. Starches, which make up a significant amount of food waste, are the primary component of non-synthetic starch-based bioplastics, which are made by converting starch into films (Zhang et al., 2014). In second-generation biorefineries, lignin extracted from biowastes is largely (98%) burnt for power generation. The complicated phenolic structure of lignin, on the other hand, has generated interest in using it as monomers, polymer grafts, or polymer additives for specific polymers (Schutyser et al., 2018; Wang et al., 2020).

The most common natural polymer is cellulose which is derived from plant biomass or particular cellulose-producing bacteria. It is employed in food packing constituents or as a nano-filler additive with additional bioplastics to increase barrier characteristics and tensile strength (Azeredo et al., 2019; Vilarinho et al., 2018). Despite the fact that cellulose is eco-friendly, about 60% of sea-bed microplastics account for regenerated cellulose, that is used to make "viscose" or "rayon" textile fibres (Henry et al., 2019). Cigarette filters, made from a similar substance called cellulose acetate, are a major source of litter. Because of its acetylation, which makes the material hydrophobic, cellulose acetate degrades very slowly (Robertson et al., 2012).

11.5.6 Biobased Polyurethanes

Polyurethanes (PU) are used commercially on a scale of more than 18 million tonnes, primarily in the form of flexible and rigid foams. Major health concerns have emerged due to the usage of poisonous phosgene and cancer-causing isocyanate monomers during the production of conventional polyurethanes (Cornille et al., 2017). Non-isocyanate bioPUs can be produced instead using cyclic carbonates and diamines from vegetable oils (Blattmann et al., 2016; Harmsen et al., 2014; Rokicki et al., 2015). The cycloaddition of epoxides with CO_2 can also form cyclic carbonates.

11.6 LIFE CYCLE ASSESSMENT OF BIOPLASTICS

The major method through which scientists and policymakers can analyse the merits and disadvantages of adopting bioplastics instead of traditional plastics is life cycle assessment (LCA). LCA is a systematic technique for analysing the environmental and socio-economic effects of the production and utilisation of a specific good (Di Bartolo et al., 2021). The general standards of LCA are ISO 14040 and ISO 14044 (Rosenboom et al., 2022). At the worldwide level, there are various LCA standards in use, as well as guidelines that are applicable in the European Union. The total LCA of bioplastics must examine a number of impact areas, including environmental, social, and economic impacts, as well as several options for end-of-life possibilities of the product. The cradle-to-gate and cradle-to-grave assessments are the two basic approaches to a product's LCA, depending on the system boundaries (Finkbeiner, 2013). Cradle-to-gate refers to a review that takes place from the beginning of the resource extraction process (cradle) to the end of production (factory gate). In the case of biobased products, this includes crop cultivation and biomass pre-processing, as well as any transportation involved (Di Bartolo et al., 2021).

The cradle-to-grave assessment considers the product's whole life cycle, from raw material extraction to end-of-life management. This comprises all parts of the cradle-to-gate evaluation, as well as the sale, storage, and usage by consumers of the product, as well as its disposal (Di Bartolo et al., 2021).

Environmental influence categories are often considered in LCA studies, but social and economic factors are also important. The social life cycle assessment examines how the extraction or production of raw materials, as well as the manufacturing, distribution, use, and disposal of items, might have negative societal consequences (Spierling et al., 2018). When used in conjunction with LCA, life cycle costing (LCC) is referred to as environmental LCC (E-LCC) (Spierling et al., 2018). E-LCC encompasses all expenses associated with a life cycle of the product, regardless of who is responsible for them. Environmental cost elements are taken into account, like ecological taxes and emissions control expenditures. The LCA of bioplastics is essential to realise the principle of circular bioeconomy in relation to bioplastics and their long-term durability. Bioplastics are made from low-cost or residual carbon sources, but economic studies are necessary to confirm the feasibility of bioplastics production procedures (Talan et al., 2020).

11.7 END-OF-LIFE TREATMENT OPTIONS OF BIOPLASTICS AND BIOBASED POLYMERS

Plastic leakage into the environment is a major problem caused by poor End-of-life management (Geyer et al., 2017; Jambeck et al., 2015). Bioplastic recycling is often recognised as the most environmentally beneficial end-of-life alternative, and it is preferred over ordinary composting. Bioplastic recycling streams, on the other hand, are less well-developed than those for conventional plastics (Hottle et al., 2017; Morão & de Bie, 2019). The presence of additives in practically every finished plastic product complicates plastic and bioplastic recycling (Geyer et al., 2017). This section examines the end-of-life alternatives for bioplastics, taking into account the existing and future recycling options.

11.8 MECHANICAL RECYCLING

The easiest, most affordable, and most popular type of recycling is mechanical recycling (Hong & Chen, 2017). It entails separating plastic waste from polymeric material, eliminating tags, rinsing, manual tearing, melting, and rebuilding into new shapes. Re-extrusion has been done in the literature; however, mechanical recycling of bioplastics is not yet commercially viable. PLA and PHA are mechanically recycled, which results in quality degradation like tensile strength and molecular weight loss (Lamberti et al., 2020; Vu et al., 2020). Due to mechanical recycling inability to efficiently remove impurities and additives from polymer waste, as well as the inherent thermal and mechanical stress, the products are typically "downcycled" into lower-quality commodities. Additional challenges include coloured or low-density substances (films, foams), as well as therapeutic pollutants, which can make products non-recyclable (Bucknall, 2020; Hong & Chen, 2017). As a result, food-grade recycled plastic is difficult to achieve (Briassoulis et al., 2013).

To improve the quality of reused polymers, virgin polymers are frequently combined with recyclates (Chanda & Roy, 2006; Hong & Chen, 2017). Mechanical recycling, on the other hand, is frequently considered as the most suitable end-of-life alternative due to its deviation from virgin resources. Mechanically recycled plastic often has a lesser impact on the environment than virgin plastic. For example, recycled PET (rPET) has a twofold lower environmental effect than fresh PET (GHG emissions from transportation and process energy usage), while recycled PE and PP (rPE and rPP, respectively) have a threefold lower environmental impact than virgin materials (Gu et al., 2017). However, there is relatively little recycling for this type of recycling: Around 10% of high-density PE and PET are recycled globally, compared to closer to zero for PP and polystyrene. Recyclable textiles and fibre products are also uncommon (Geyer et al., 2017). PET from soft drink bottles is the most widely processed recycled plastic. Its performance as a polycondensation polymer can be enhanced through solid-state post-polymerisation (essentially, eliminating volatile by-products of polymerisation by warming recycled particles under pressure), which raises the molecular weight of recyclates for industrial applications.

11.9 CHEMICAL RECYCLING

Chemical recycling, as opposed to mechanical recycling, can produce high-quality polymers from waste, a process known as "upcycling." Depolymerisation of plastic products yields their monomeric components, which can subsequently be repolymerised using methods that allow for regulated polymerisation to yield preferred polymers (like those with regulated molecular weight). Low-molecular-weight fibre polyesters, for example, can be easily broken down into monomers and subsequently polymerised into the longer-chain polyesters needed for bottles (Park & Kim, 2014; Pudack et al., 2020). Impurities and colour can be eliminated as well. Solvolysis and thermolysis are the most common methods for chemical recycling.

Solvolysis is a solvent-based depolymerisation process in which polymers containing cleavable groups throughout their backbone, like ester linkages in PET, PEF, and PLA, are exposed to hydrolysis, glycolysis, or methanolysis (Demarteau et al., 2020; Hong & Chen, 2017; Pudack et al., 2020). Aromatic polyesters are more hydrolysable than aliphatic polyesters like PLA, PBS, or PHAs. PLA can be hydrolysed to 95% lactic acid production devoid of a catalyst around 160°C–180°C for two hours with a four-times lower energy requirement (Lamberti et al., 2020). It can be depolymerised to 90% cyclic lactide monomers after six hours using Zn transesterification catalysts (Alberti & Enthaler, 2020). The resultant monomers can be used to make high-quality polymers as a feedstock. Chemical recycling, on the other hand, is more costly and less economically viable than mechanical recycling due to the requirement for chemicals and more complicated separation machines. Chemical recycling makes up less than 1% of all recyclable plastic. Several significant chemical companies are working on ways to make chemcycled polymers globally viable with virgin polymers. In plastic applications, chemically recoverable polymers can be configured and used to address enduring end-of-life issues and support a circular materials economy (Coates & Getzler, 2020; Hong & Chen, 2017). With this method, monomers for high-quality condensation polymers, like polyesters and polyamides, can be produced.

Thermolysis is the process of pyrolyzing polyolefins that do not have hydrolysable functional groups at temperatures ranging from 200°C to 800°C (based on the catalyst utilised and also, polymer used) in lack of oxygen. In these circumstances, the C–C linkages disintegrate, converting the polymer either back into hydrocarbon oil or gas as feedstock or straight into olefin monomers.Conventional refineries and polymerisation companies can then use this feedstock (Anuar Sharuddin et al., 2016; Hees et al., 2019; Zhao et al., 2020).

11.10 COMPOSTING AND BIODEGRADATION

Composting and biodegradation is referred as the digestion of compounds by different microorganisms and physiological transition of polymeric substances into water, carbon dioxide, and other inorganic compounds by many recognised species (Künkel et al., 2016). Physical techniques are frequently utilised to support this process, particularly those that help with particle size reduction and fragmentation. Micronization or extrusion, for example, can amorphise crystalline structures into semi-crystalline plastics, making them more prone to enzymatic degradation (Austin et al., 2018). Microbial enzymes, acids, or bases can accelerate hydrolysis, which breaks vulnerable linkages in permeable amorphous portions of a polymer, usually aliphatic esters. UV radiation photodegrades tertiary and aromatic C–C bonds, often resulting in brittle and discoloured material. Metallic catalysts embedded in the polymer can improve this reaction (Sivan, 2011). Metals can also cause oxo-degradation (decomposition by oxidation); however, this can result in fragmentation into microplastics and inadequate digestion. As a result, oxo-degradation has been prohibited in the European Union and Switzerland.

Because the pace of biodegradation is strongly influenced by the elemental composition of a polymer, stabilising additives, environmental factors (such as the availability of H_2O and O_2), and any microorganisms employed, biodegradation is not an easy process (Lambert & Wagner, 2017). In industrial composting facilities, home compost, and even in open water, these parameters are frequently not met. Biodegradable plastics, like PLA shopping bags and cutlery, are frequently rejected by composters because the required decomposition timeframes are longer than the average composting duration of six to eight weeks (X. Chen & Yan, 2020).

11.11 INCINERATION

In the United States, 20% of end-of-life plastic waste is burnt; in Europe, 40% of end-of-life plastic garbage is incinerated. Emission of carbon dioxide are total zero when solely carbon/hydrogen/oxygen-containing renewable substance is burnt, while a portion of the generated thermal energy can be reclaimed for energy generation. Nevertheless, the burning of N-, S-, and Cl-containing polymers, on the other hand, creates poisonous NOx, SOx, and HCl. Similarly, additives in polymers can produce a variety of harmful chemicals when they are burnt, necessitating potentially costly capture and treatment (Chanda & Roy, 2006; Harding et al., 2007). Moreover, there are issues about a "locking-in" effect, in which high incineration facility investment costs and the necessity for ongoing garbage intake may hinder the development of recycling technologies.

11.12 BIOLOGICAL RECYCLING

Alike chemical recycling and as an alternative to complete biodegradation, microbes and their hydrolytic enzymes can be utilised to break down condensation polymers into smaller sub-units, that is, monomers (Jarerat et al., 2006). These biological mechanisms are still being researched, but they have the potential to be greener than the synthetic approach. Aromatic polyesters are often resistant to enzymatic hydrolysis, whereas aliphatic esters are easily hydrolysed. At ambient temperatures, however, *Ideonellasakaiensis 201-F6*, a bacterium identified in a Japanese recycling facility, can depolymerise PET in 40 days (Yoshida et al., 2016). The PETase enzyme, interestingly, is only effective in the degradation of aromatic polyesters and unsuccessful in the breakdown of aliphatic polyesters (Austin et al., 2018). To improve substrate specificity and thermal stability, leaf compost cutinase can be genetically engineered. At temperatures around the glass transition of PET (75°C), the improved enzyme can depolymerise 90% of micronised, amorphous PET into monomers in ten hours (Tournier et al., 2020). The amorphous chain mobility increases at this temperature, making it more susceptible to microbial destruction. The terephthalic acid monomer obtained can be utilised to make bottle-grade PET (Wei et al., 2019). PEF has also been depolymerised using this approach (Pellis et al., 2016; Weinberger et al., 2017).

Polyurethanes are substantially less biodegradable than polyesters due to the strength of the urethane linkages. Fungi and soil bacteria, on the other hand, can assist hydrolyse the ester groups in polyester-containing polyurethane (Espinosa et al., 2020; Russell et al., 2011). Improved understanding of enzyme activity and gene editing to improve microorganism selectivity could actually boost polyurethane biorecycling.

Polyolefin materials are very hard to degrade biologically because they lack breakable functional groups throughout their backbones. They are very hydrophobic in nature, have a high molecular weight, and contain stabilising additives (C. C. Chen et al., 2020; Gewert et al., 2015). Some organisms are thought to metabolise small fragments of less than 5,000 Da; but the molecular weight of most polyolefin polymers exceeds millions of daltons. Waxworm bacteria and *Pseudomonas* strains have been found to partially biodegrade PE films (5%–20%) over a period of one to two months (Gajendiran et al., 2016; Jaiswal et al., 2020; Santo et al., 2013; Yang et al., 2000).

11.13 LANDFILL

Landfills are still the most common method of garbage disposal in many countries: in the United States, 58% of waste is disposed of in landfills (2014), and in Europe, 27.3% of waste is disposed of in landfills (2017) (Geyer et al., 2017). Landfills that are poorly maintained and permeable are a major source of pollution in the environment. Biodegradable polymers also should be restricted out of landfills since they can decompose anaerobically into CH_4, which has a greenhouse gas impact >20 times greater than CO_2. In 2006, it was projected that just 10% of CH_4 produced in landfills was trapped worldwide, which is a strategy that provides the option for energy efficiency while also enhancing the climate and public health (Anenberg et al., 2012; Themelis & Ulloa, 2007). Landfilling taxes, according to the UN, might make recycling more cost-effective.

11.14 ANAEROBIC DIGESTION

In a methanisation "biogas" factory, regulated fermentation (which takes place without oxygen) converts biodegradable polymer waste into methane. Following the capture and burning of the methane, which results in the production of carbon dioxide and water, heat and energy can be retrieved for use. This technique produces energy and generates a total zero-carbon residue for the bioplastic waste (Hobbs et al., 2019; Stagner, 2015). By including components like a "bioreactor landfill," which cycles water to increase microbial activity for methane yield, anaerobic digestion efficacy can be enhanced (Themelis & Ulloa, 2007). At rising temperatures, numerous polymers, including PLA, polycaprolactones, PHAs, and thermoplastic starch, can be digested anaerobically.

11.15 CONCEPT OF BIOPLASTICS AND BIOBASED POLYMER IN A CIRCULAR ECONOMY

Bioplastics, or materials that are biobased, biodegradable, or both biobased and biodegradable, have several advantages: they benefit our economy, society, and environment. They represent an economically creative sector that is developing between 20% and 100% each year, with a current share of roughly 1% of the worldwide plastics market (European Bioplastics, 2016). In Europe, investment in bioplastics production and R&D is contingent on a long-term legislative framework that encourages the use, reuse, and recycling of these materials. The development of a strong and resilient bioplastic requires a favourable legislative framework for the circular and biobased economy throughout the value chain.

Achieving circular economy in case of plastics presents substantial obstacles. This is because our existing techniques for plastic generation, usage and disposal generally fails to meet most of the circular economy's concepts. The prevalence of fossil fuels as polymeric feedstocks, for example, obviously defies a core circular economy premise of solely using renewable resources. Other inconsistencies can be noticed in the present state of plastics after they have been used. Despite recent improvements in recycling processes, the majority of end-of-life plastics are still consigned to landfill or increasingly burnt for energy recovery, both of which contribute to environmental damage in different ways and constitute a significant loss of a valuable resource.

Significant changes to current practices will be required to achieve the circular economy of plastics. This changes include new and sustainable approaches to eco-design, reuse, maintenance and repair, renting and sharing, recycling, and chemical conversion of plastics. (Iaquaniello et al., 2018). In addition to this, necessary social and economic changes are also required to attain circular economy of plastics. Several governments, including the European Union, local governments, and private businesses, have implemented laws to support the circular economy in an attempt to balance environmental concerns with economic growth. If completely implemented, a circular plastics economy would not only keep plastics in use for a wide range of applications, but it would also lessen the environmental harm caused by plastic waste. Current estimates suggest that the circular economy might contribute US$ 1 trillion annually to the global economy, suggesting that the economic benefits could be considerable (Korhonen et al., 2018).

11.16 CONCLUSION AND FUTURE PERSPECTIVES

The utilisation of renewable resources does not guarantee long-term viability. Sustainability is mostly determined by the process of making the material, the place to be used and the recycling process, rather than the substance's constituent parts. Bioplastics, on the other hand, have the potential to transform various plastic-intensive sectors towards a circular economy as technology progresses. Almost every fossil-based application has a biobased equivalent; however, they are typically in limited and expensive quantities, and they do not necessarily have significant environmental benefits.The ability to assess, analyse, and compare the long-term viability and environmental impact of fossil- and biobased materials is critical. Although there are hazards (consumption of fertilisers, societal risks, etc.) associated with the adoption of biobased plastics, it is clear that it provides an alternative to fossil-based production and may thus become a requirement in the future.

To offer polymer building blocks in a cost-effective and sustainable way, biorefinery processes must, however, become more efficient and follow concepts of "green" chemistry, such as employing non-toxic chemicals and conserving energy. A possible method for boosting microorganism productivity in the consumption of biomass, the polymerisation of bioplastics (particularly PHAs), and biological depolymerisation for recycling is gene editing.

LCA is the primary tool; however, to make LCAs increasingly transparent, uniform, and comparative, methodology principles need to be standardised. Existing bioplastic labels must be updated for use on both global and local levels. The labels should be able to convey globally accepted quality standards and yet, still reflect the end-of-life of plastics in the domestic market.

If compostable or biodegradable plastics are not separated from recycling streams, they may cause difficulties. Nonetheless, applications such as compost bags have advantages because the bag and contents can be co-digested, avoiding the need for separation and generating energy and compost. During the "usage" phase, materials should be created to ensure rapid degradation without producing technological concerns while maintaining mechanical qualities. Another claimed benefit is the use of biodegradable plastics to reduce pollution caused by leaks in open environments; however, this appears to be unduly hopeful at the moment. The size and density of the plastic waste, as well as agglomeration with other materials, might influence the outcome in natural ecosystems, which differ dramatically between geographic regions and seasons. Furthermore, employing biodegradable alternatives does not eliminate the economic loss caused by plastic trash. In this regard, stronger laws, the promotion of an environmentally friendly culture, and early investment in sustainability-focused education could prove to be a more efficient use of resources.

ACKNOWLEDGEMENTS

Authors acknowledge the INSPIRE Fellowship received by the Research Student under the Innovation in Science Pursuit for Inspired Research (INSPIRE) scheme (DST/INSPIRE Fellowship/ [IF190230]; dated: November 21, 2020); Technology Mission Division (Energy, Water & all Others), Department of Science & Technology, Ministry of Science & Technology, Government of India (Reference Number: DST/TMD/IC-MAP/2K20/02 Project Title: DST-IIT Hyderabad Integrated Clean Energy Material Acceleration Platform on Bioenergy and Hydrogen); Department of Science and Technology-Science and Engineering Research Board (DST-SERB-No. SB/YS/LS-47/2013), India; Department of Science and Technology-Science and Engineering Research Board (DST-SERB-No. SB/YS/LS-47/2013), India; Department of Science and Technology-Promotion of University Research and Scientific Excellence (DST-PURSE) [DST letter No. SR/PURSE phase 2/38 (G); dated: February 21, 2017], India; RUSA – Phase 2.0 grant [Letter No. F.24-51/ 2014-U, Policy (TN Multi-Gen), Department of Education Government of India; dated: October 09, 2018]; Scheme for Promotion of Academic and Research Collaboration (SPARC) (No. SPARC/2018–2019/P485/SL; dated: March 15, 2019) and University Science Instrumentation Centre (USIC), Alagappa University, Karaikudi, Tamil Nadu, India.

REFERENCES

Alberti, C., & Enthaler, S. (2020). Depolymerization of end-of-life poly(lactide) to lactide via zinc-catalysis. *ChemistrySelect*, *5*(46), 14759–14763. https://doi.org/10.1002/SLCT.202003979

Andrady, A. L. (2011). Microplastics in the marine environment. *Marine Pollution Bulletin*, *62*(8), 1596–1605. https://doi.org/10.1016/J.MARPOLBUL.2011.05.030

Anenberg, S. C., Schwartz, J., Shindell, D., Amann, M., Faluvegi, G., Klimont, Z., Janssens-Maenhout, G., Pozzoli, L., van Dingenen, R., Vignati, E., Emberson, L., Muller, N. Z., Jason West, J., Williams, M., Demkine, V., Kevin Hicks, W., Kuylenstierna, J., Raes, F., & Ramanathan, V. (2012). Global air quality and health co-benefits of mitigating near-term climate change through methane and black carbon emission controls. *Environmental Health Perspectives*, *120*(6), 831–839. https://doi.org/10.1289/EHP.1104301

Anuar Sharuddin, S. D., Abnisa, F., Wan Daud, W. M. A., & Aroua, M. K. (2016). A review on pyrolysis of plastic wastes. *Energy Conversion and Management*, *115*, 308–326. https://doi.org/10.1016/J.ENCONMAN.2016.02.037

Aslam, M., Kalyar, M. A., & Raza, Z. A. (2018). Polyvinyl alcohol: A review of research status and use of polyvinyl alcohol based nanocomposites. *Polymer Engineering & Science*, *58*(12), 2119–2132. https://doi.org/10.1002/PEN.24855

Austin, H. P., Allen, M. D., Donohoe, B. S., Rorrer, N. A., Kearns, F. L., Silveira, R. L., Pollard, B. C., Dominick, G., Duman, R., Omari, K. El, Mykhaylyk, V., Wagner, A., Michener, W. E., Amore, A., Skaf, M. S., Crowley, M. F., Thorne, A. W., Johnson, C. W., Lee

Woodcock, H., ... Beckham, G. T. (2018). Characterization and engineering of a plastic-degrading aromatic polyesterase. *Proceedings of the National Academy of Sciences of the United States of America*, *115*(19), E4350–E4357. https://doi.org/10.1073/PNAS.1718804115/SUPPL_FILE/PNAS.1718804115.SAPP.PDF

Azeredo, H. M. C., Barud, H., Farinas, C. S., Vasconcellos, V. M., & Claro, A. M. (2019). Bacterial cellulose as a raw material for food and food packaging applications. *Frontiers in Sustainable Food Systems*, *3*, 7. https://doi.org/10.3389/FSUFS.2019.00007/BIBTEX

Ben Halima, N. (2016). Poly(vinyl alcohol): Review of its promising applications and insights into biodegradation. *RSC Advances*, *6*(46), 39823–39832. https://doi.org/10.1039/C6RA05742J

Blattmann, H., Lauth, M., Mülhaupt, R., Blattmann, H., Mülhaupt, R., & Lauth, M. (2016). Flexible and bio-based nonisocyanate polyurethane (NIPU) foams. *Macromolecular Materials and Engineering*, *301*(8), 944–952. https://doi.org/10.1002/MAME.201600141

Boucher, J., & Friot, D. (n.d.). *Primary Microplastics in the Oceans: A Global Evaluation of Sources*. International Union for Conservation of Nature.

Briassoulis, D., Hiskakis, M., & Babou, E. (2013). Technical specifications for mechanical recycling of agricultural plastic waste. *Waste Management*, *33*(6), 1516–1530. https://doi.org/10.1016/J.WASMAN.2013.03.004

Bucknall, D. G. (2020). Plastics as a materials system in a circular economy. *Philosophical Transactions of the Royal Society A*, *378*(2176). https://doi.org/10.1098/RSTA.2019.0268

Burgess, S. K., Leisen, J. E., Kraftschik, B. E., Mubarak, C. R., Kriegel, R. M., & Koros, W. J. (2014). Chain mobility, thermal, and mechanical properties of poly(ethylene furanoate) compared to poly(ethylene terephthalate). *Macromolecules*, *47*(4), 1383–1391. https://doi.org/10.1021/MA5000199/SUPPL_FILE/MA5000199_SI_001.PDF

Carbery, M., O'Connor, W., & Palanisami, T. (2018). Trophic transfer of microplastics and mixed contaminants in the marine food web and implications for human health. *Environment International*, *115*, 400–409. https://doi.org/10.1016/J.ENVINT.2018.03.007

Carlos Morales-Huerta, J., Martínez De Ilarduya, A., & Muñoz-Guerra, S. (2016). Poly(alkylene 2,5-furandicarboxylate)s (PEF and PBF) by ring opening polymerization. *Polymer*, *87*, 148–158. https://doi.org/10.1016/J.POLYMER.2016.02.003

Chafran, L. S., Paiva, M. F., Freitas, J. O. C., Sales, M. J. A., Dias, S. C. L., & Dias, J. A. (2019). Preparation of PLA blends by polycondensation of D,L-lactic acid using supported 12-tungstophosphoric acid as a heterogeneous catalyst. *Heliyon*, *5*(5), e01810. https://doi.org/10.1016/J.HELIYON.2019.E01810

Chanda, M., & Roy, S. K. (2006). *Plastics Technology Handbook*. https://doi.org/10.1201/9781420006360

Chanprateep, S. (2010). Current trends in biodegradable polyhydroxyalkanoates. *Journal of Bioscience and Bioengineering*, *110*(6), 621–632. https://doi.org/10.1016/J.JBIOSC.2010.07.014

Chen, C. C., Dai, L., Ma, L., & Guo, R. T. (2020). Enzymatic degradation of plant biomass and synthetic polymers. *Nature Reviews Chemistry*, *4*(3), 114–126. https://doi.org/10.1038/s41570-020-0163-6

Chen, X., & Yan, N. (2020). A brief overview of renewable plastics. *Materials Today Sustainability*, *7–8*, 100031. https://doi.org/10.1016/J.MTSUST.2019.100031

Chia, W. Y., Ying Tang, D. Y., Khoo, K. S., Kay Lup, A. N., & Chew, K. W. (2020). Nature's fight against plastic pollution: Algae for plastic biodegradation and bioplastics production. *Environmental Science and Ecotechnology*, *4*, 100065. https://doi.org/10.1016/J.ESE.2020.100065

Coates, G. W., & Getzler, Y. D. Y. L. (2020). Chemical recycling to monomer for an ideal, circular polymer economy. *Nature Reviews Materials*, *5*(7), 501–516. https://doi.org/10.1038/s41578-020-0190-4

Cornille, A., Auvergne, R., Figovsky, O., Boutevin, B., & Caillol, S. (2017). A perspective approach to sustainable routes for non-isocyanate polyurethanes. *European Polymer Journal*, *87*, 535–552. https://doi.org/10.1016/J.EURPOLYMJ.2016.11.027

Crump, A., Mullens, C., Bethell, E. J., Cunningham, E. M., & Arnott, G. (2020). Microplastics disrupt hermit crab shell selection. *Biology Letters*, *16*(4). https://doi.org/10.1098/RSBL.2020.0030

Dechent, S. E., Kleij, A. W., & Luinstra, G. A. (2020). Fully bio-derived CO_2 polymers for non-isocyanate based polyurethane synthesis. *Green Chemistry*, *22*(3), 969–978. https://doi.org/10.1039/C9GC03488A

Demarteau, J., Olazabal, I., Jehanno, C., & Sardon, H. (2020). Aminolytic upcycling of poly(ethylene terephthalate) wastes using a thermally-stable organocatalyst. *Polymer Chemistry*, *11*(30), 4875–4882. https://doi.org/10.1039/D0PY00067A

Di Bartolo, A., Infurna, G., & Dintcheva, N. T. (2021). A review of bioplastics and their adoption in the circular economy. *Polymers*, *13*(8), 1229. https://doi.org/10.3390/POLYM13081229

Dusselier, M., Van Wouwe, P., Dewaele, A., Jacobs, P. A., & Sels, B. F. (2015). Shape-selective zeolite catalysis for bioplastics production. *Science*, *349*(6243), 78–80. https://doi.org/10.1126/SCIENCE.AAA7169/SUPPL_FILE/DUSSELIER.SM.PDF

Espinosa, M. J. C., Blanco, A. C., Schmidgall, T., Atanasoff-Kardjalieff, A. K., Kappelmeyer, U., Tischler, D., Pieper, D. H., Heipieper, H. J., & Eberlein, C. (2020). Toward biorecycling: Isolation of a soil bacterium that grows on a polyurethane oligomer and monomer. *Frontiers in Microbiology*, *11*, 404. https://doi.org/10.3389/FMICB.2020.00404/BIBTEX

European Bioplastics (n.d.) *Global bioplastics production capacities continue to grow despite low oil price - European bioplastics e.V.* Retrieved August 15, 2022, from https://www.european-bioplastics.org/market-data-update-2016/

Ferreira, F. V., Cividanes, L. S., Gouveia, R. F., & Lona, L. M. F. (2019). An overview on properties and applications of poly(butylene adipate-co-terephthalate)-PBAT based composites. *Polymer Engineering & Science*, *59*(s2), E7–E15. https://doi.org/10.1002/PEN.24770

Finkbeiner, M. (2013). Product environmental footprint-breakthrough or breakdown for policy implementation of life cycle assessment? *The International Journal of Life Cycle Assessment*, *19*(2), 266–271. https://doi.org/10.1007/S11367-013-0678-X

Fleckenstein, P., Rosenboom, J. G., Storti, G., & Morbidelli, M. (2018). Synthesis of cyclic (ethylene furanoate) oligomers via cyclodepolymerization. *Macromolecular Reaction Engineering*, *12*(4), 1800018. https://doi.org/10.1002/MREN.201800018

Gajendiran, A., Krishnamoorthy, S., & Abraham, J. (2016). Microbial degradation of low-density polyethylene (LDPE) by Aspergillus clavatus strain JASK1 isolated from landfill soil. *3 Biotech*, *6*(1), 1–6. https://doi.org/10.1007/S13205-016-0394-X

Gallo, F., Fossi, C., Weber, R., Santillo, D., Sousa, J., Ingram, I., Nadal, A., & Romano, D. (2018). Marine litter plastics and microplastics and their toxic chemicals components: The need for urgent preventive measures. *Environmental Sciences Europe*, *30*(1), 1–14. https://doi.org/10.1186/S12302-018-0139-Z/FIGURES/1

Garlotta, D. (2001). A literature review of poly(lactic acid). *Journal of Polymers and the Environment*, *9*(2), 63–84. https://doi.org/10.1023/A:1020200822435

Gewert, B., Plassmann, M. M., & Macleod, M. (2015). Pathways for degradation of plastic polymers floating in the marine environment. *Environmental Science: Processes & Impacts*, *17*(9), 1513–1521. https://doi.org/10.1039/C5EM00207A

Geyer, R., Jambeck, J. R., & Law, K. L. (2017). Production, use, and fate of all plastics ever made. *Science Advances*, *3*(7). https://doi.org/10.1126/SCIADV.1700782

Gregory, M. R. (2009). Environmental implications of plastic debris in marine settingsentanglement, ingestion, smothering, hangers-on, hitch-hiking and alien invasions. *Philosophical Transactions of the Royal Society B: Biological Sciences*, *364*(1526), 2013–2025. https://doi.org/10.1098/RSTB.2008.0265

Gu, F., Guo, J., Zhang, W., Summers, P. A., & Hall, P. (2017). From waste plastics to industrial raw materials: A life cycle assessment of mechanical plastic recycling practice based on a real-world case study. *The Science of the Total Environment, 601–602*, 1192–1207. https://doi.org/10.1016/J.SCITOTENV.2017.05.278

Harding, K. G., Dennis, J. S., von Blottnitz, H., & Harrison, S. T. L. (2007). Environmental analysis of plastic production processes: Comparing petroleum-based polypropylene and polyethylene with biologically-based poly-β-hydroxybutyric acid using life cycle analysis. *Journal of Biotechnology, 130*(1), 57–66. https://doi.org/10.1016/J.JBIOTEC.2007.02.012

Harmsen, P. F. H., Hackmann, M. M., & Bos, H. L. (2014). Green building blocks for bio-based plastics. *Biofuels, Bioproducts and Biorefining, 8*(3), 306–324. https://doi.org/10.1002/BBB.1468

Hees, T., Zhong, F., Stürzel, M., & Mülhaupt, R. (2019). Tailoring hydrocarbon polymers and all-hydrocarbon composites for circular economy. *Macromolecular Rapid Communications, 40*(1), 1800608. https://doi.org/10.1002/MARC.201800608

Henry, B., Laitala, K., & Klepp, I. G. (2019). Microfibres from apparel and home textiles: Prospects for including microplastics in environmental sustainability assessment. *Science of The Total Environment, 652*, 483–494. https://doi.org/10.1016/J.SCITOTENV.2018.10.166

Hobbs, S. R., Parameswaran, P., Astmann, B., Devkota, J. P., & Landis, A. E. (2019). Anaerobic codigestion of food waste and polylactic acid: Effect of pretreatment on methane yield and solid reduction. *Advances in Materials Science and Engineering, 2019*. https://doi.org/10.1155/2019/4715904

Hong, M., & Chen, E. Y. X. (2017). Chemically recyclable polymers: A circular economy approach to sustainability. *Green Chemistry, 19*(16), 3692–3706. https://doi.org/10.1039/C7GC01496A

Hottle, T. A., Bilec, M. M., & Landis, A. E. (2017). Biopolymer production and end of life comparisons using life cycle assessment. *Resources, Conservation and Recycling, 122*, 295–306. https://doi.org/10.1016/j.resconrec.2017.03.002

Iaquaniello, G., Centi, G., Salladini, A., Palo, E., & Perathoner, S. (2018). Waste to chemicals for a circular economy. *Chemistry - A European Journal, 24*(46), 11831–11839. https://doi.org/10.1002/CHEM.201802903

Jaiswal, S., Sharma, B., & Shukla, P. (2020). Integrated approaches in microbial degradation of plastics. *Environmental Technology and Innovation, 17*. https://doi.org/10.1016/J.ETI.2019.100567

Jambeck, J. R., Geyer, R., Wilcox, C., Siegler, T. R., Perryman, M., Andrady, A., Narayan, R., & Law, K. L. (2015). Plastic waste inputs from land into the ocean. *Science, 347*(6223), 768–771. https://doi.org/10.1126/SCIENCE.1260352/SUPPL_FILE/JAMBECK.SM.PDF

Jan Kole, P., Löhr, A. J., Van Belleghem, F. G. A. J., & Ragas, A. M. J. (2017). Wear and tear of tyres: A stealthy source of microplastics in the environment. *International Journal of Environmental Research and Public Health, 14*(10). https://doi.org/10.3390/IJERPH14101265

Jarerat, A., Tokiwa, Y., & Tanaka, H. (2006). Production of poly(L-lactide)-degrading enzyme by Amycolatopsis orientalis for biological recycling of poly(L-lactide). *Applied Microbiology and Biotechnology, 72*(4), 726–731. https://doi.org/10.1007/S00253-006-0343-4

Jem, K. J., & Tan, B. (2020). The development and challenges of poly (lactic acid) and poly (glycolic acid). *Advanced Industrial and Engineering Polymer Research, 3*(2), 60–70. https://doi.org/10.1016/J.AIEPR.2020.01.002

Jiang, Y., Marang, L., Tamis, J., van Loosdrecht, M. C. M., Dijkman, H., & Kleerebezem, R. (2012). Waste to resource: Converting paper mill wastewater to bioplastic. *Water Research, 46*(17), 5517–5530. https://doi.org/10.1016/J.WATRES.2012.07.028

Kalia, V. C., Raizada, N., & Sonakya, V. (2000). Bioplastics.

Kane, I. A., Clare, M. A., Miramontes, E., Wogelius, R., Rothwell, J. J., Garreau, P., & Pohl, F. (2020). Seafloor microplastic hotspots controlled by deep-sea circulation. *Science*, *368*(6495), 1140–1145. https://doi.org/10.1126/SCIENCE.ABA5899/SUPPL_FILE/ABA5899_KANE_SM.PDF

Kawai, F., & Hu, X. (2009). Biochemistry of microbial polyvinyl alcohol degradation. *Applied Microbiology and Biotechnology*, *84*(2), 227–237. https://doi.org/10.1007/S00253-009-2113-6

Korhonen, J., Honkasalo, A., & Seppälä, J. (2018). Circular economy: The concept and its limitations. *Ecological Economics*, *143*, 37–46. https://doi.org/10.1016/J.ECOLECON.2017.06.041

Künkel, A., Becker, J., Börger, L., Hamprecht, J., Koltzenburg, S., Loos, R., Schick, M. B., Schlegel, K., Sinkel, C., Skupin, G., & Yamamoto, M. (2016). Polymers, Biodegradable. *Ullmann's Encyclopedia of Industrial Chemistry*, 1–29. https://doi.org/10.1002/14356007.N21_N01.PUB2

Labet, M., & Thielemans, W. (2009). Synthesis of polycaprolactone: A review. *Chemical Society Reviews*, *38*(12), 3484–3504. https://doi.org/10.1039/B820162P

Lambert, S., & Wagner, M. (2017). Environmental performance of bio-based and biodegradable plastics: The road ahead. *Chemical Society Reviews*, *46*(22), 6855–6871. https://doi.org/10.1039/C7CS00149E

Lamberti, F. M., Román-Ramírez, L. A., & Wood, J. (2020). Recycling of bioplastics: Routes and benefits. *Journal of Polymers and the Environment*, *28*(10), 2551–2571. https://doi.org/10.1007/S10924-020-01795-8

Lebreton, L. C. M., Van Der Zwet, J., Damsteeg, J. W., Slat, B., Andrady, A., & Reisser, J. (2017). River plastic emissions to the world's oceans. *Nature Communications*, *8*(1), 1–10. https://doi.org/10.1038/ncomms15611

Leong, H. Y., Chang, C. K., Khoo, K. S., Chew, K. W., Chia, S. R., Lim, J. W., Chang, J. S., & Show, P. L. (2021). Waste biorefinery towards a sustainable circular bioeconomy: A solution to global issues. *Biotechnology for Biofuels*, *14*(1), 1–15. https://doi.org/10.1186/S13068-021-01939-5/TABLES/3

Loos, K., Zhang, R., Pereira, I., Agostinho, B., Hu, H., Maniar, D., Sbirrazzuoli, N., Silvestre, A. J. D., Guigo, N., & Sousa, A. F. (2020). A perspective on PEF synthesis, properties, and end-life. *Frontiers in Chemistry*, *8*, 585. https://doi.org/10.3389/FCHEM.2020.00585/BIBTEX

Madhavan Nampoothiri, K., Nair, N. R., & John, R. P. (2010). An overview of the recent developments in polylactide (PLA) research. *Bioresource Technology*, *101*(22), 8493–8501. https://doi.org/10.1016/J.BIORTECH.2010.05.092

Matsumura, S., Kurita, H., & Shimokobe, H. (1993). Anaerobic biodegradability of polyvinyl alcohol. *Biotechnology Letters*, *15*(7), 749–754. https://doi.org/10.1007/BF01080150

Medeiros Garcia Alcântara, J., Distante, F., Storti, G., Moscatelli, D., Morbidelli, M., & Sponchioni, M. (2020). Current trends in the production of biodegradable bioplastics: The case of polyhydroxyalkanoates. *Biotechnology Advances*, *42*. https://doi.org/10.1016/J.BIOTECHADV.2020.107582

Mendhulkar, V. D., & Shetye, L. A. (2017). Synthesis of biodegradable polymer polyhydroxyalkanoate (PHA) in cyanobacteria synechococcus elongates under mixotrophic nitrogen- and phosphate-mediated stress conditions. *Industrial Biotechnology*, *13*(2), 85–93. https://doi.org/10.1089/IND.2016.0021

Mohanrasu, K., Guru Raj Rao, R., Dinesh, G. H., Zhang, K., Sudhakar, M., Pugazhendhi, A., Jeyakanthan, J., Ponnuchamy, K., Govarthanan, M., & Arun, A. (2021). Production and characterization of biodegradable polyhydroxybutyrate by Micrococcus luteus isolated from marine environment. *International Journal of Biological Macromolecules*, *186*, 125–134. https://doi.org/10.1016/J.IJBIOMAC.2021.07.029

Mohanrasu, K., Premnath, N., Siva Prakash, G., Sudhakar, M., Boobalan, T., & Arun, A. (2018). Exploring multi potential uses of marine bacteria; an integrated approach for PHB production, PAHs and polyethylene biodegradation. *Journal of Photochemistry and Photobiology B: Biology*, *185*, 55–65. https://doi.org/10.1016/J.JPHOTOBIOL.2018.05.014

Mohanrasu, K., Rao, R. G. R., Dinesh, G. H., Zhang, K., Prakash, G. S., Song, D. P., Muniyasamy, S., Pugazhendhi, A., Jeyakanthan, J., & Arun, A. (2020). Optimization of media components and culture conditions for polyhydroxyalkanoates production by Bacillus megaterium. *Fuel*, *271*, 117522. https://doi.org/10.1016/J.FUEL.2020.117522

Morão, A., & de Bie, F. (2019). Life cycle impact assessment of polylactic acid (PLA) produced from sugarcane in Thailand. *Journal of Polymers and the Environment*, *27*(11), 2523–2539. https://doi.org/10.1007/S10924-019-01525-9/TABLES/7

Nikodinovic-Runic, J., Guzik, M., Kenny, S. T., Babu, R., Werker, A., & O'Connor, K. E. (2013). Carbon-rich wastes as feedstocks for biodegradable polymer (polyhydroxyalkanoate) production using bacteria. *Advances in Applied Microbiology*, *84*, 139–200. https://doi.org/10.1016/B978-0-12-407673-0.00004-7

Park, S. H., & Kim, S. H. (2014). Poly (ethylene terephthalate) recycling for high value added textiles. *Fashion and Textiles*, *1*(1), 1–17. https://doi.org/10.1186/S40691-014-0001-X/TABLES/3

Pellis, A., Haernvall, K., Pichler, C. M., Ghazaryan, G., Breinbauer, R., & Guebitz, G. M. (2016). Enzymatic hydrolysis of poly(ethylene furanoate). *Journal of Biotechnology*, *235*, 47–53. https://doi.org/10.1016/J.JBIOTEC.2016.02.006

Penczek, S., Szymanski, R., Duda, A., & Baran, J. (2003). Living polymerization of cyclic esters - a route to (bio)degradable polymers. Influence of chain transfer to polymer on livingness. *Macromolecular Symposia*, *201*(1), 261–270. https://doi.org/10.1002/MASY.200351129

Plastics Europe (n.d.). *Plastics - the facts 2020*. Retrieved August 15, 2022, from https://plasticseurope.org/knowledge-hub/plastics-the-facts-2020/

Pudack, C., Stepanski, M., & Fässler, P. (2020). PET recycling - contributions of crystallization to sustainability. *Chemie Ingenieur Technik*, *92*(4), 452–458. https://doi.org/10.1002/CITE.201900085

Righetti, M. C., Marchese, P., Vannini, M., Celli, A., Lorenzetti, C., Cavallo, D., Ocando, C., Müller, A. J., & Androsch, R. (2020). Polymorphism and multiple melting behavior of bio-based poly(propylene 2,5–furandicarboxylate). *Biomacromolecules*, *21*(7), 2622–2634. https://doi.org/10.1021/ACS.BIOMAC.0C00039/SUPPL_FILE/BM0C00039_SI_001.PDF

Robertson, R. M., Thomas, W. C., Suthar, J. N., & Brown, D. M. (2012). Accelerated degradation of cellulose acetate cigarette filters using controlled-release acid catalysis. *Green Chemistry*, *14*(8), 2266–2272. https://doi.org/10.1039/C2GC16635F

Rochman, C. M., Tahir, A., Williams, S. L., Baxa, D. V., Lam, R., Miller, J. T., Teh, F. C., Werorilangi, S., & Teh, S. J. (2015). Anthropogenic debris in seafood: Plastic debris and fibers from textiles in fish and bivalves sold for human consumption. *Scientific Reports*, *5*(1), 1–10. https://doi.org/10.1038/srep14340

Rokicki, G., Parzuchowski, P. G., & Mazurek, M. (2015). Non-isocyanate polyurethanes: Synthesis, properties, and applications. *Polymers for Advanced Technologies*, *26*(7), 707–761. https://doi.org/10.1002/PAT.3522

Rosenboom, J. G., Hohl, D. K., Fleckenstein, P., Storti, G., & Morbidelli, M. (2018). Bottle-grade polyethylene furanoate from ring-opening polymerisation of cyclic oligomers. *Nature Communications*, *9*(1), 1–7. https://doi.org/10.1038/s41467-018-05147-y

Rosenboom, J. G., Langer, R., & Traverso, G. (2022). Bioplastics for a circular economy. *Nature Reviews Materials*, *7*(2), 117–137. https://doi.org/10.1038/s41578-021-00407-8

Russell, J. R., Huang, J., Anand, P., Kucera, K., Sandoval, A. G., Dantzler, K. W., Hickman, D. S., Jee, J., Kimovec, F. M., Koppstein, D., Marks, D. H., Mittermiller, P. A., Núñez, S. J., Santiago, M., Townes, M. A., Vishnevetsky, M., Williams, N. E., Vargas, M. P. N., Boulanger, L. A., ... Strobel, S. A. (2011). Biodegradation of polyester polyurethane by endophytic fungi. *Applied and Environmental Microbiology*, *77*(17), 6076–6084. https://doi.org/10.1128/AEM.00521-11

Sajid, M., Zhao, X., & Liu, D. (2018). Production of 2,5-furandicarboxylic acid (FDCA) from 5-hydroxymethylfurfural (HMF): Recent progress focusing on the chemical-catalytic routes. *Green Chemistry*, *20*(24), 5427–5453. https://doi.org/10.1039/C8GC02680G

Santo, M., Weitsman, R., & Sivan, A. (2013). The role of the copper-binding enzyme - laccase - in the biodegradation of polyethylene by the actinomycete Rhodococcus ruber. *International Biodeterioration and Biodegradation*, *84*, 204–210. https://doi.org/10.1016/J.IBIOD.2012.03.001

Schmidt, C., Krauth, T., & Wagner, S. (2017). Export of plastic debris by rivers into the sea. *Environmental Science and Technology*, *51*(21), 12246–12253. https://doi.org/10.1021/ACS.EST.7B02368/SUPPL_FILE/ES7B02368_SI_002.ZIP

Schutyser, W., Renders, T., Van Den Bosch, S., Koelewijn, S. F., Beckham, G. T., & Sels, B. F. (2018). Chemicals from lignin: An interplay of lignocellulose fractionation, depolymerisation, and upgrading. *Chemical Society Reviews*, *47*(3), 852–908. https://doi.org/10.1039/C7CS00566K

Siracusa, V., & Blanco, I. (2020). Bio-polyethylene (Bio-PE), bio-polypropylene (Bio-PP) and bio-poly(ethylene terephthalate) (Bio-PET): Recent developments in bio-based polymers analogous to petroleum-derived ones for packaging and engineering applications. *Polymers*, *12*(8), 1641. https://doi.org/10.3390/POLYM12081641

Sivan, A. (2011). New perspectives in plastic biodegradation. *Current Opinion in Biotechnology*, *22*(3), 422–426. https://doi.org/10.1016/J.COPBIO.2011.01.013

Smith, M., Love, D. C., Rochman, C. M., & Neff, R. A. (2018). Microplastics in seafood and the implications for human health. *Current Environmental Health Reports*, *5*(3), 375. https://doi.org/10.1007/S40572-018-0206-Z

Spierling, S., Röttger, C., Venkatachalam, V., Mudersbach, M., Herrmann, C., & Endres, H. J. (2018). Bio-based plastics - a building block for the circular economy? *Procedia CIRP*, *69*, 573–578. https://doi.org/10.1016/J.PROCIR.2017.11.017

Stagner, J. (2015). Methane generation from anaerobic digestion of biodegradable plastics - a review. *International Journal of Environmental Studies*, *73*(3), 462–468. https://doi.org/10.1080/00207233.2015.1108607

Straub, S., Hirsch, P. E., & Burkhardt-Holm, P. (2017). Biodegradable and petroleum-based microplastics do not differ in their ingestion and excretion but in their biological effects in a freshwater invertebrate gammarus fossarum. *International Journal of Environmental Research and Public Health*, *14*(7). https://doi.org/10.3390/IJERPH14070774

Talan, A., Kaur, R., Tyagi, R. D., & Drogui, P. (2020). Bioconversion of oily waste to polyhydroxyalkanoates: Sustainable technology with circular bioeconomy approach and multidimensional impacts. *Bioresource Technology Reports*, *11*. https://doi.org/10.1016/J.BITEB.2020.100496

Talan, A., & Tyagi, R. D. (2020). Education and human resource development for sustainability. *Sustainability*, 413–438. https://doi.org/10.1002/9781119434016.CH20

Terzopoulou, Z., Karakatsianopoulou, E., Kasmi, N., Majdoub, M., Papageorgiou, G. Z., & Bikiaris, D. N. (2017). Effect of catalyst type on recyclability and decomposition mechanism of poly(ethylene furanoate) biobased polyester. *Journal of Analytical and Applied Pyrolysis*, *126*, 357–370. https://doi.org/10.1016/J.JAAP.2017.05.010

Themelis, N. J., & Ulloa, P. A. (2007). Methane generation in landfills. *Renewable Energy*, *32*(7), 1243–1257. https://doi.org/10.1016/J.RENENE.2006.04.020

Tokiwa, Y., Calabia, B. P., Ugwu, C. U., & Aiba, S. (2009). Biodegradability of plastics. *International Journal of Molecular Sciences, 10*(9), 3722–3742. https://doi.org/10.3390/ijms10093722

Tournier, V., Topham, C. M., Gilles, A., David, B., Folgoas, C., Moya-Leclair, E., Kamionka, E., Desrousseaux, M. L., Texier, H., Gavalda, S., Cot, M., Guémard, E., Dalibey, M., Nomme, J., Cioci, G., Barbe, S., Chateau, M., André, I., Duquesne, S., & Marty, A. (2020). An engineered PET depolymerase to break down and recycle plastic bottles. *Nature, 580*(7802), 216–219. https://doi.org/10.1038/s41586-020-2149-4

Tullo, A. (2021). PHA makers move on production plans.

UNEP. (2016). *UNEP-marine litter vital graphics. United Nations environment programme and GRID-Arendal. Nairobi and Arendal.* www.unep.org, www.grida.no., www.unep.org

Vilarinho, F., Sanches Silva, A., Vaz, M. F., & Farinha, J. P. (2018). Nanocellulose in green food packaging. *Critical Reviews in Food Science and Nutrition, 58*(9), 1526–1537. https://doi.org/10.1080/10408398.2016.1270254

Von Moos, N., Burkhardt-Holm, P., & Köhler, A. (2012). Uptake and effects of microplastics on cells and tissue of the blue mussel Mytilus edulis L. after an experimental exposure. *Environmental Science and Technology, 46*(20), 11327–11335. https://doi.org/10.1021/ES302332W/SUPPL_FILE/ES302332W_SI_001.PDF

Vu, D. H., Åkesson, D., Taherzadeh, M. J., & Ferreira, J. A. (2020). Recycling strategies for polyhydroxyalkanoate-based waste materials: An overview. *Bioresource Technology, 298*. https://doi.org/10.1016/J.BIORTECH.2019.122393

Wang, Z., Ganewatta, M. S., & Tang, C. (2020). Sustainable polymers from biomass: Bridging chemistry with materials and processing. *Progress in Polymer Science, 101*, 101197. https://doi.org/10.1016/J.PROGPOLYMSCI.2019.101197

Wei, R., Breite, D., Song, C., Gräsing, D., Ploss, T., Hille, P., Schwerdtfeger, R., Matysik, J., Schulze, A., Zimmermann, W., Wei, R., Hille, P., Zimmermann, W., Breite, D., Schulze, A., Song, C., Gräsing, D., Matysik, J., Ploss, T., & Schwerdtfeger, R. (2019). Biocatalytic degradation efficiency of postconsumer polyethylene terephthalate packaging determined by their polymer microstructures. *Advanced Science, 6*(14), 1900491. https://doi.org/10.1002/ADVS.201900491

Weinberger, S., Canadell, J., Quartinello, F., Yeniad, B., Arias, A., Pellis, A., & Guebitz, G. M. (2017). Enzymatic degradation of poly(ethylene 2,5-furanoate) powders and amorphous films. *Catalysts, 7*(11), 318. https://doi.org/10.3390/CATAL7110318

Woodruff, M. A., & Hutmacher, D. W. (2010). The return of a forgotten polymer-polycaprolactone in the 21st century. *Progress in Polymer Science, 35*(10), 1217–1256. https://doi.org/10.1016/J.PROGPOLYMSCI.2010.04.002

Xu, J., & Guo, B. H. (2010). Poly(butylene succinate) and its copolymers: Research, development and industrialization. *Biotechnology Journal, 5*(11), 1149–1163. https://doi.org/10.1002/BIOT.201000136

Yadav, B., Pandey, A., Kumar, L. R., & Tyagi, R. D. (2020). Bioconversion of waste (water)/residues to bioplastics- A circular bioeconomy approach. *Bioresource Technology, 298*, 122584. https://doi.org/10.1016/j.biortech.2019.122584

Yang, C., Hua, Q., & Shimizu, K. (2000). Energetics and carbon metabolism during growth of microalgal cells under photoautotrophic, mixotrophic and cyclic light-autotrophic/dark-heterotrophic conditions. *Biochemical Engineering Journal, 6*(2), 87–102. https://doi.org/10.1016/S1369-703X(00)00080-2

Yoshida, S., Hiraga, K., Takehana, T., Taniguchi, I., Yamaji, H., Maeda, Y., Toyohara, K., Miyamoto, K., Kimura, Y., & Oda, K. (2016). A bacterium that degrades and assimilates poly(ethylene terephthalate). *Science, 351*(6278), 1196–1199. https://doi.org/10.1126/SCIENCE.AAD6359/SUPPL_FILE/AAD6359-YOSHIDA-SM.PDF

Zhang, Y., Rempel, C., & Liu, Q. (2014). Thermoplastic starch processing and characteristics-a review. *Critical Reviews in Food Science and Nutrition*, *54*(10), 1353–1370. https://doi.org/10.1080/10408398.2011.636156

Zhao, D., Wang, X., Miller, J. B., & Huber, G. W. (2020). The chemistry and kinetics of polyethylene pyrolysis: A process to produce fuels and chemicals. *ChemSusChem*, *13*(7), 1764–1774. https://doi.org/10.1002/CSSC.201903434

Ziccardi, L. M., Edgington, A., Hentz, K., Kulacki, K. J., & Kane Driscoll, S. (2016). Microplastics as vectors for bioaccumulation of hydrophobic organic chemicals in the marine environment: A state-of-the-science review. *Environmental Toxicology and Chemistry*, *35*(7), 1667–1676. https://doi.org/10.1002/ETC.3461

Zumstein, M. T., Schintlmeister, A., Nelson, T. F., Baumgartner, R., Woebken, D., Wagner, M., Kohler, H. P. E., McNeill, K., & Sander, M. (2018). Biodegradation of synthetic polymers in soils: Tracking carbon into CO_2 and microbial biomass. *Science Advances*, *4*(7). https://doi.org/10.1126/SCIADV.AAS9024

12 Bioplastics and Its Bio-nanocomposites for a Circular Economy

Standards, Labelling and Regulations

Orebotse Joseph Botlhoko, Rakgoshi Lekalakala and Sudhakar Muniyasamy

12.1 INTRODUCTION

The public concern about the presence of waste pollutants in the environment is from plastic materials, which are non-biodegradable, and the very same plastics are used in food packaging, manufacturing, agricultural, building materials, medical and pharmaceutical applications and other areas, which make them important to our lives. Plastics are a need in today's economy, as they are ubiquitous in practically all industrial and domestic sectors (Ncube et al., 2020; Kibria et al., 2023). Plastic products are widely used and are in high demand because of their high functionality, low cost, light weight, ease of production, and good physical and mechanical feature (Ncube et al., 2021). Plastics, on the other hand, pollute or harm the environment because they are not readily biodegradable and resist to microbial decomposition (Vroman & Tighzert, 2009). The global production of plastic was projected to be around 370 million tons in 2019, and 390.7 million tons of plastic in 2021, with only 9% of it being recycled, 12% being incinerated, and the remaining being discharged into the environment or landfills. This waste disposal substance has a negative influence on the environment and distract the food chain, hence poses a risk to animals and human health (Kumar et al., 2021; Al-Khairy et al., 2022). Furthermore, these materials emit obnoxious gases when combusted, thus causing atmospheric pollution. It should not also be forgotten that the disposal of plastics is becoming difficult as a result of diminishing landfill space. Meanwhile, several approaches are proposed to reduce plastic pollution. Where conventional plastics are concerned, the implementation of a circular economy is proposed, whereby the use, reuse, and recycling of plastics is promoted to establish a closed-loop system.

Equally so, looking at the severity of problems caused by conventional polymers and their composites, there is a necessity to discuss an alternative sustainable, eco-friendly plastic for possible applications such as bioplastics in detail (Zuccarello

DOI: 10.1201/9781003304142-12

et al., 2018). Bioplastics are eco-friendly type of biological materials made from renewable biomass and can be biodegradable, making them feasible alternatives for reducing non-biodegradable plastic waste while avoiding the risks posed by petroleum-based polymers made from fossil fuels (Bakar & Othman, 2019). Importantly, the word bioplastic is normally used interchangeably with the word biodegradable. However, the word bioplastics define polymers that fulfil one or both of the following criteria: those that are both biobased and biodegradable, those that are wholly biobased, and those that are wholly biodegradable as shown in Figure 12.1 (Moshood et al., 2022). Biobased polymers are those that are manufactured entirely or partially from biomass, which includes any sort of renewable organic material of biological origin, as well as organic waste (Di Bartolo et al., 2021). While biodegradable polymers are those materials that can degrade into natural components such as carbon dioxide, water, and biomass through the activity of microorganisms (Rahman & Bhoi, 2021). In a more particular sense, biodegradable plastic must meet certain biodegradation requirements in industrial composting facilities and must adhere to strict guidelines, such as specific length of time, temperature, and humidity of particular conditions in the phase propagation approach as illustrated in Figure 12.2 (Moshood et al., 2022). Herein, the test method identifies microbial growth by releasing CO_2. The adaptation of the microbial population to the substrate is known as the lag phase, which is followed by the biodegradation phase. This phase of biodegradation occurs when the microbial population consumes the carbon substrate to sustain its cellular life. Final stage is the plateau phase, when the substrate is entirely consumed by microorganisms (Goel et al., 2021). The biodegradable phases are supported by the European standard EN 134325:2002, and it is explained that 90% of the material must degrade within six months by breaking down into simple molecules such as water, CO_2, and methane (Oberti & Paciello, 2022).

Some of the interesting examples of bioplastics/biopolymers on the market include protein, starch, cellulose, DNA, RNA, lipids, collagen, and carbohydrates. On the other hand, polylactic acid (PLA) and polyhydroxyalkanoates (PHAs) account for

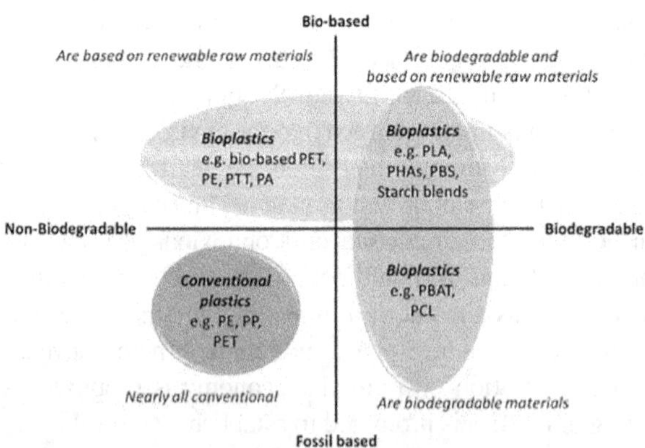

FIGURE 12.1 Classification of bioplastics, both biodegradable and non-biodegradable. Redesigned on the basis of information available in (Moshood et al., 2022).

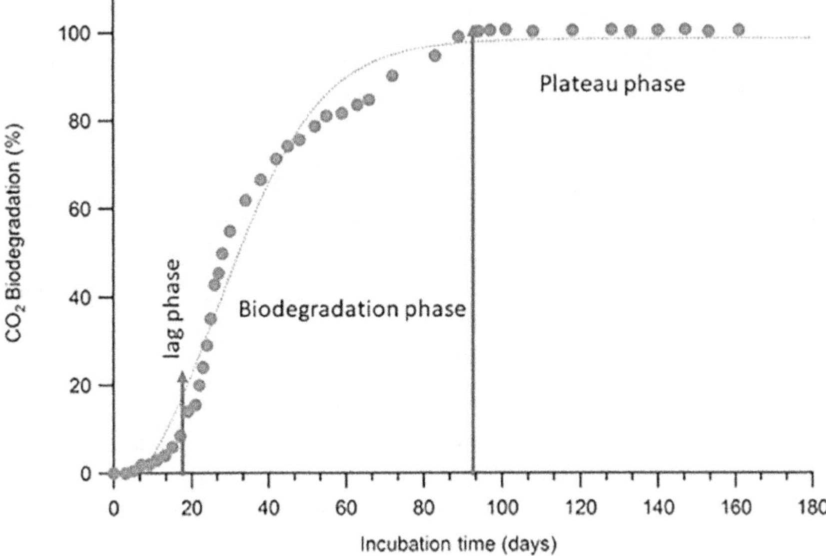

FIGURE 12.2 CO_2 release curve of typical lag phase, biodegradation phase and plateau phase.

over 1 million tons of the global production capacities of bioplastics or biodegradable plastics polymers (Naser et al., 2021). Therefore, PLA and PHAs are the most dominant bioplastics in research and industry-standard materials for different applications subjected to their unique properties. For example, these bioplastic materials are widely used in packaging, including starch which is inexpensive and readily available and is perfect for this task. Moreover, bioplastics are incorporated into cement during concrete mixtures and dry premixed mortars to increase the desired properties or as additive that can optimize the product. They are also used in the construction industry of interior decoration (Oberti & Paciello, 2022). Also, bioplastics play a critical role in medical and agricultural industry. In some instances, bioplastics are combined with nanomaterials to reinforce or enhance their desired properties and practical applications.

More importantly, the use of natural materials, bioplastics and bio-nanocomposites provides not only the solution for waste disposal but also provides reduction of carbon and energy footprint. It is apparent that sustainable low-carbon societies require biomass-derived polymers rather than petroleum-based plastics (Lee & Na, 2013; Atiwesh et al., 2021). Hence, the research & development based on the paradigm shift towards the application of biopolymers, bio-nanocomposites sustainability, and eco-friendly materials is supported by the waste management legislation. The use of biopolymers and their bio-nanocomposites can make a critical contribution to the implementation of circular economy model. In this regard, a safe and clean environment for living beings can be achieved. However, these bioplastics have some drawbacks that limit their use as individual polymers. For example, PLA is a three-carbon membered thermoplastic with hydroxyl and carbonyl at the end, and it is brittle. On the other hand, PLA has low heat distortion temperature and slow crystallization rate, is not readily biodegradable at or below 30°C temperature profiles, and is prone to high degradation rate during processing (Ecker et al., 2019; Wasti & Adhikari, 2020; Farah et al., 2016; Su et al., 2019). It can rapidly deform, especially if under stress. In this context, it is important to resolve the PLA limitation by blending it with a ductility biodegradable polymer or incorporating nanomaterials. Incorporation of nanomaterials into biopolymers is a practical and economical way to develop new materials with unique properties such as biodegradability, thermal, mechanical, barrier, and antimicrobial properties. In addition, natural fibres can also be used to reinforce the biopolymers nanocomposites since they possess many advantages such as renewability, biodegradability, low cost, low density, sustainability, and environmentally benign.

12.2 CIRCULAR ECONOMY AND BIOPLASTICS

Biopolymers and bio-nanocomposite materials play an important role in implementing a functional circular economy. The circular economy principles and technologies emphasizes reduced raw material use, product reuse without losing its performance, waste discharges, and product recycling approach (Oberti & Paciello, 2021; John & Thomas, 2022). Herein, it is where waste materials are converted into a valuable source so they can stay within the loop (John & Thomas, 2022). Recycling can be of three types: mechanical, chemical, and organic (Oberti & Paciello, 2021). It is critical to enact the circular economy approach at every step along the plastic supply chain to minimize environmental impact and plastic waste at large. The robust plastic collection, circular recycling, composting, and improved process energy efficiency in plastic and bioplastic manufacturing combined with the use of renewable power is also critical and ideal for green environment and sustainability (Figure 12.3) (Rosenboom et al., 2022). In contrast to a linear economy based on a manufacture, use, and disposal model, the combined circular economy and biopolymers effect will ensure that materials are retained in the loop. A circular approach to plastics will solve the problem of plastic waste pollution in landfills and oceans, as well as the negative health consequences of microplastics on marine and human life. This can be accomplished

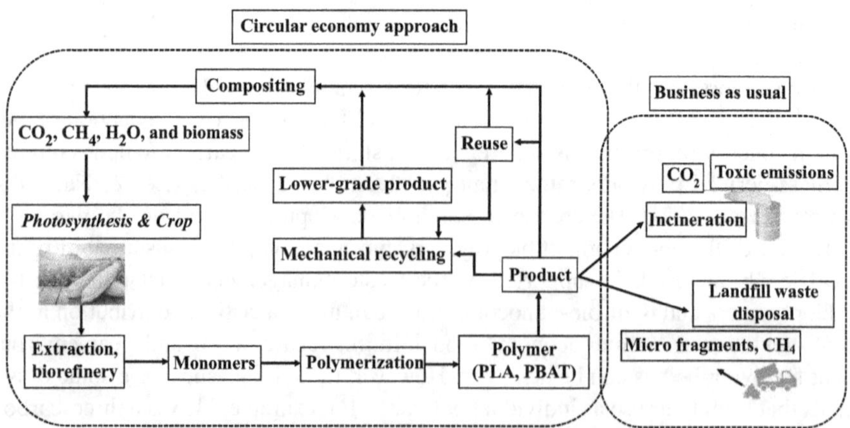

FIGURE 12.3 Ideal circular economy of plastic.

by improving recycling processes and producing biopolymer and bio-nanocomposite materials as an alternative to petroleum-based feedstock. The circular economy and the biopolymer application model are comparable in that they both lower the demand for fossil carbon while increasing the usage of waste and by-products (John & Thomas, 2022).

The industry and consumers can be forced to use the bioplastic through political legislation and regulations as well as the development of functional biodegradable standard test methods; this approach will fully facilitate the implementation of the circular economy model. In this regard, the European Commission issued Directive 904/2019, known as single-use plastic, banning plastic disposable products in all EU countries from January 2022. Equally so, Europeans have developed test standard methods such as EN 134325:2002. This paves the way for the implementation of a circular economy and the use of bioplastics (Oberti & Paciello, 2021).

12.3 BIOPOLYMERS AND NANOMATERIALS: A SHORT DESCRIPTION

12.3.1 Description of Biopolymers

Biopolymers are the organic substances present in natural sources. The term biopolymer originates from the Greek words bio and polymer, representing nature and living organisms like plants and microbes. These are made up of monomeric units bonded together by covalent bonds, which form larger molecules. As per the IUPAC definition, a macromolecule is a single molecule made up of repeating units (Ezeoha & Ezenwanne, 2013; Jenkins et al., 1996). They are a renewable resource, unlike most of the polymers which are petroleum-based polymers. The biopolymers are found to be biocompatible and biodegradable, making them useful in distinct applications, such as edible films, emulsions, manufacturing, packaging materials in the food industry, and medical and pharmaceutical industries (Udayakumar et al., 2021).

The benefit of biopolymers over petroleum-based polymers is mostly due to their cost effectiveness, biocompatible and biodegradable, thus purely eco-friendliness, and user-friendly materials. In addition, biopolymers are abundant materials and have unique properties and structures. Biopolymers have a distinctive helical structure, stiffness, charge-free chains, and strong resistance to salt and cold; as a result, they thicken and stabilize better under harsh pool conditions (Das et al., 2022). Biopolymers are slowly becoming the industry standard for plastic packaging that is incorporated into society and encourages a healthy and sustainable lifestyle. According to Market studies, the biopolymers/biodegradable plastics market is estimated to reach $6.12 billion by 2023 (Baranwal et al., 2022). Biopolymers are being explored for use in new ways, using nanomaterial reinforcements to improve their characteristics and practical applications.

12.3.2 Description of Nanomaterials

Nanomaterials are generally regarded as fairly new materials possessing at least one dimension less than 100 nm (1–100 nm range) (Lin et al., 2014; He et al., 2019). Laterally, nanomaterials are a bridge between atomic or molecular structures and bulk materials. Nanomaterials are known to occur naturally or are produced as the by-products of combustion reactions. However, due to their specific role and dynamic physicochemical properties, they are frequently engineered for various research purpose and industrial applications (Lin et al., 2014; Tiede et al., 2008). The influence of scientific research based on nanomaterials is drawn from their particle size, shape, high surface-to-volume ratio characteristics, physicochemical properties, and reinforcement capability. Properties of nanomaterials are superior or different form their bulk counterparts (Mourdikoudis et al., 2018). Nanomaterials are present in a large variety of industries and consumer products, including building materials, cosmetics products, packaging materials, food additives, agriculture, and electronics, and recently, it was estimated that their market value reached about $3 trillion with 6 million jobs (Vasile, 2018; He et al., 2019). Some examples of nanomaterials on the market include clay, carbon nanorods, carbon nanotubes (CNTs), fullerenes, graphene and its derivatives, quantum dots, titanium oxides, zinc oxides, liposomes, dendrimers, gadolinium nitride nanowires, silver nanoparticle (NPs), gold NPs, platinum NPs, and magnetic NPs (He et al., 2019).

The benefit of nanoparticles over bulk materials is mostly due to their small size. Because of their high porosity and surface-to-volume ratio, they are in high demand for industrial applications. Their high surface-to-volume ratio, for example, is extremely important in the medical area since it influences the likelihood of efficiently fighting various diseases. This effect is ascribed to the nanomaterial-enabled bonding of cells and active substances. Nanomaterials are the most used reinforcement materials for polymeric systems at a very low content. They practically improve the mechanical, thermal, and barrier properties of composites to some degree due to their size, low density and shape characteristics (Bhattacharya, 2016).

12.4 APPLICATIONS OF BIOPOLYMERS AND BIO-NANOCOMPOSITES

Biopolymers and their bio-nanocomposites manufactured from biomass have wider applications in various fields including food packaging, biomedical, automotive, construction, agriculture, water treatment and others (Figure 12.4) (Rhim et al., 2013; Bilal et al., 2020; Dogra et al., 2022). Biobased materials have been widely accepted in many industries due to their eco-friendly, non-toxic, reduced carbon footprints, and renewable resources, providing opportunities for recycling including thermal, mechanical and organic recycling. Apart from that biobased materials provide good physical-chemical properties in certain applications and cost competitiveness as compared to petroleum-based conventional materials. In certain cases, modification of biopolymers can be used to improve certain properties to meet their application specifications. For example, PLA is a very brittle material and improves the ductile properties, starch can be blended with PLA to modify the inherent brittleness behaviours. Similar cases, bio-nanocomposites can be prepared by using ductile polymers such as PBAT and thermoplastic starch (TPS) to improve the physical, mechanical and thermal properties by reinforcement with nanomaterials such as cellulose nanocrystals, cellulose nanofibers, nanoclay and others (Bilal et al., 2020; Anstey et al., 2016; Bastioli, 2020; Bhagwat et al., 2020; Godfrey, 2020).

Application of Biopolymers and Bio-nanocomposites

- Synthesis of Nanomaterials
- Biomedical
- Agriculture
- Food packaging
- Water treatment
- Construction
- Automotive
- Environmental
- Fire retardant
- Smart materials

FIGURE 12.4 Applications of biopolymers and its bio-nanocomposites.

12.5 OPPORTUNITIES AND CHALLENGES

12.5.1 Biodegradable Testing Standards, Labelling, and Regulations

Due to environmental concerns induced by plastic waste disposal and its accumulation, such as leaving a high carbon footprint and contributions towards greenhouse gas emissions. Plastic materials are made from petroleum-based synthetic polymers and are non-biodegradable in landfill, compost, fresh waste and marine environments (Muniyasamy et al., 2013; Muniyasamy & John, 2017). It becomes a serious environmental problem of these disposal plastic products causing terrestrial and aquatic habitats. In environmental consciousness, polymer materials should be aligned with the concept of recirculation in terms of reuse, recycling, and renewability (Figure 12.5).

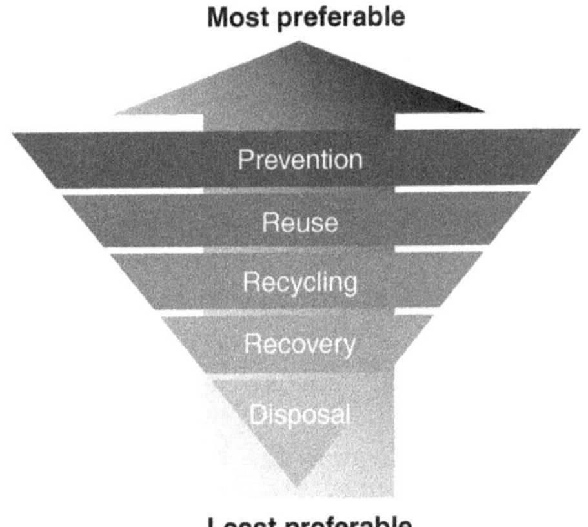

FIGURE 12.5 Plastic waste management hierarchy.

Developing of environmentally friendly plastics has been one major way towards this goal and reason being, to ensure that plastics, in their end-of-life use, do not contribute negatively towards the environment. As biobased plastics and biodegradable plastics are not easily distinguishable to traditional plastics, it is necessary to provide their environmental claims ensuring quality and labelling and this can only be done through standardization and certification systems such as biodegradability evaluation and toxicity (Muniyasamy et al., 2013, 2019).

Biodegradation is a biochemical interaction between a material/product and aerobic microorganisms. A product's ability to biodegrade depends on the amount of carbon available for microbial consumption. It is generally measured by CO_2 and CH_4 production as explained in the above section. Challenges with biodegradation testing include the complexity of the biochemical interaction, the composition of the materials tested, and the specific needs of each biodegradation test. Materials that are made up of components 'known' to biodegrade sometimes do not 'pass', and materials made of inorganic components do not necessarily 'fail' the various assessments of biodegradability so, knowing which biodegradation test method to use for a specific material is key (Narayan, 2006; Nazareth et al., 2019).

Since different polymers degrade in different environments (landfill, compost, fresh water, marine water, aerobic and anaerobic conditions), it is necessary that each industrial bioproduct come with a biodegradation claim coupled to a specific well-defined environment. And it is for this reason that specific biodegradation test in terms of carbon mineralization methods have been designed for each environmental compartment by various International Standards Organization (ISO), European committee on Normalization (CEN) and the American Society of Testing Materials (ASTM) (EN13432, 2000).

To claim biodegradable materials, it must be defined by,

- The disposal system – composting, anaerobic digestor, landfill, soil and marine water.
- Time required for complete microbial utilization in the selected disposal environment defined time frame or less.
- Complete utilization of the substrate carbon by the microorganisms as measured by the evolved CO_2 (aerobic) and $CO_2 + CH_4$ (anaerobic) leaving no residues.
- No negative effects on the disposal system, including the presence of high levels of regulated metals and other harmful components.
- Disintegration, fragmentation, primary degradability, partial biodegradability, or will eventually biodegrade is not an option – serious health and environmental consequences can occur.

Successful biodegradation demonstration can be managed through the selection of the appropriate method and preparation of the samples to fit the requirements of the specific test method (Table 12.1; Figure 12.6).

TABLE 12.1
Major Standards Available for Biodegradation Testing of Polymeric Materials

Standards	Title
OECD 301B	OECD 301B is a frequently requested aerobic biodegradation test for liquids. OECD 301B is a 28-day respirometry test that analyses CO_2 evolution and is suitable for both poorly soluble and absorbing material samples
OECD 302	This test is used for the determination of the biodegradability of a solution typically not readily biodegradable. Often used for materials that are known to be insoluble
OECD 306	OECD 306 is used when a material needs to be specifically shown to biodegrade in seawater. It requires the use of a marine water inoculum. It is often used in conjunction with the OECD 301/302 Sludge Inoculum protocol
ISO 14852	Determination of the ultimate aerobic biodegradability of plastic materials in an aqueous medium – Method by analysis of evolved carbon dioxide
ISO 17088 & 14855:1	Determination of the ultimate aerobic biodegradability and disintegration of plastic materials under controlled composting conditions – Method by analysis of evolved carbon dioxide; Amendment 1: Use of activated vermiculite instead of mature compost
ISO 17088 & 14855:2	Determination of the ultimate aerobic biodegradability and disintegration of plastic materials under controlled composting conditions – Part 2: Gravimetric measurement of carbon dioxide evolved in a laboratory-scale test
ISO 14593	ISO 14593 evaluates the ultimate aerobic biodegradability of organic compounds in an aqueous medium. ISO 14593 is a minimum 28-day biodegradation test
ISO 17556	Plastics – Determination of the ultimate aerobic biodegradability in soil by measuring the oxygen demand in a respirometer or the amount of carbon dioxide evolved
ISO 20200	Plastics – Determination of the degree of disintegration of plastic materials under simulated composting conditions in a laboratory-scale test
EN 13432	Packaging – Evaluation of the ultimate aerobic biodegradability of packaging materials under controlled composting conditions – method by analysis of released carbon dioxide
EN 14047	Packaging – Determination of the ultimate aerobic biodegradability of packaging materials in an aqueous medium – Method by analysis of evolved carbon dioxide
EN 17033	Biodegradable mulch films for use in agriculture and horticulture. Requirements and test methods
prCEN/TR 15822	Plastics – Biodegradable plastics in or on soil – Recovery, disposal and related environmental issues
ASTM D5209	Standard test method for determining the aerobic biodegradation of plastic materials in the presence of municipal sewage sludge
ASTM D5272	Standard practice for outdoor exposure testing of photodegradable plastics
ASTM D5510	Standard practice for heat ageing of oxidatively degradable plastics
ASTM D5511	Standard test method for determining anaerobic biodegradation of plastic materials under high-solids anaerobic-digestion conditions
ASTM D5526	Anaerobic biodegradation test of plastic materials under accelerated landfill conditions
ASTM D5864	Determining aerobic aquatic biodegradation of lubricants or their components

(*Continued*)

TABLE 12.1 (*Continued*)
Major Standards Available for Biodegradation Testing of Polymeric Materials

Standards	Title
ASTM D5988	Standard test method for determining aerobic biodegradation in soil of plastic materials or residual plastic materials after composting
ASTM D6400	Standard specification for labelling of plastics designed to be aerobically composted in municipal or industrial facilities
ASTM D5338	Standard test method for determining the degree and rate of aerobic biodegradation of plastic materials in a controlled composting environment
ASTM D6954	Standard guide for exposing and testing plastics that degrade in the environment by a combination of oxidation and biodegradation
PAS 9017-2018	Biodegradation of polyolefins in an open-air terrestrial environment – Specification
ASTM D6691	Standard test method for determining aerobic biodegradation of plastic materials in the marine environment by a defined microbial consortium
NF T51–800	Plastics — Specifications for plastics suitable for home composting
AS 5810	Biodegradable plastics suitable for home composting

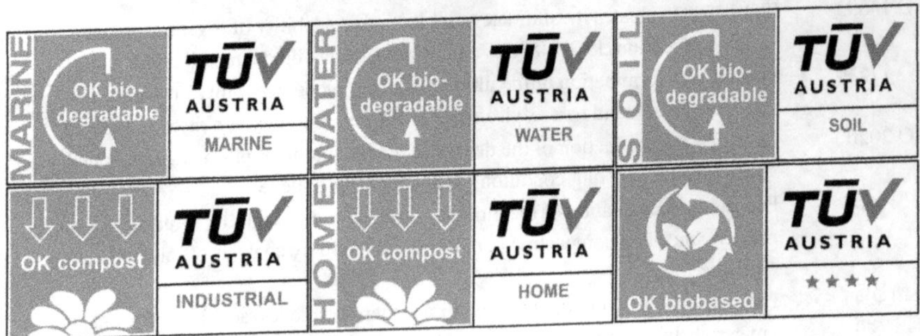

FIGURE 12.6 Major international biodegradation labelling and certification used for biodegradable materials and its products.

12.6 COMPOST BIODEGRADATION

ASTM D6400 standard is equivalent of the ISO 14855 and EN 13432, for biodegradation testing under aerobic composting in municipal and industrial facilities, which is recognized by many municipalities, regulatory agencies, for making biodegradability claims about a product or material (EN13432, 2000).

This standard specifies procedures and requirements for materials and products (finished products) that are suitable for composting (home or industrial). Materials and products are considered as suitable for home composting (at 23°C ± 2°C) and industrial composting (at 58°C ± 2°C) as a complete product including all the individual components meet the requirements. The following aspects need meet the below aspects to claim as compostable plastics and its products.

i. Mineralization – CO_2 emission during composting.
ii. Disintegration during composting.
iii. Ecotoxicity test on the quality of the resulting compost, including the presence of high levels of regulated metals and other harmful components.

12.6.1 Soil Biodegradation

Currently, the main international standards are ASTM D5988 and ISO 17556 for determining aerobic biodegradation of plastic materials in or on soil and deals with the extent of biodegradation over a time (period) by the microorganisms present in the soil-mineralized plastics. This standard was recently superseded by the new standard specification EN17033 for biodegradable mulch films for use in agriculture and horticulture, which specifies the necessary requirements and test methods.

EN 17033 specifies test methods and evaluation criteria regarding the biodegradation, ecotoxicity, film properties, and constituents of the biodegradable mulch films. To use the pre-existing and well-established certification "OK Biodegradable soil", the test materials and products require 90% CO_2 conversion within 24 months in a soil biodegradation test. Additionally, the standard includes a new, more comprehensive ecotoxicity testing and evaluation scheme taking into account relevant terrestrial organism groups such as plants, invertebrates (e.g. earthworm), and microorganisms (e.g. nitrification inhibition test); important ecological processes that are critical due to their role in maintaining soil functions by breaking down organic matter and formulating soil structure and ecologically recycling of materials; and relevant exposure pathways of degradation products such as soil pore water, soil pore air and soil material. Moreover, the standard strictly defines restrictions regarding different potentially harmful constituents, such as regulated metals and substances of very high concern.

12.6.2 Aqueous Biodegradation Test

OECD 301B and OCED 306 are the most used readily and inherent biodegradability methods in a liquid environment (fresh water and marine water). The OECD 301B and OCED 306 are an aerobic biodegradation test that introduces a material to an inoculum in a closed environment and measures biodegradation of the material by CO_2 evolution. The OECD 301B test method can be used for highly soluble, poorly soluble, and even for materials with certain concentrations known to be insoluble. Common materials tested with OECD 301B include fuels, lubricants, oil, surfactants, and personal care products. Formulations and other solutions can also be tested with the 301B.

12.6.3 Anaerobic Biodegradation Test

ASTM D5511 is the most used standard test method for determining anaerobic biodegradation of plastic materials under high-solids anaerobic-digestion conditions. The anaerobic process operated temperature at 50°C–52°C conditions that yield conversion of carbon in the sample to gaseous (CO_2/CH_4) under these conditions.

According to this standard method, we monitor the biodegradation in three steps,

i. Disintegration/fragmentation of the test sample over time (primary degradation).
ii. % Biodegradation based on CO_2/CH_4 gases evolved.
iii. Ecotoxicity after biodegradation.

12.7 CONCLUSION

The use of biopolymers and bio-nanocomposites has opportunities to address the environmental footprint and addressing the post-consumer plastic waste pollution issues. The trend of plastic production, consumption and mounting of environmental concern caused by plastic waste, are in favour of an increasing acceptance and diffusion of environmentally friendly polymeric materials. The improved utilization of agricultural undervalued by-products in biomaterial formulations by green technology research (renewable, recyclable and reusable) has opportunities to create new biomaterials for uses in packaging, agriculture, biomedical, and other sectors. Renewable resources can be used as cost-effective feedstocks for the production of environmentally friendly polymers and have the potential to replace conventional commodity plastics in those segments where recycling or feedstock recovery is difficult and heavily penalized from an economic standpoint. Biodegradation end-of-life options of environmentally friendly polymers are necessary to be followed and validated as per international standards, regulations, and certification. It is worth to noting that biobased polymer and bio-nanocomposite materials from biomass have a potential role to play in the development of the bioeconomy due to their potential to address environmental concerns regarding plastic waste and economy challenges.

REFERENCES

Al-Khairy, D., Fu, W., Alzahmi, A. S., Twizere, J. C., Amin, S. A., Salehi-Ashtiani, K., & Mystikou, A. (2022). Closing the gap between bio-based and petroleum-based plastic through bioengineering. *Microorganisms*, *10*(12), 2320.

Anstey, A., Muniyasamy, S., Reddy, M. M., Misra, M., & Mohanty, A. (2014). Processability and biodegradability evaluation of composites from poly (butylene succinate) (PBS) bioplastic and biofuel co-products from Ontario. *Journal of Polymers and the Environment*, *22*, 209–218.

Atiwesh, G., Mikhael, A., Parrish, C. C., Banoub, J., & Le, T. A. T. (2021). Environmental impact of bioplastic use: A review. *Heliyon*, *7*(9), e07918.

Bakar, N. F. A., & Othman, S. A. (2019). Corn bio-plastics for packaging application. *Journal of Design for Sustainable and Environment*, *1*(1), 1–3.

Baranwal, J., Barse, B., Fais, A., Delogu, G. L., & Kumar, A. (2022). Biopolymer: A sustainable material for food and medical applications. *Polymers*, *14*(5), 983.

Bastioli, C. (Ed.). (2020). *Handbook of Biodegradable Polymers*. Walter de Gruyter GmbH & Co KG.

Bhagwat, G., Gray, K., Wilson, S. P., Muniyasamy, S., Vincent, S. G. T., Bush, R., & Palanisami, T. (2020). Benchmarking bioplastics: A natural step towards a sustainable future. *Journal of Polymers and the Environment*, *28*, 3055–3075.

Bhattacharya, M. (2016). Polymer nanocomposites-a comparison between carbon nanotubes, graphene, and clay as nanofillers. *Materials*, *9*(4), 262.

Bilal, M., Rasheed, T., Nabeel, F., & Iqbal, H. M. (2020). Bionanocomposites from biofibers and biopolymers. In *Biofibers and Biopolymers for Biocomposites: Synthesis, Characterization and Properties.* (pp. 135–157). Springer, Cham.

Das, A., Ringu, T., Ghosh, S., & Pramanik, N. (2022). A comprehensive review on recent advances in preparation, physicochemical characterization, and bioengineering applications of biopolymers. *Polymer Bulletin*, *80*, 1–66.

Di Bartolo, A., Infurna, G., & Dintcheva, N. T. (2021). A review of bioplastics and their adoption in the circular economy. *Polymers*, *13*(8), 1229.

Dogra, V., Verma, D., & Fortunati, E. (2022). Biopolymers and nanomaterials in food packaging and applications. *In Nanotechnology-Based Sustainable Alternatives for the Management of Plant Diseases* (pp. 355–374). Elsevier.

Ecker, J. V., Burzic, I., Haider, A., Hild, S., & Rennhofer, H. (2019). Improving the impact strength of PLA and its blends with PHA in fused layer modelling. *Polymer Testing*, *78*, 105929.

EN13432, C. D. N. (2000). *Packaging-Requirements for Packaging Recoverable through Composting and Biodegradation-Test Scheme and Evaluation Criteria for the Final Acceptance of Packaging*. The European Committee for Standardization, Brussels.

Ezeoha, S. L., & Ezenwanne, J. N. (2013). Production of biodegradable plastic packaging film from cassava starch. *IOSR Journal of Engineering*, *3*(10), 14–20.

Farah, S., Anderson, D. G., & Langer, R. (2016). Physical and mechanical properties of PLA, and their functions in widespread applications-A comprehensive review. *Advanced Drug Delivery Reviews*, *107*, 367–392.

Godfrey, L., Görgens, J. F., & Roman, H. (Eds.). (2020). *Opportunities for Biomass and Organic Waste Valorisation: Finding Alternative Solutions to Disposal in South Africa.* Routledge.

Goel, V., Luthra, P., Kapur, G. S., & Ramakumar, S. S. V. (2021). Biodegradable/bio-plastics: myths and realities. *Journal of Polymers and the Environment*, *29*, 3079–3104.

He, X., Deng, H., & Hwang, H. M. (2019). The current application of nanotechnology in food and agriculture. *Journal of Food and Drug Analysis*, *27*(1), 1–21.

Jenkins, A. D., Kratochvíl, P., Stepto, R. F. T., & Suter, U. W. (1996). Glossary of basic terms in polymer science (IUPAC recommendations 1996). *Pure and Applied Chemistry*, *68*(12), 2287–2311.

John, M. J., & Thomas, S. (2022). Bio-based materials: Contribution to advancing circular economy. *Polymers*, *14*(18), 3887.

Kibria, M. G., Masuk, N. I., Safayet, R., Nguyen, H. Q., & Mourshed, M. (2023). Plastic waste: Challenges and opportunities to mitigate pollution and effective management. *International Journal of Environmental Research*, *17*(1), 20.

Kumar, R., Verma, A., Shome, A., Sinha, R., Sinha, S., Jha, P. K., ... & Vara Prasad, P. V. (2021). Impacts of plastic pollution on ecosystem services, sustainable development goals, and need to focus on circular economy and policy interventions. *Sustainability*, *13*(17), 9963.

Lee, G. N., & Na, J. (2013). Future of microbial polyesters. *Microbial Cell Factories*, *12*(1), 54.

Lin, P. C., Lin, S., Wang, P. C., & Sridhar, R. (2014). Techniques for physicochemical characterization of nanomaterials. *Biotechnology Advances*, *32*(4), 711–726.

Moshood, T. D., Nawanir, G., Mahmud, F., Mohamad, F., Ahmad, M. H., & AbdulGhani, A. (2022). Sustainability of biodegradable plastics: New problem or solution to solve the global plastic pollution? *Current Research in Green and Sustainable Chemistry*, *5*, 100273.

Mourdikoudis, S., Pallares, R. M., & Thanh, N. T. (2018). Characterization techniques for nanoparticles: Comparison and complementarity upon studying nanoparticle properties. *Nanoscale*, *10*(27), 12871–12934.

Muniyasamy, S., & John, M. J. (2017). Biodegradability of biobased polymeric materials in natural environments. In Thakur, V. K., Thakur, M. K., Kessler, M. R. (Eds.), *Handbook of Composites from Renewable Materials*. (pp. 625–653). Wiley.

Muniyasamy, S., Anstey, A., Reddy, M. M., Misra, M., & Mohanty, A. (2013). Biodegradability and compostability of lignocellulosic based composite materials. *Journal of Renewable Materials*, *1*(4), 253–272.

Muniyasamy, S., Mohanrasu, K., Gada, A., Mokhena, T. C., Mtibe, A., Boobalan, T., ... & Arun, A. (2019). Biobased biodegradable polymers for ecological applications: A move towards manufacturing sustainable biodegradable plastic products. *Integrating Green Chemistry and Sustainable Engineering*, *8*, 215–253.

Narayan, R. (2006). Biobased and biodegradable polymer materials: Rationale, drivers, and technology exemplars. *Degradable Polymers and Materials*, *939*, 282–306.

Naser, A. Z., Deiab, I., & Darras, B. M. (2021). Poly (lactic acid) (PLA) and polyhydroxyalkanoates (PHAs), green alternatives to petroleum-based plastics: A review. *RSC Advances*, *11*(28), 17151–17196.

Nazareth, M., Marques, M. R., Leite, M. C., & Castro, Í. B. (2019). Commercial plastics claiming biodegradable status: Is this also accurate for marine environments? *Journal of Hazardous Materials*, *366*, 714–722.

Ncube, L. K., Ude, A. U., Ogunmuyiwa, E. N., Zulkifli, R., & Beas, I. N. (2020). Environmental impact of food packaging materials: A review of contemporary development from conventional plastics to polylactic acid based materials. *Materials*, *13*(21), 4994.

Ncube, L. K., Ude, A. U., Ogunmuyiwa, E. N., Zulkifli, R., & Beas, I. N. (2021). An overview of plastic waste generation and management in food packaging industries. *Recycling*, *6*(1), 12.

Oberti, I., & Paciello, A. (2022). Bioplastic as a substitute for plastic in construction industry. *Encyclopedia*, *2*(3), 1408–1420.

Rahman, M. H., & Bhoi, P. R. (2021). An overview of non-biodegradable bioplastics. *Journal of Cleaner Production*, *294*, 126218.

Rhim, J. W., Park, H. M., & Ha, C. S. (2013). Bio-nanocomposites for food packaging applications. *Progress in Polymer Science*, *38*(10–11), 1629–1652.

Rosenboom, J. G., Langer, R., & Traverso, G. (2022). Bioplastics for a circular economy. *Nature Reviews Materials*, *7*(2), 117–137.

Su, S., Kopitzky, R., Tolga, S., & Kabasci, S. (2019). Polylactide (PLA) and its blends with poly (butylene succinate)(PBS): A brief review. *Polymers*, *11*(7), 1193.

Tiede, K., Boxall, A. B., Tear, S. P., Lewis, J., David, H., & Hassellöv, M. (2008). Detection and characterization of engineered nanoparticles in food and the environment. *Food Additives and Contaminants*, *25*(7), 795–821.

Udayakumar, G. P., Muthusamy, S., Selvaganesh, B., Sivarajasekar, N., Rambabu, K., Banat, F., ... & Show, P. L. (2021). Biopolymers and composites: Properties, characterization and their applications in food, medical and pharmaceutical industries. *Journal of Environmental Chemical Engineering*, *9*(4), 105322.

Vasile, C. (2018). Polymeric nanocomposites and nanocoatings for food packaging: A review. *Materials*, *11*(10), 1834.

Vroman, I., & Tighzert, L. (2009). Biodegradable polymers. *Materials*, *2*(2), 307–344.

Wasti, S., & Adhikari, S. (2020). Use of biomaterials for 3D printing by fused deposition modeling technique: A review. *Frontiers in Chemistry*, *8*, 315.

Zuccarello, B., Marannano, G., & Mancino, A. (2018). Optimal manufacturing and mechanical characterization of high performance biocomposites reinforced by sisal fibers. *Composite Structures*, *194*, 575–583.

13 Life Cycle Assessment (LCA) Study of Bioplastic Biopolymers

Rakgoshi Lekalakala, Orebotse Joseph Botlhoko and Sudhakar Muniyasamy

13.1 INTRODUCTION

Environmental impact is one of the key areas of concern when it comes to the use of plastics. Petro-based plastics, being made from non-renewable resources, have a high carbon footprint, are major in land and oceanic pollution, and contribute to global warming (Zheng & Suh 2019). On the other hand, biopolymers are made from renewable resources and have a significantly lower carbon footprint. Additionally, biopolymers are biodegradable, which means that they do not contribute to the build-up of plastic waste in the environment. This makes biopolymers a more sustainable option, in terms of environmental impact, compared to petro-based plastics (Bishop et al., 2021; Dormer et al., 2013; Zheng & Suh, 2019).

Service life performance is an important factor that must be considered when assessing the value proposition of biopolymers. In terms of strength and durability, petro-based plastics outperform biopolymers. However, biopolymers have the advantage of being biodegradable, which means that they do not pose a threat to the environment even if they are not disposed of properly. Additionally, biopolymers can be designed to have specific properties, such as being compostable, which makes them suitable for specific applications where conventional plastics are not a viable option (Cadenazzi et al., 2019; Favi et al., 2022; Lizin et al., 2013).

End of life analysis is also an important consideration when assessing the value proposition of biopolymers (Amato et al., 2017; Chen et al., 2019; Funazaki et al., 2003; Senan-Salinas et al., 2019). Petro-based plastics can persist in the environment for hundreds of years, which poses a significant threat to wildlife and ecosystems (Dormer et al., 2013; Sangale et al., 2012). In contrast, biopolymers are biodegradable and can be broken down by natural means, which means that they do not pose a long-term threat to the environment. Additionally, biopolymers can be composted, which provides an environmentally friendly means of disposal (Chen et al., 2019; Ciriminna & Pagliaro, 2020).

The use of biopolymers offers several advantages over petro-based plastics, especially in terms of environmental impact, end-of-life analysis and service life

DOI: 10.1201/9781003304142-13

performance. Biopolymers are made from renewable resources, have a lower carbon footprint, and are biodegradable, which makes them a more sustainable option than petro-based plastics. However, petro-based plastics have the advantage of being stronger and more durable, which makes them suitable for specific applications. A comprehensive assessment of the value proposition of biopolymers can help to crystallise the benefits of biopolymers for a wider audience, inform resource allocation decisions, and pre-empt best use case scenarios for bioplastics (Goonoo et al., 2013; Gowthaman et al., 2021; Hargreaves et al., 2018).

Given the exodus from petro-based non-degradable plastic towards biobased biodegradable polymers, there arises a need to qualitatively and quantitatively assess the chasm of value proposition with regards to environmental impact, service life performance and end-of-life analysis between the two categories of polymer materials. This can help crystallise the value proposition of biopolymers for a wider spectrum of audiences, back resource allocation decisions, and pre-empt best-use case scenarios for bioplastics (Figure 13.1).

Life cycle analysis (LCA) is a powerful tool that can be used to evaluate the overall sustainability of different materials, including biopolymers. It takes into account various factors such as environmental footprint, land-use impacts, biogenic carbon and end-of-life management. The results of an LCA can help to provide a comprehensive understanding of the sustainability of different materials and help inform decisions regarding their use (Das et al., 2021; Ramesh & Vinodh, 2020; Tabone et al., 2010).

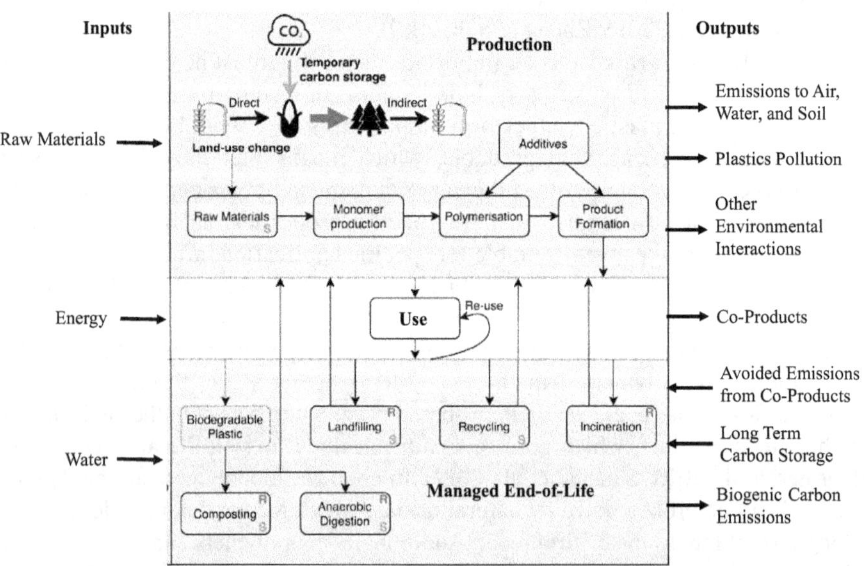

FIGURE 13.1 A simplified schematic of a plastic value chain represented in LCA. The diagram also gives an indication of carbon flows throughout the system, with S representing carbon storage, and R representing the release of the carbon (Bishop et al., 2021).

13.2 LCA OF BIOCOMPOSITES

LCA helps us to understand the various aspects of materials production, including the synthesis of raw materials, usage of the finished product and waste management (Chen & Patel, 2012). Therefore, it is very important to include LCA when designing materials from renewable resources. In our case, biocomposite designing should consider the LCA for understanding their true "green-ness". The biggest challenge in designing structurally and functionally stable sustainable biocomposites is to devise an environmentally benign synthesis and modification process with appropriate disposal or waste management methods. Recycling and composting are the major waste management procedures that can be adopted in handling biocomposites (Vilaplana et al., 2010). However, due to the complexity and sensitivity of biocomposites towards recycling, composting can be the preferred route of disposal. Earlier sections of this chapter deal with the various aspects of composting and the biodegradation mechanisms of these biocomposites. The introduction of LCA into the design of biocomposites will give a more exact picture of their sustainability and whether they are suitable alternatives for their petro-based counterparts.

LCA is a procedural outline that is used to assess and estimate the environmental impacts of a product during its life cycle. The environmental impacts assessed include climate change, eutrophication, acidification, ozone depletion, greenhouse gas emissions, toxicological stress on human health and ecosystems, the depletion of resources, water use, land use, noise, and others. LCA is a very useful tool for evaluating environmental sustainability and identifying objectives for improvement. LCA methodology is a widely accepted approach for characterising environmental impacts and has been standardised under the ISO 14040 series (ISO 14040-ISO14043). LCA can be used to measure the environmental impacts of products globally, regionally, and locally. LCA consist of the following steps, as described in Figure 13.2.

i. **Goal and scope definition**: The goal and scope definition of an LCA provides a clear picture of the product system in terms of the system boundaries and a functional unit (Rebitzer et al., 2004).
ii. **Inventory analysis**: The inventory analysis comprises a thorough anthology of all of the environmental inputs and outputs of each stage of the life cycle (Murphy & Bartle, 2004).
iii. **Impact assessment**: A process whereby environmental impacts from the inventory are reviewed, and generally, the overall environmental performance of the product is determined.
iv. **Interpretation**: Serves to evaluate the study in order to derive recommendations and conclusions, and is conducted at every stage.

It is important to include waste management routes into the LCA of biocomposites when comparing them with their petroleum counterparts. A feature article doubted the sustainability of polyhydroxyalkanoates (PHAs) in terms of higher nonrenewable energy usage, as they did not consider the waste management procedures for these materials (Gerngross, 1999). The end-life management of biocomposites is very important as it relates to energy recovery and the possible emission or

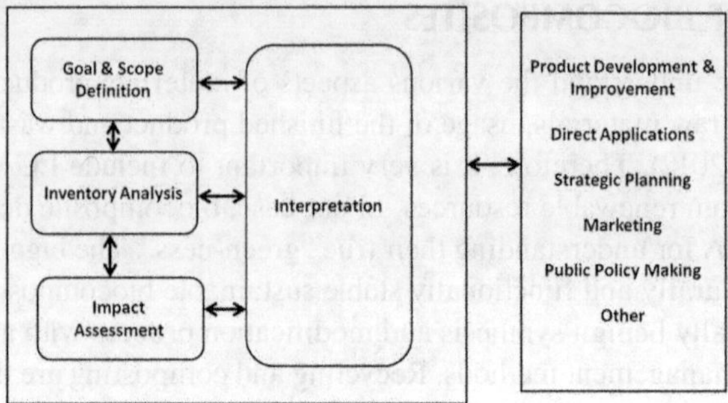

FIGURE 13.2 Life cycle analysis: stages involved.

sequestration of carbon compounds. Most of the LCA studies concerning polylactic acid (PLAs) compared their energy consumption and greenhouse gas emissions with their petro-based counterparts.

It is well known that natural fibres offer better results in terms of sustainability compared to their synthetic counterparts. In a review by Joshi et al. (2004), natural fibres were compared with glass fibre composites. It was found that natural fibres were more sustainable due to four categories: (i) lesser impact on the environment in producing natural fibres, (ii) natural fibres have lower density, (iii) higher loading of natural fibres for comparable performance, and (iv) better waste management routes for natural fibres. This study also revealed the negative impacts of natural fibres were influenced by pesticides and the usage of other chemicals. However, the larger environmental benefits of natural fibres during their service life and end-life phase have outweighed the negative impacts. Added advantages of using natural fibre composites include low density, leading to a reduction in the weight of automotives, which reduces overall fuel consumption and greenhouse gas emissions. The study has also concluded that the longer service life of natural fibre composite-based products provided higher environmental benefits. Also, additional energy can be recovered by using incineration as the end-life waste management procedure. This study compared biocomposites based on different natural fibres for automotive applications and revealed that the natural fibres showed better environmental performance than clay-based nanocomposites of poly(hydroxyl butyrate) (PHB) for electronics and automotive applications.

However, it is important to note that LCA can also be a source of error due to the inherent limitations and weaknesses in the analysis. One of the biggest challenges in conducting an LCA is defining the boundary conditions. This includes deciding which aspects of the life cycle to include in the analysis, such as the production, use, and disposal of the material. If the boundary conditions are not well defined, the results of the LCA may not provide a complete or accurate picture of the sustainability of the material in question. Another weakness of LCA is the lack of accurate data and information regarding the environmental impact of different materials. This can make it difficult to compare the sustainability of different materials, as the data may not be available or may be unreliable. This can also lead to incorrect conclusions being drawn from the LCA (Bishop et al., 2021; Bjørn et al., 2018; Curran, 2014; Hottle et al., 2013; Wittmaier et al., 2009).

Additionally, LCA can be subject to bias, as the results of the analysis may be influenced by the specific assumptions and methodologies used. For example, if the LCA is conducted using a certain set of assumptions regarding the environmental impact of a material, the results may not reflect the true sustainability of the material in question. This can lead to inaccurate conclusions being drawn from the LCA, which can have a significant impact on resource allocation decisions. While LCA is a useful tool for evaluating the overall sustainability of different materials, it is important to be aware of its limitations and weaknesses. Careful consideration must be given to defining the boundary conditions, ensuring the availability of accurate data, and avoiding bias in the analysis in order to ensure that the results of the LCA are reliable and provide a comprehensive picture of the sustainability of the material in question.

13.3 RESOURCE SUSTAINABILITY

The efficient allocation of resources is becoming increasingly important as the world's population continues to grow and the depletion of resources accelerates. This is particularly true for polymer materials, where biobased and biodegradable polymers provide an attractive alternative to petro-based polymers. Biobased polymers have several advantages over petro-based polymers, including renewable sourcing of raw materials, circularity of bioresources, and more efficient extraction through farming rather than mining (Devadas et al., 2021; Rosenboom et al., 2022).

Biobased polymers are sourced from renewable sources, such as plants and microorganisms, as opposed to petro-based polymers, which are derived from fossil fuels. This means that biobased polymers have a smaller carbon footprint and are more environmentally friendly than petro-based polymers. Additionally, biobased polymers can be sourced from a variety of crops, such as corn, sugarcane, and cassava, which can be grown and harvested on a regular basis, making them a sustainable and renewable source of raw materials. Another advantage of biobased polymers is that they support a circular resource economy, as opposed to a linear fossil resource economy. This means that after a biobased polymer has reached the end of its useful life, it can be decomposed and broken down into its constituent parts, which can then be used as a source of energy or raw materials for others. In contrast, petro-based polymers are not biodegradable and contribute to the growing problem of plastic waste in the world's oceans and landfills (Kim & Dale, 2005; Klemm et al., 2005; Nanni et al., 2021; Rommi et al., 2016).

Furthermore, the extraction of biobased raw materials is often more efficient than the extraction of fossil fuels. For example, biobased raw materials can be obtained through farming, which is a less energy-intensive process than oil drilling. In addition, the conversion of biobased raw materials into biodegradable polymer monomers is often more efficient and less expensive than the process of converting fossil fuels into petro-based polymer monomers when fermentation is contrasted with the operations of an oil refinery. Biopolymers have become increasingly popular in recent years due to concerns over resource sustainability and the depletion of fossil fuels. However, the production of biopolymers is not without its challenges. One important aspect to consider is the role of agriculture and the competition for resources between biopolymers and other forms of agriculture (Chia et al., 2018).

As previously mentioned, biopolymers are typically made from plant matter such as corn, sugarcane, or potatoes. This presents a challenge for agriculture, as the demand for raw materials for biopolymer production competes with the demand for fresh produce. The global demand for polymer materials is estimated to be in the range of millions of tons, and it is unlikely that this demand can be satisfied agriculturally without disrupting the food value chain. This competition for resources is further exacerbated by the increasing global demand for food and the limited availability of arable land. In addition, the production of biopolymers can have a significant impact on land use and soil quality. The cultivation of crops for biopolymer production can lead to the degradation of fertile soil and the loss of biodiversity. Furthermore, the use of pesticides and fertilisers in the cultivation of crops for biopolymer production can lead to environmental pollution and the contamination of water sources. Due to the limitations of resource availability and the competition for resources between biopolymers and other forms of agriculture, it is unlikely that biopolymers will completely replace petro-based polymers. Instead, the two types of polymers will likely coexist in the global economy, each with its own unique set of advantages and disadvantages (Bolaji et al., 2021; Kasapis et al., 2009).

Petro-based polymers are sometimes cheaper and more easily produced than biopolymers. They are also more flexible and have better mechanical properties. However, petro-based polymers are non-renewable and contribute to greenhouse gas emissions and other forms of environmental pollution. In contrast, biopolymers are renewable and have a lower environmental impact, but they are typically more expensive and less flexible than petro-based polymers. Thus, the production of biopolymers presents a number of challenges for agriculture and resource sustainability. The competition for resources between biopolymers and other forms of agriculture is a major concern, as the demand for biopolymer materials is estimated to reach record volumes with a double-digit compounded annual growth rate.

13.4 ENERGY USAGE

Energy is a critical aspect when evaluating the sustainability and environmental impact of polymer materials. Biopolymers and fossil-based polymers, have varying energy demands during production and processing, and small variations in energy consumption exist between the two types. Injection moulding, a commonly used processing method, is one example where these variations can occur, and energy consumption depends on factors such as the preconditioning steps, such as drying and machine warm-up, as well as the processing temperature, drying time, and temperature. Biopolymers, such as PLA and PHA, are renewable and biodegradable, making them more sustainable alternatives to traditional fossil-based polymers. However, the processing of biopolymers often requires higher temperatures and longer drying times, leading to higher energy consumption than fossil-based polymers. For example, a study showed that energy consumption during the injection moulding of PHA was approximately 20% higher than that of polypropylene (PP), a fossil-based polymer (Schulze et al., 2017).

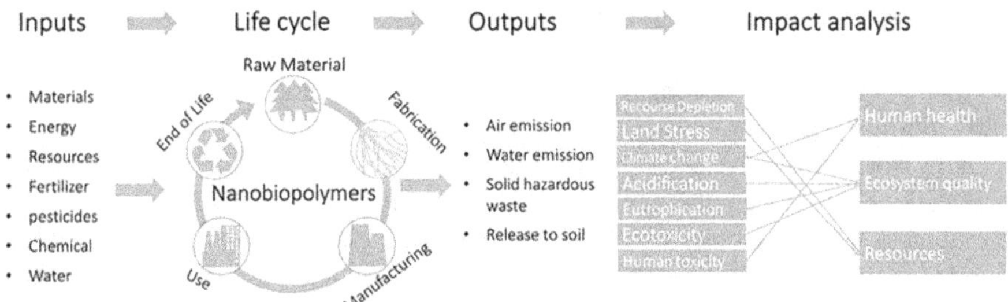

FIGURE 13.3 Life cycle assessment process for nanobiopolymers.

In contrast, fossil-based polymers tend to require lower processing temperatures and shorter drying times, leading to lower energy consumption. For example, a study by Benavides et al. found that energy consumption during the processing of polyethylene (PE) was approximately 10% lower compared to that of PLA. However, it is important to note that the exact values for energy consumption will vary from case to case, and the differences between biopolymers and fossil-based polymers tend to be negligible in the grand scheme of things. In addition, the overall energy consumption of polymer processing depends on a range of factors, including processing conditions, material properties, and equipment efficiency. In conclusion, energy consumption is an important aspect to consider when evaluating the sustainability and environmental impact of polymer materials. While small variations in energy demand exist between biopolymers and fossil-based polymers, the exact values will vary from case to case, and the overall energy consumption will depend on a range of factors (Benavides et al., 2020) (Figure 13.3).

13.5 IMPACT OF APPLICATION

When evaluating the sustainability and environmental impact of biopolymers, application suitability is a significant parameter that is relevant to LCA but difficult to measure. Biopolymers have several applications where they are highly suitable, thus increasing their overall positive impact on sustainability.

Food packaging, medical devices, and agricultural products are examples of applications where biopolymers can make a significant difference. Conventional petroleum-based materials used in these applications often result in excessively contaminated waste, making recycling costly and labour-intensive. Biopolymers, on the other hand, being biodegradable, are thus exceptionally suitable for these applications where their biodegradability can be used to mitigate the disposal issue. As an example, products made from biopolymers can be designed to biodegrade in the farm environment, thus reducing environmental contamination caused by plastics, the labour cost required for waste collection and the need for landfills (Udayakumar et al., 2021).

The biodegradable nature of biopolymers also makes them suitable for applications in the medical industry, such as single-use medical devices. These products can be disposed of in an industrial composting facility, where they will biodegrade and not contribute to environmental contamination nor take up space in the environment for lengthy periods with landfilling. The same can be said for direct food contact packaging articles where the value proposition of recycling diminishes with the necessity of excessive washing and drying while the properties of the material inherently diminish (Baranwal et al., 2022; Rebelo et al., 2017).

It is worth noting that optimising the suitability of biopolymers for specific applications will depend on a range of factors, including the design of the product, the processing conditions, and the properties of the material. However, the overall trend towards increasing the use of biopolymers in applications where conventional materials are problematic and unrecyclable is clear. Application suitability is a crucial parameter when evaluating the sustainability and environmental impact of biopolymers. Biopolymers have several applications, including food packaging, medical devices, and agricultural products, where recyclability is inherently challenging. Their biodegradable nature makes them highly suitable, reducing environmental contamination and the need for landfills. This, in turn, increases their overall positive impact on sustainability (Niaounakis, 2015; Rebelo et al., 2017) (Figure 13.4).

However, conventional plastics still have the advantage over biopolymers in applications that require long service life due to their longevity. Additionally, recycling conventional plastics makes sense in applications where the product can be easily collected and separated from other waste streams, and the plastic is not contaminated. This section will further discuss the contrast of advantages of conventional plastics

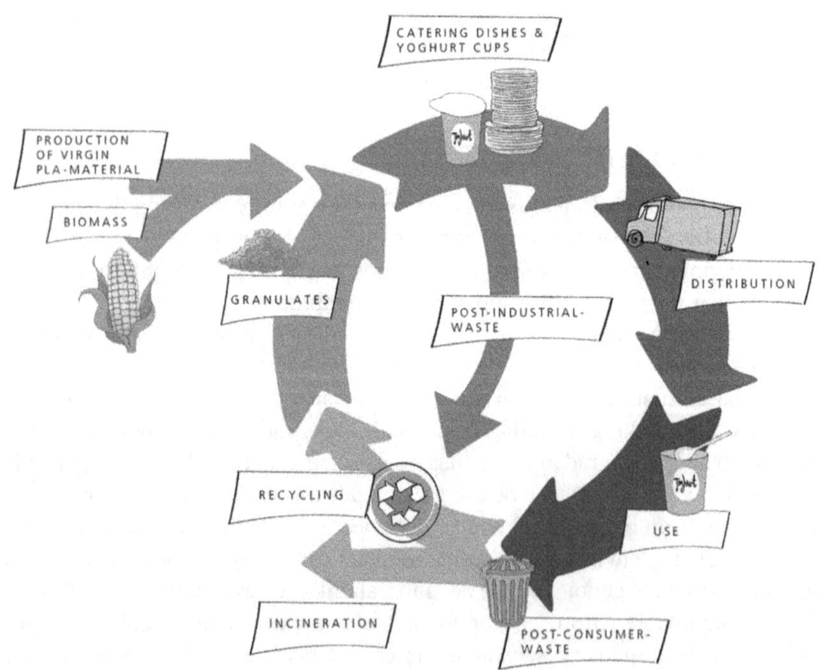

FIGURE 13.4 LCA of recycled polymer.

over biopolymers in certain applications, the availability of additives that reinforce the recyclability of conventional plastics, and the suitability of biopolymers for food contact packaging.

Conventional plastics have been widely used for decades due to their durability, low cost, and ease of production. In applications that require long service life, such as construction materials and automotive parts, conventional plastics retain the advantage over biopolymers due to their longevity and recyclability. Biopolymers may degrade over time, reducing their mechanical properties and leading to a shorter service life, thus reducing the level of extractable value per unit. Additionally, biopolymers are typically more expensive than conventional plastics, making them less attractive for long-term applications. In many applications, recycling of conventional plastics makes sense due to the availability of well-established waste management infrastructure and the ease of collecting and separating the plastic from other waste streams, increasing circularity of resource economy for conventional plastics. Additionally, there are no additives available that reinforce the mechanical properties of recycled plastics, increasing the service life of the material and adding to resource sustainability, especially where such conventional polymers are biobased. For example, the use of recycled high-density polyethylene (HDPE) in construction materials has been shown to have a lower carbon footprint than virgin HDPE (COWI, 2018).

A further example of where biopolymers may be well suited is in food contact packaging. The contamination of conventional plastics with foodstuff can make them unrecyclable, and multi-layered and/or metalised films used in packaging can also make recycling difficult. Biopolymers are biodegradable, which means that they can break down naturally without causing harm to the environment. Additionally, the weight of the article in food contact packaging is typically not a significant factor, as the volume of waste generated is relatively small. This means that waste collectors may be more willing to accept biopolymers for recycling. Overall, a well-suited application can make all the difference when evaluating the sustainability of biopolymers versus conventional petroleum-based polymers (COWI, 2018).

13.6 GLOBAL WARMING POTENTIAL OF BIOPOLYMERS

The global warming potential of biopolymers is one of their key selling points ahead of conventional plastic materials. This is another metric in LCA that can be difficult to measure. There is a myriad of impact categories under this umbrella, and transparency and unbiased reporting are crucial to inform policy decisions in the right direction. Some of the vital impact categories in global warming potential include acidification potential, eutrophication potential, resource depletion, photochemical oxidant formation, ozone depletion, ecotoxicity, human toxicity, particulate matter formation, energy, land-use and water consumption where it is generally accepted that biopolymers have better environmental sustainability over petro-based polymers. Petro-based polymers do, however, have an advantage over biopolymers with regard to greenhouse gas emissions at end-of-life since the carbon remains immobilised while it is released into the atmosphere when biopolymers biodegrade, which may exacerbate the planetary greenhouse effect at scale. This may be difficult to juxtapose with land pollution, microplastic pollution and persistent organic pollutants resulting from the ill-managed non-degradable plastic waste (Imtiaz et al., 2021).

In general, biopolymers have a better environmental sustainability profile than petro-based polymers in most of these global warming potential categories. For instance, biopolymers are produced from renewable resources, which reduces the depletion of finite petroleum reserves and reduces the associated greenhouse gas emissions.

However, petro-based polymers have an advantage over biopolymers with regard to greenhouse gas emissions at end-of-life. The carbon in petro-based polymers remains immobilised in landfills, while biopolymers biodegrade and release GHGs into the atmosphere. This trade-off between the greenhouse gas emissions associated with biodegradation and the land pollution, microplastic pollution, and persistent organic pollutants resulting from non-degradable plastics can be difficult to resolve. Each class of material may be advantageous to use when this is considered in concert with application and ease of recyclability.

Transparency and unbiased reporting are crucial in informing policy decisions related to biopolymers and their overall impact on the environment. Accurate and reliable data on the global warming potential of biopolymers can help policymakers make informed decisions about the use and regulation of biopolymers and other sustainable materials (Figure 13.5).

The global warming potential of biopolymers is a critical metric in LCA that depends on several impact categories. Biopolymers have a better environmental sustainability profile than petro-based polymers in most categories, but the trade-off between the GHG emissions associated with biodegradation and the land pollution, microplastic pollution, and persistent organic pollutants resulting from

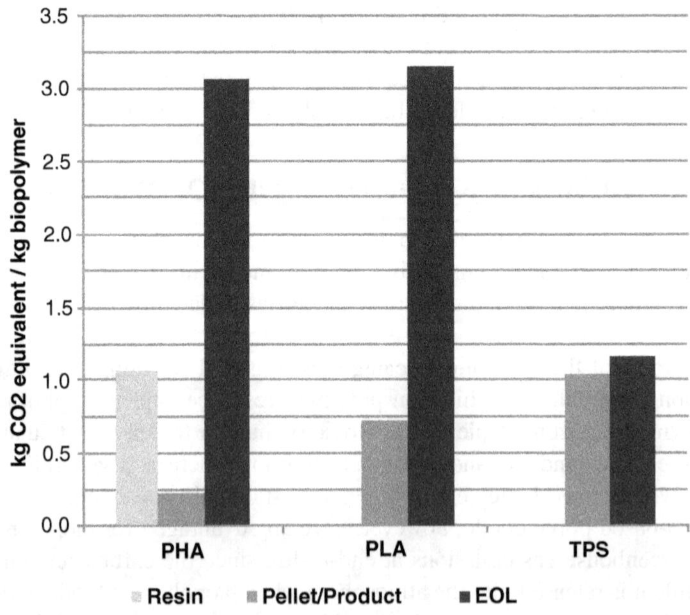

FIGURE 13.5 Average global warming potential of biopolymers based on the extent of the system boundaries of the 15 different LCAs (Hottle et al., 2013).

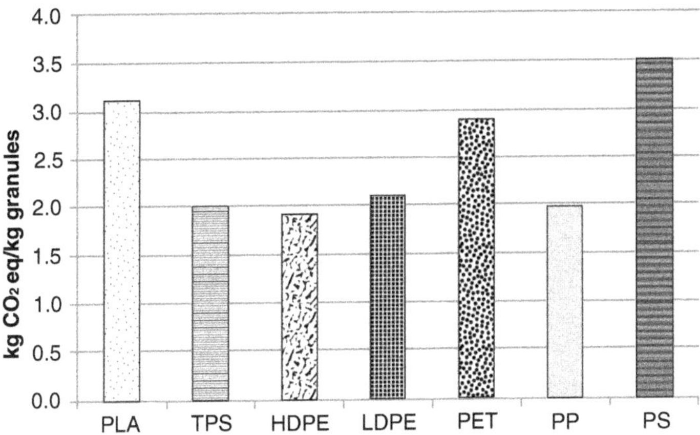

FIGURE 13.6 Life cycle environmental impacts of PLA and TPS compared to petroleum-based polymers per kg of granule (starch) (Data taken from ecoinvent v2.2 and TRACI v2.00. PLA = polylactic acid, TPS = thermoplastic starch, HDPE = high-density polyethylene, LDPE = low-density polyethylene, PET = polyethylene terephthalate, PP = polypropylene, PS = polystyrene. CTUe = Comparative Toxic Unit ecosystem, CTUh = Comparative Toxic Unit human health. (Hottle et al., 2013).)

non-degradable plastics is difficult to resolve. Transparency and unbiased reporting are crucial in informing policy decisions related to biopolymers and their impact on the environment.

For example, a recent study compared the environmental impact of different types of biopolymer films for packaging with conventional polyethylene film. The study found that PLA and thermoplastic starch (TPS) had lower environmental impact than conventional polyethylene film in terms of global warming potential, acidification potential, and eutrophication potential. However, the study also found that TPS had higher impacts in terms of land use and water depletion than PLA and conventional polyethylene film. Such comparative LCA studies can provide valuable information on the environmental sustainability of biopolymers and guide their application in specific areas (González-García et al., 2016) (Figure 13.6).

13.7 FUTURE PERSPECTIVES

The future of biopolymers is bright when used hand in hand with LCA to refine their applicability and better communicate their value proposition. More LCA studies comparing individual biopolymer with conventional polymers intended for replacement in specific application can be valuable towards inform policy, product design, research, development, and innovation. More development towards refining LCA methodologies and principles can also improve its functionality in analysis of resource economies.

To further enhance the functionality of LCA, more development towards refining LCA methodologies and principles can improve its analysis of resource economies. This can include expanding the boundaries of LCA to include upstream impacts, such as land use and water consumption, and incorporating circular economy principles to evaluate the potential for biopolymer waste to be recycled or biodegraded. Such development can provide a more comprehensive and accurate assessment of the environmental sustainability of biopolymers and inform policy decisions and research towards the creation of more sustainable biopolymers.

Summarily, more comparative LCA studies and development towards refining LCA methodologies and principles can improve the functionality of LCA and guide the deployment, mass adoption, product design, research, development, and innovation of biopolymers towards more sustainable and environmentally friendly solutions.

13.8 CONCLUSION

While biopolymers will continue to supplant conventional plastics with remediation of the environment. An application and material-specific case-by-case LCA study would be prudent to validate and quantify biopolymer value proposition and highlight the areas of highest impact. LCA has proven to be a valuable tool for assessing the environmental sustainability of biopolymers and can help emphasise their impact and inform policy for efficient deployment and mass adoption in synergy with conventional polymers. Going forward, it can positively add to the utility of LCA by implementing scales of quantification. It will also clarify the qualitative analysis of LCA if the overarching principles are unified into a few key concepts that embody the concepts and make it easy to convey the message of sustainability. This can also make comparative analysis simple and compact.

The environmental sustainability of biopolymers is complex, and a comprehensive LCA can help quantify and qualify their environmental impact and inform policy for efficient deployment and mass adoption. In contrast, LCA evaluates the environmental impact of a product or process throughout its entire life cycle, from raw material extraction and processing to production, transportation, use, and end-of-life. The use of LCA for biopolymers can help to identify the areas of highest environmental impact and quantify their environmental benefits compared to conventional polymers.

To further enhance the utility of LCA for biopolymers, scales of quantification can be implemented. These scales can provide a numerical score for the environmental performance of a biopolymer, making it easier to compare its environmental impact with other materials. This can help to guide the development of more environmentally friendly biopolymers and inform policy and regulatory decisions that can promote their efficient deployment and mass adoption.

In addition, unifying the overarching principles of LCA into a few key concepts can help to clarify the qualitative analysis of LCA and make it easier to convey the message to the masses. This can enable comparative analysis to be simple and compact, allowing consumers to make informed decisions about the environmental impact of different products. By unifying the principles of LCA, the environmental sustainability of biopolymers can be communicated more effectively, making it easier to inform and educate the general public about the benefits and limitations of biopolymers.

REFERENCES

Amato, A., Rocchetti, L., & Beolchini, F. (2017). Environmental impact assessment of different end-of-life LCD management strategies. *Waste Management*, 59, 432–441.

Baranwal, J., Barse, B., Fais, A., Delogu, G. L., & Kumar, A. (2022). Biopolymer: a sustainable material for food and medical applications. *Polymers*, 14(5), 983.

Benavides, P. T., Lee, U., & Zarè-Mehrjerdi, O. (2020). Life cycle greenhouse gas emissions and energy use of polylactic acid, bio-derived polyethylene, and fossil-derived polyethylene. *Journal of Cleaner Production*, 277, 124010.

Bishop, G., Styles, D., & Lens, P. N. (2021). Environmental performance comparison of bioplastics and petrochemical plastics: a review of life cycle assessment (LCA) methodological decisions. *Resources, Conservation and Recycling*, 168, 105451.

Bjørn, A., Owsianiak, M., Molin, C., & Laurent, A. (2018). Main characteristics of LCA. In Hauschild, M. Z., Rosenbaum, R. K., & Olsen, S. I. (Eds.), *Life Cycle Assessment: Theory and Practice* (pp. 9–16). Cham: Springer.

Bolaji, I., Nejad, B., Billham, M., Mehta, N., Smyth, B., & Cunningham, E. (2021). Multi-criteria decision analysis of agri-food waste as a feedstock for biopolymer production. *Resources, Conservation and Recycling*, 172, 105671.

Cadenazzi, T., Dotelli, G., Rossini, M., Nolan, S., & Nanni, A. (2019). Life-cycle cost and life-cycle assessment analysis at the design stage of a fiber-reinforced polymer-reinforced concrete bridge in Florida. *Advances in Civil Engineering Materials*, 8(2), 128–151.

Chen, G. Q., & Patel, M. K. (2012). Plastics derived from biological sources: present and future: a technical and environmental review. *Chemical Reviews*, 112(4), 2082–2099.

Chen, Y., Cui, Z., Cui, X., Liu, W., Wang, X., Li, X., & Li, S. (2019). Life cycle assessment of end-of-life treatments of waste plastics in China. *Resources, Conservation and Recycling*, 146, 348–357.

Chia, S. R., Chew, K. W., Show, P. L., Yap, Y. J., Ong, H. C., Ling, T. C., & Chang, J. S. (2018). Analysis of economic and environmental aspects of microalgae biorefinery for biofuels production: a review. *Biotechnology Journal*, 13(6), 1700618.

Ciriminna, R., & Pagliaro, M. (2020). Biodegradable and compostable plastics: a critical perspective on the dawn of their global adoption. *ChemistryOpen*, 9(1), 8–13.

COWI. (2018). Carbon footprint of plastics waste recycling. *European Commission*. Retrieved from https://ec.europa.eu/environment/integration/research/newsalert/pdf/carbon_footprint_of_plastics_waste_recycling_535na2_en.pdf

Curran, M. (2014). Strengths and Limitations of Life Cycle Assessment. In: Klöpffer, W. (eds) *Background and Future Prospects in Life Cycle Assessment*. LCA Compendium – The Complete World of Life Cycle Assessment. Springer, Dordrecht., 189–206. https://doi.org/10.1007/978-94-017-8697-3_6

Das, S., Liang, C., & Dunn, J. B. (2021). Life cycle assessment of polymers and their recycling. In *Circular Economy of Polymers: Topics in Recycling Technologies* (pp. 143–170). American Chemical Society.

Devadas, V. V., Khoo, K. S., Chia, W. Y., Chew, K. W., Munawaroh, H. S. H., Lam, M. K., ... & Show, P. L. (2021). Algae biopolymer towards sustainable circular economy. *Bioresource Technology*, 325, 124702.

Dormer, A., Finn, D. P., Ward, P., & Cullen, J. (2013). Carbon footprint analysis in plastics manufacturing. *Journal of Cleaner Production*, 51, 133–141.

Favi, C., Moroni, F., Lutey, A. H., & Rodríguez, N. B. (2022). Life cycle analysis of engineering polymer joining methods using adhesive bonding: fatigue performance and environmental implications. *Procedia CIRP*, 105, 565–570.

Funazaki, A., Taneda, K., Tahara, K., & Inaba, A. (2003). Automobile life cycle assessment issues at end-of-life and recycling. *JSAE Review*, 24(4), 381–386.

Gerngross, T. U. (1999). Can biotechnology move us toward a sustainable society? *Nature Biotechnology*, 17(6), 541–544.

González-García, S., Moreira, M. T., Feijoo, G., & Pena, N. (2016). Comparative environmental assessment of biopolymer films for food packaging through life cycle thinking. *Journal of Cleaner Production*, 112, 4373–4383.

Goonoo, N., Bhaw-Luximon, A., Bowlin, G. L., & Jhurry, D. (2013). An assessment of biopolymer-and synthetic polymer-based scaffolds for bone and vascular tissue engineering. *Polymer International*, 62(4), 523–533.

Gowthaman, N. S. K., Lim, H. N., Sreeraj, T. R., Amalraj, A., & Gopi, S. (2021). Advantages of biopolymers over synthetic polymers: social, economic, and environmental aspects. In *Biopolymers and Their Industrial Applications* (pp. 351–372). Elsevier.

Hargreaves, A. J., Vale, P., Whelan, J., Alibardi, L., Constantino, C., Dotro, G., ... & Campo, P. (2018). Coagulation-flocculation process with metal salts, synthetic polymers and biopolymers for the removal of trace metals (Cu, Pb, Ni, Zn) from municipal wastewater. *Clean Technologies and Environmental Policy*, 20, 393–402.

Hottle, T. A., Bilec, M. M., & Landis, A. E. (2013). Sustainability assessments of bio-based polymers. *Polymer Degradation and Stability*, 98(9), 1898–1907.

Imtiaz, L., Kashif-ur-Rehman, S., Alaloul, W. S., Nazir, K., Javed, M. F., Aslam, F., & Musarat, M. A. (2021). Life cycle impact assessment of recycled aggregate concrete, geopolymer concrete, and recycled aggregate-based geopolymer concrete. *Sustainability*, 13(24), 13515.

ISO-14044, Environmental management – Life cycle assessment – principles and framework. 2006. https://www.iso.org/standard/37456.html.

ISO-14041, Environmental management – Life cycle assessment – goal and scope definition and inventory analysis. 1998.

ISO-14042, Environmental management – Life cycle assessment – life cycle impact assessment. 2000.

ISO-14044, Environmental management – Life cycle assessment – life cycle interpretation. 2006. https://www.iso.org/standard/38498.html.

Joshi, S. V., Drzal, L. T., Mohanty, A. K., & Arora, S. (2004). Are natural fiber composites environmentally superior to glass fiber reinforced composites? *Composites Part A: Applied Science and Manufacturing*, 35(3), 371–376.

Kasapis, S., Norton, I. T., & Johan, B. (Eds.). (2009). *Modern Biopolymer Science: Bridging the Divide between Fundamental Treatise and Industrial Application*. Academic Press.

Kim, S., & Dale, B. (2005). Life cycle assessment study of biopolymers (polyhydroxyalkanoates)-derived from no-tilled corn (11 pp). *The International Journal of Life Cycle Assessment*, 10, 200–210.

Klemm, D., Heublein, B., Fink, H. P., & Bohn, A. (2005). Cellulose: fascinating biopolymer and sustainable raw material. *Angewandte Chemie International Edition*, 44(22), 3358–3393.

Lizin, S., Passel, S. V., Schepper, E.D., Maes, W., Lutsen, L., Manca, J., & Vanderzande, D. (2013). Life cycle analyses of organic photovoltaics: a review. *Energy & Environmental Science*, 6, 3136–3146.

Murphy, R., & Bartle, I. (2004). *Summary Report, Biodegradable Polymers and Sustainability: Insight from Life Cycle Assessment*. National Non Food Crops Centre, UK.

Nanni, A., Parisi, M., & Colonna, M. (2021). Wine by-products as raw materials for the production of biopolymers and of natural reinforcing fillers: a critical review. *Polymers*, 13(3), 381.

Niaounakis, M. (2015). *Biopolymers: Applications and Trends*. William Andrew.

Ramesh, P., & Vinodh, S. (2020). State of art review on life cycle assessment of polymers. *International Journal of Sustainable Engineering*, 13(6), 411–422.

Rebelo, R., Fernandes, M., & Fangueiro, R. (2017). Biopolymers in medical implants: a brief review. *Procedia Engineering*, 200, 236–243.

Rebitzer, G., Ekvall, T., Frischknecht, R., Hunkeler, D., Norris, G., Rydberg, T., ... & Pennington, D. W. (2004). Life cycle assessment: part 1: framework, goal and scope definition, inventory analysis, and applications. *Environment International*, 30(5), 701–720.

Rommi, K., Rahikainen, J., Vartiainen, J., Holopainen, U., Lahtinen, P., Honkapää, K., & Lantto, R. (2016). Potato peeling costreams as raw materials for biopolymer film preparation. *Journal of Applied Polymer Science*, 133(5), 42862.

Rosenboom, J-G., Langer, R., & Traverso, G. (2022). Bioplastics for a circular economy. *Nature Reviews Materials* 7, 117–137.

Sangale, M. K., Shahnawaz, M., & Ade, A. B. (2012). A review on biodegradation of polythene: the microbial approach. *Journal of Bioremediation and Biodegradation*, 3(10), 1–9.

Schulze, C., Juraschek, M., Herrmann, C., & Thiede, S. (2017). Energy analysis of bioplastics processing. *Procedia CIRP*, 61, 600–605.

Senan-Salinas, J., Garcia-Pacheco, R., Landaburu-Aguirre, J., & García-Calvo, E. (2019). Recycling of end-of-life reverse osmosis membranes: comparative LCA and cost-effectiveness analysis at pilot scale. *Resources, Conservation and Recycling*, 150, 104423.

Tabone, M. D., Cregg, J. J., Beckman, E. J., & Landis, A. E. (2010). Sustainability metrics: life cycle assessment and green design in polymers. *Environmental Science & Technology*, 44(21), 8264–8269.

Udayakumar, G. P., Muthusamy, S., Selvaganesh, B., Sivarajasekar, N., Rambabu, K., Banat, F., ... & Show, P. L. (2021). Biopolymers and composites: properties, characterization and their applications in food, medical and pharmaceutical industries. *Journal of Environmental Chemical Engineering*, 9(4), 105322.

Vilaplana, F., Strömberg, E., & Karlsson, S. (2010). Environmental and resource aspects of sustainable biocomposites. *Polymer Degradation and Stability*, 95(11), 2147–2161.

Wittmaier, M., Langer, S., & Sawilla, B. (2009). Possibilities and limitations of life cycle assessment (LCA) in the development of waste utilization systems-applied examples for a region in Northern Germany. *Waste Management*, 29(5), 1732–1738.

Zheng, J., & Suh, S. (2019). Strategies to reduce the global carbon footprint of plastics. *Nature Climate Change*, 9(5), 374–378.

14 A Global Perspective of Beginning to the End of the Microplastics in the Aquatic Environment

Tracing the Missing Link – Bioplastics

Durgadevi Sabapathi, Selvaraj Anitha, Anathanarayanan Yuvaraj, Muthusamy Govarthanan, Nachimuthu Karmegam, Muniyandi Biruntha, Gnanasekaran Sivakumar, A. Arun and Ponnuchamy Kumar

14.1 INTRODUCTION

It is well-known that plastics are synthetic or semi-synthetic polymers derived from hydrocarbons (Ahmed et al., 2018). In other words, plastics are high molecular-weight polymeric substances generated from the remains of fossil fuels such as oil, natural gas, and coal (Geyer et al., 2017; Lear et al., 2021). Numerous chemical modifications have been devised to produce a vast array of plastic polymers (Munno et al., 2020). Therefore, their incorporation into everyday household items, from electronics to pharmaceuticals, is inevitable (Rahimi and García, 2017). Based on the ingredients used in their manufacture, plastics are classified into two categories: thermoplastics and thermosets (Desidery and Lanotte, 2022; Kazemi et al., 2021). Firstly, the production of thermoplastics involves a series of melting and solidification (Nanda and Berruti, 2021). According to reports, the degradation of thermoplastics in the environment is gradual (Chamas et al., 2020). Meanwhile, a few known microplastics used in our routine day-to-day life are polyethylene terephthalate (PET), low-density polyethylene (LDPE), polyvinyl chloride (PVC), high-density polyethylene (HDPE), polypropylene (PP), and polystyrene (PS) (Evode et al., 2021; Padervand et al., 2020; Valerio et al., 2020). Similarly, thermoset plastics employ high-molecular-weight polymers that undergo a chemical reaction to generate an

DOI: 10.1201/9781003304142-14

irreversible 3D matrix (Zhang et al., 2020). Thermosets are more challenging to recycle than thermoplastics due to the base materials employed in their production (Jehanno and Sardon, 2019). Epoxy, polyester, silicone, polyurethane, and vulcanized rubber are a few examples of thermoset plastics (Yue et al., 2019).

Plastics can be found in virtually everything, from water bottles to clothing (Akanyange et al., 2022; Browne et al., 2011). The versatility, low cost, safety, lightweight, and durability of plastics and other synthetic organic polymers make it unavoidable (Hossain et al., 2020; Ivleva, 2021). In addition to the above, plastics are utilized in many items (Andrady and Neal, 2009). This includes refrigerators, computers, automobile components, and textiles (Heller et al., 2020). In the same way, plastics play a pivotal role in medical items, namely disposable syringes, intravenous sets, glucose bottles, disposable plastic aprons, catheters, and cannulas (Varnava and Patrickios, 2021). As a result, they have overtaken their predecessors, paper, glass, and cardboard (Hou et al., 2018). However, its management after release into the environment is the primary concern (Geyer et al., 2017). Upon exposure to sun, plastics are transformed into numerous forms by physical degradation and reach the aquatic environment in the form of tiny plastic fragments, "microplastics" (Sobhani et al., 2020).

14.2 SOURCES OF MICROPLASTICS IN THE ENVIRONMENT

The National Oceanic and Atmospheric Administration (NOAA) defines microplastics as particles in the water that are <5 mm in size (Campanale et al., 2019; Dyachenko et al., 2017; Lenaker et al., 2021) and different categorization as presented in Figure 14.1. Microplastics pollution has recently been exposed largely to concern all life forms (Huang et al., 2021). As the ocean's big plastic waste continues to break down into millions of tiny bits, its concentration will rise (Cressey,

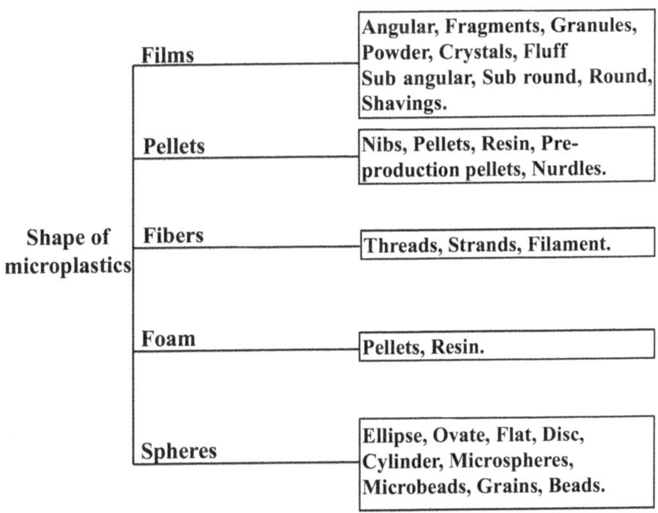

FIGURE 14.1 Categorization of microplastics based on their shapes. (Adapted from Yuan and Nag (2022), Patra et al. (2022), and Lusher et al. (2017).)

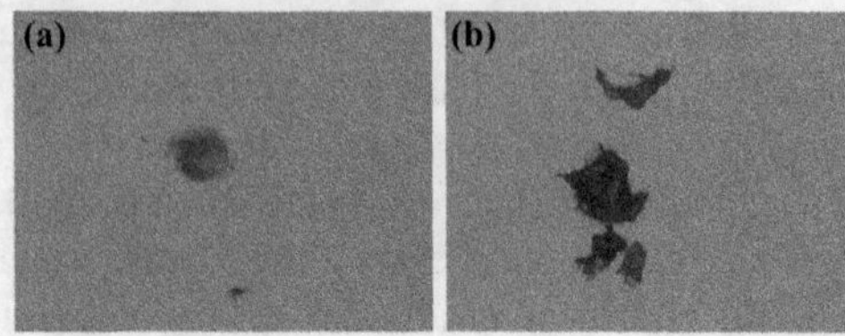

FIGURE 14.2 (a) Phase contrast microscope micrographs of MPs-PST and (b) MPs-PET. Magnifications 40x.

2016; Jambeck et al., 2015) and depicted in Figure 14.2. This deterioration results from environmental dangers such as river run-off, building debris, public littering, and natural disasters, all of which contribute to the direct paths of plastics into the waterways (De Matteis et al., 2022). There are many primary and secondary sources from which microplastics are generated. The primary sources include that the plastics that are introduced intentionally into the environment (El Hadri et al., 2020; Eriksen et al., 2014; Xu et al., 2020). For example, the "microbeads", which are small plastic particles used in personal care products (Ammala, 2013; King et al., 2017). However, due to their diminutive size, they will invade water systems and the natural environment (Eriksen et al., 2014; Sussarellu et al., 2016). Secondary sources of microplastics are accidentally introduced into the water due to harsh solar radiation (Malankowska et al., 2021). As a result, they eventually broke into fragments, leading to the formation of microplastics through weathering and degradation (Ekvall et al., 2019; Gigault et al., 2016). For example, synthetic fabrics that decompose into microfibers reach the atmosphere, rivers, lakes, and bigger bodies of water (Cutroneo et al., 2020; Stanton et al., 2020).

14.3 MICROPLASTICS IN THE AQUATIC ECOSYSTEM

As a result of their widespread emission, microplastics pose various obstacles to economic growth and have adverse effects on tourism (Chaudhry and Sachdeva, 2021). Consequently, deterioration in marine ecosystems that endanger human health is also observed. An estimated 700 species ingest microplastics, from microscopic zooplankton to giant whales (Choi et al., 2020; Teichert et al., 2021; Thushari and Senevirathna, 2020). Therefore, the occurrence of microplastics in diverse sources was explored in this section. The microplastics in freshwater are the least studied upon comparison to various aquatic environments (Lu et al., 2021). Taken this into consideration, the occurrence of microplastics in the freshwater ecosystem is summarized in Table 14.1. This limited information revealed that the abundance of microplastics is comparable to sea contamination levels (Lusher, 2015). Altogether, the primary source of microplastics in freshwater could be due to anthropogenic inputs (Tong et al., 2020).

Microplastics are often ubiquitous, yet their presence in marine ecosystems is rather modest (Avio et al., 2017; Parton et al., 2020). Researchers are searching for cost-effective methods of microplastic detection, collection, and removal from marine ecosystems. This is one of the largest barriers in the fight against microplastic pollution (Kristanti et al., 2022). There are currently no cleanup methods in place,

TABLE 14.1
Existence of Microplastics in Freshwater

S.No	Sample Type	MPs Range/Average	Size	Distribution (%)	Spectroscopy	Polymer Types	References
1	Water	288 pieces/m^3	1–5 and 0.3 mm	86.51	FTIR-ATR	PE; PET; PP; PVC	Amrutha and Warrier (2020)
	Sediment	96 pieces/kg					
	Soil	84.45 pieces/kg					
2	sediment	0.68–148.31 and 11.48–63.79 ng/g	5–10 mm	70	FTIR	PET; PE; PP; PS	Sarkar et al. (2019)
3	Sediment	40.7 particles/m^2	0.3–0.6; 0.6–1.18; 1.18–2.36 and 2.36–4.75 mm	96.10	FTIR-ATR, FP-XRF	PE; PP; PA; PS; PET; PUR; alkyd; CE; ABS; PVC, PVFM	Robin et al. (2020)
4	Water	1.25 particles/m^3				PE; PP; RY; PS	
5	Sediment	496 items/m^2	>5 mm	91	Raman spectra	Polymers; PP; PE; PS	Sruthy and Ramasamy (2017)
6	Sediment	5500 pellets	-	-	FTIR-ATR	PE; PP	Veerasingam et al. (2016)
7	Sediment	309 items/kg	−1 to 0.3 mm	80	ATR-FTIR	NYL; PE; PS; PP; PVC	Bharath K et al. (2021)
	Water	28 items/km^2	0.3–2 mm				
8	Water	5.9 items/L	0.33–2 mm	99	ATR-FTIR, SEM	HDPE; LDPE; PP; PS	Gopinath et al. (2020)
	Sediment	27 items/kg	2 mm				
9	Water	0.038 particles/L	2459 ± 209 μm	91	ATR-FTIR	RYN; CL; A; PET; PVC, PE; NYL	Napper et al. (2021)
10	Water	288 pieces/m^3	1–5 mm and 0.3–1 mm	80	FTIR -ATR	PE; PET; PVC; PP	Amrutha and Warrier (2020)
	Sediment	96 pieces/kg					
	Soil	84.45 pieces/kg					
11	Sediment	5.9 particles/L and 27 particles/kg	0.3 mm, 1 mm, and 2 mm	100	ATR-FTIR, SEM-EDAX	HDPE, LDPE, PP, PS,	Gopinath et al. (2020)
12	Water and sediment	28 items per square km^2 and 309 items/kg	0.3 mm, 1 mm, and 2 mm	90	ATR FT-IR	NYL; PE; PS; PP; PVP	Bharath et al. (2021)

nor have the sources that continue to add microplastics to ocean waters been shut down (Alimi et al., 2018). Table 14.2 illustrates the presence of microplastics in seawater. Due to substantial anthropogenic activities, the prevalence of microplastics in sediment has grown since its inception (Wong et al., 2020). The buildup of microplastics in deep-sea sediments is poorly known due to a number of practical challenges (Woodall et al., 2014; Zhang et al., 2020). In particular, the ocean is vast and deep, and obtaining samples from deep-sea environments is prohibitively expensive and fraught with

TABLE 14.2
Existence of Microplastics in Marine Environments

S. No	MPs Range/ Average	Size	Distribution (%)	Spectroscopy	Polymer Types	References
1	19.87 items/L	500 μm–1 mm	58	FTIR-ATR, SEM & EDAX	PE; PP; PP-PE; PA; PET; PVC; PS; RA; PV	Narmatha Sathish et al. (2020)
2	13.4 items/L	1–5 mm	64	FTIR-ATR and SEM- EDAX	PE, PE, PA, PS, PP, PE:PP, PP-PE, PVA	Sathish et al. (2020)
3	21.60 items/L	0.5–1 mm	100	FTIR-ATR analysis	PE; PP	Patterson et al. (2019)
4	0.93/m^2	35.29 to 5010 μm	–	ATR-FTIR	IS; PEI; A (AC fiber); PPS; EVA; A; EVAc; PIPU; PVC	Goswami et al. (2020)
5	1.25$\pm$0.88 particles/m^3	0.3–0.6; 0.6–1.18; 1.18–2.36 and 2.36–4.75 mm	70	FTIR-ATR, FP-XRF	PE, CL, RY, PE, PP	Robin et al. (2020)
6	0.80–6.99 items/m^3	0.29–2 mm and 0.08–0.8 mm	93.8	Micro-Raman spectra and SEM	PP, PBR, PA, PE, PEVA, PP	Anu Pavithran (2021)
7	54$\pm$41 to 619$\pm$377 items/L	0.5–1 mm	40.89	ATR-FTIR	E, PP, PS, PET and PA	Keerthika et al. (2022)
8	550,000 L (550 m^3)	150–500, 501–750, 751–1000 and 1001–5000 μm	71	BX53 Olympus microscope	Plastic fibers	Alfaro-Núñez et al. (2021)
9	0.014–12.51 items/m^3	25–250 μm	–	FT-IR	PA-NYL; PE, uPVC fibre, PE-fibres, PP	Cole et al. (2015)

logistical challenges (Barrett et al., 2020). According to the literature, microplastics accumulate in sediments on the seabed either directly by sinking through the water column or indirectly via currents and sediments carried over continental slopes (Kane and Clare, 2019). Nevertheless, it is crucial to determine the presence of microplastics in sediments. Table 14.3 illustrates the sediments with a large amount of microplastics.

Microplastic investigations on wastewater treatment plants (WWTP) and sludge are included in Table 14.4. It has been stated that modern wastewater treatment facilities can remove up to 95% of microplastics (Frehland et al., 2020; Xu et al., 2018). Unfortunately, significant quantities of microplastics are being dumped into natural waterways via WWTPs (Akarsu et al., 2020). Similarly, the obtained sludge is regarded as the predominant microplastic source (Li et al., 2018). In reality, sludge considerably affects the microplastic concentrations in soils (Corradini et al., 2019). Microplastics inputs to agricultural lands from wastewater and sewage sludge reuse have been estimated to be between 65,000 and 230,000 tonnes per year, making it one of the potential reserves for microplastics (Nguyen et al., 2022; Zilinskaite et al., 2022).

14.4 IDENTIFICATION AND CHARACTERIZATION OF MICROPLASTICS IN WATER

Studies tracking microplastics need consistent, standardized techniques of measurement to ensure accuracy and comparability (Lu et al., 2021). Research into the buildup of microplastics in tissues, meantime, necessitates the development of suitable analytical tools (Dong et al., 2023). Current efforts to detect microplastics include a wide variety of analytical approaches (Möller et al., 2020). Most of them, however, are still in their infant stages and need more standardization (Guo et al., 2022). Sampling, extraction, separation, identification, and quantification are the fundamental steps in evaluating environmental microplastics (Duis and Coors, 2016; Hanvey et al., 2017). A few microplastics are visible to the naked eye after extraction. However, cutting-edge equipment and methods are needed to identify tiny microplastics (Kovač Viršek et al., 2016). Therefore, techniques for identifying and characterizing microplastics are presented here.

The most common method for detecting microplastics in the environment, which are typically fibers, pieces, and beads, is light microscopy (Wang et al., 2021). However, it is extremely difficult to characterize plastic particles smaller than 100 μm that are colorless or amorphous (Woo et al., 2021). The identification of microplastics may also be hampered by inadequate sample particle separation (Nguyen et al., 2019). As a result, the scanning electron microscope can create remarkable, highly magnified images of microplastics (Shi et al., 2022). High-resolution images can distinguish between organic and microplastic particles by examining at surface roughness (Khoironi et al., 2020; Sait et al., 2021). In a few specific circumstances, transmission electron microscopy can be used to characterize microplastics (Yang et al., 2021). An electron microscope and energy dispersive spectroscopy can be used to map the constituent elements on the surface of microplastics (Jiang et al., 2020; Xiong et al., 2021). The crystallinity of microplastics can occasionally be assessed using polarized light microscopy as well (Rodríguez Chialanza et al., 2018). Polarized microscopy, however, can only be used with clear microplastics (Iñiguez et al., 2017; Shim et al., 2017).

TABLE 14.3
Existence of Microplastics in Sediment

S.No	MPs Range/Average	Size	Distribution (%)	Spectroscopy	Polymer Types	Reference
1	134.29 items/kg	1–5 mm	44	FTIR-ATR, SEM & EDAX	PE; PP; PP-PE; PA;PET; PEST; PVC; PS; RA; PVA	Narmatha Sathish et al. (2020)
2	12.75 items/kg	1–3 mm	100	FTIR-ATR analysis	PE, PP, PE, PA; paint	Patterson et al. (2019)
3	414.35 items/kg	100–1000 μm	100	Raman spectral poly(dimer acid-co-al polyamine), polypropylene, melami polyvinyl form polybutadiene	butadieneacrylonitrileacrylic acid polysulfide, poly(perfluoroethylene oxide), polyvinyl benzoa polyvinyl chloride, nylon epoxy epichlorhydr acrylonitrile butadiene	Patchaiyappan et al. (2020)
4	22.4 kg dry weight/km^2	–	40.6	–	Plastic bags, Fo wrappers, Personal ca products, Cu (Styrofoam), Bevera bottles, Plastic rope/ n pieces, Bottle or contain caps	Selvam et al. (2021)
5	385 items/kg	0.5–3 mm	100	FTIR-ATR SEM-EDAX	PE; PP; NY; PES, PS	Sathish et al. (2019)
6	220 MPs/kg; 181 MPs/kg; 45 MPs/kg	>1 mm	100	Fluorescence microscopy, SEM-EDS and FTIR	PET, PE, PVC, PP, PS, polyester, polyamides	Tiwari et al. (2019)
7	0.55 kg/100m^2	–	73.8	–	Plastic	Daniel et al. (2020)
8	191 items/kg	5–1 mm	70	FTIR and SEM	PE; PE + PP; PP; PS; PC	Ashwini and Varghese (2020)

(Continued)

TABLE 14.3 (*Continued*)
Existence of Microplastics in Sediment

S.No	MPs Range/Average	Size	Distribution (%)	Spectroscopy	Polymer Types	Reference
9	8.96 kg/m	–	56.42	–	Food wrappers; cups; bottle and caps; thermocol/Styrofoam; food wrappers	Arun Kumar et al. (2019)
10	72.03 ± 19.16 microplastic particles/100 g	300 µm–1 mm	56.32	Raman spectroscopy	Polypropylene, H—LDPE, polystyrene polyurethane	Dowarah and Devipriya (2019)
11	9233 mg/m²	1–2.5 mm	60.1	FTIR-ATR	PE; PP; PS; NY	Karthik et al. (2018)
12	403 pieces	>1.01–200 mm	60.8	FTIR spectroscopy	PE, PS, NY, PVC	Vidyasakar et al. (2018)
13	505 pieces	–	–	FTIR	Polystyrene, polyethylene, polycarbonate, polyvinyl chloride; nylon	Krishnakumar et al. (2018)
14	3.24 g/m²	5–100 mm	80	–	plastic	Jayasiri et al. (2013)
15	277.90 items/kg	0.01–3 mm	49	Digital Microscope	Polyethylene; polypropylene; polystyrene	Perumal and Muthuramalingam (2021)
16	45.17 particles/kg	46.72–5024 µm	–	ATR-FTIR	IS; PEM, A (AC fiber); PPS; EVA, AN-NYL; EVAc; PIPU, PVC	Goswami et al. (2020)
17	1029 items/m²	>1 mm	96	ATR-FTIR spectroscopy	PE, PP, PS, PA, PA, PVC	Imhof et al. (2017)
18	40.7 ± 33.2 particles/m²	0.3–0.6; 0.6–1.18; 1.18–2.36 and 2.36–4.75 mm	99.99	SMZ, FTIR-ATR, FP-XRF	PE; CE; RY; PL; PP	Robin et al. (2020)

TABLE 14.4
Existence of Microplastics in Wastewater Treatment Plant

S.No	MPs Range/Average	Size	Distribution (%)	Spectroscopy	Polymer Types	Reference
1	1.5 MP/L	392 μm (± 27 μm SE)	76.9	Micro FTIR	PE; PP; PS; PVC; NYL	Tagg et al. (2020)
2	1.0–1.143 g/mL	400 and 45 mm	95	FTIR and microscope	PE microbeads	Carr et al. (2016)
3	0.21 to 1.5 microplastics/L	25–500 μm	92–99	ATR-FTIR	PE; PET; PS; PP	Ziajahromi et al. (2017)
4	15.70 (±5.23) MP/L	1.342 mm (±0.519)	98.41	FT-IR	Alkyd fragment; PP fiber, PE microbeads	Murphy et al. (2016)
5	12.43 (±2.70) ML/L 204, 9.73 (±3.04) ML/L, 3.21 (±0.50) ML/L, and 1.23 (±0.15) ML/L	400–600 μm	90.1	FTIR	LDPE; HDPE; Acryl (acrylate); PP; PEP; PS; BPL; NYL; PUR; PET; MCR; PTFE (Teflon); MMF (melamine); PES (polyester); PVI; PIB; RBB	Bayo et al. (2020)
6	8.1×10^8 items/day	60–2800 μm	95	FTIR-ATR	PP; PE; PVS; POM	Blair et al. (2019)
7	37.7–97.2 no of microplastics/g of sludge	100–500 μm	67.1	FT-IR	PEL; PE; PP; PVC	Harley-Nyang et al. (2022)

(Continued)

TABLE 14.4 (*Continued*)
Existence of Microplastics in Wastewater Treatment Plant

S. No	MPs Range/Average	Size	Distribution (%)	Spectroscopy	Polymer Types	Reference
8	240.3 ± 31.4 n /g	222.6 µm	54.8	Raman	MPs; PA; PP; PE; PC; PVC	Liu et al (2019)
9	0–5 × 101 m^{-3} MP > 500 µm and 1 × 101 to 9 × $10^3 m^{-3}$ MP	<500 µm	–	ATR-FT-IR, (FPA)-micro-FT-IR imaging	PE; PP; PA; PVC; PS; IR-PUR; silicone; PUR-based coatings	Mintenig et al. (2017)
10	103.4 pieces/L	300–4750 µm	32–57	Raman and FTIR spectra	PE; PES; PVC; PP; NYL; PS	Hongprasith et al. (2020)
11	545.9 and 87.6 items/kg	> 0.5 mm	–	Micro-Fourier-transform infrared (µ-FTIR) spectrometer	PP; PP/PE; PET; PEB; EVA	Zhang et al. (2020)

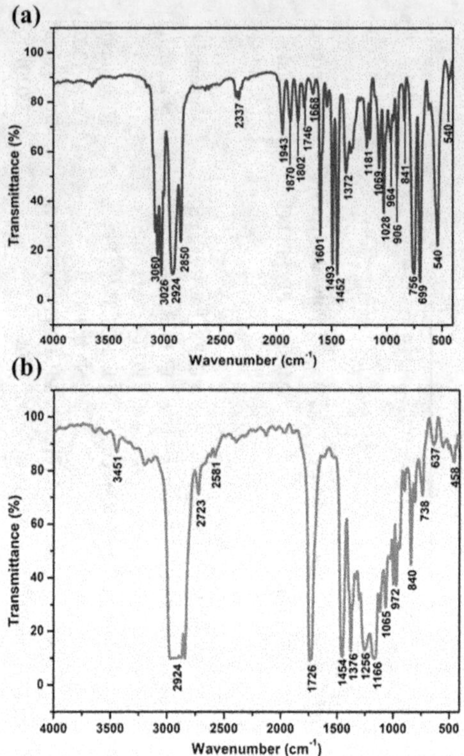

FIGURE 14.3 FT-IR spectra of MPs polystyrene (a), cellophane (b).

Fourier transform infrared (FTIR) spectroscopy displays the chemical bonding of particles by identifying carbon molecules inside polymers (Song et al., 2015) and is presented in Figure 14.3.

Moreover, a spectrum library is employed to identify the polymer (De Frond et al., 2021). Similarly, micro-FTIR may be used to identify microplastics on a single platform due to the fact that it possesses both microscopic and spectroscopic features with transmittance, reflectance, and ATR modes (Qiu et al., 2016). Microplastics that are opaque and thick do not need sample preparation for reflection and ATR modes (Veerasingam et al., 2021). ATR is capable of producing stable spectra on microplastic surfaces (Möller et al., 2022). In contrast, Raman spectroscopy uses a laser to bombard an object, and the frequency of the backscattered light changes according to the molecular structure and atoms of the target, resulting in a spectrum that is unique for each polymer (Chaisrikhwun et al., 2023). Raman spectroscopy, like FTIR, can detect plastics and provide polymer composition profiles by understanding both data libraries and algorithms (Becucci et al., 2022). Micro-Raman spectroscopy, like FTIR, combines microscopy and chemical analysis (Käppler et al., 2018; Renner et al., 2018). In micro-Raman spectroscopy, only materials with a Raman reference spectrum in the database may be identified (Trainic et al., 2020).

Thermal analysis methods, which look at the physical and chemical changes in thermally stable polymers, can be used to detect microplastics. Differential scanning calorimetric (DSC) is used to measure the rate of heat transfer (Dierkes et al., 2021; Shabaka et al., 2020). Dissolution, crystallization, transition temperatures, enthalpy, and entropy are all ways to investigate the polymer's characteristics (Samanta et al., 2022; Zamboulis et al., 2019). Due to variations in DSC characteristics among various plastic goods, DSC may be used to identify distinct kinds of polymers (Dierkes et al., 2021). Other forms of heat analysis include thermo gravimetric analysis spectroscopy (TGA) (Majewsky et al., 2016). By monitoring the sample's weight loss while it is heated at a predetermined pace in a well-controlled atmosphere, we can verify the accuracy and completeness of the data (Gomiero et al., 2021). Analysis of pyrolyzed polymer gas is performed by means of pyrolysis gas chromatography-mass spectrometry (Py-GC-MS) (Dierkes et al., 2019; Li et al., 2021). A sample's plasticity may be determined by comparing its pyrogram to that of a standard polymer (Sun et al., 2019). Unlike spectroscopy, heat analysis may be used to find polymers (Luo et al., 2021). Heat analysis is damaging therefore microplastic materials cannot be analyzed retrospectively (Woo et al., 2021). The merits of each method are weighed, and if necessary, further analytical approaches are employed to fill in the gaps. Thermal analysis and GC-MS can help find out how much microplastic is in a large sample (Mintenig et al., 2018). TGA-FTIR can only be used for qualitative research; however, TGA-FTIR-GC-MS can identify microplastics in real-time and can be used for quantitative research as well (Liu et al., 2021). Some of the byproducts of a thermal breakdown can be investigated with mass spectrometry (Albignac et al., 2022; Toapanta et al., 2021). The execution of the instruments and the processing of the data take longer duration when utilizing FTIR or Raman spectroscopy (Brandt et al., 2021).

14.5 TYPES OF MICROPLASTICS

The different types of microplastics such as PET, PVC, PP, PE, and PS have been detected in marine environments (Andrady, 2017). In addition, the microplastics contain several significant polymers, as presented in Table 14.5 (Crawford and Quinn, 2017). The amount of crystallinity in plastic polymers is determined by the number

TABLE 14.5
Types and Characterizes of Various Microplastics

Types of Microplastics	Degree of Crystallinity	Chemical Formula	Density (g/cm³)
PE	Semi-crystalline	$(C_2H_4)_n$	0.92–0.97
PA66	Semi-crystalline	$(C_{12}H_{22}N_2O_2)_n$	1.13–1.38
PS	Amorphous	$(C_8H_8)_n$	1.04–1.50
PVC	Amorphous (atactic)	$(C_2H_3Cl)_n$	1.15–1.70
PP	Semi-crystalline (isotactic, syndiotactic)	$(C_3H_6)_n$	0.88–1.23
PA6	Semi-crystalline	$(C_6H_{11}NO)_n$	1.12–1.14
PET	Amorphous	$(C_{10}H_8O_4)_n$	1.30–1.50

Source: Crawford and Quinn (2017)

of crystalline areas, and the critical chains of the polymers are bound with each other. The amount of crystallinity directly contributes to the polymer resources. Based on the crystallinity levels of the polymers, semi-crystalline plastic polymers consist of great resistance and strength. The amorphous plastic polymers appear to be highly flexible, and soft, but in reality, they have a very poor resistance capacity and strength. In particular, the amorphous plastics polymers such as rubber and glass-based materials (Guo and Wang, 2019).

Polymers actively convert glassy into rubbery form when increasing the temperature (glass conversion heat). The amount of microplastics that are released into the water column can be controlled by the microplastics' thickness. Accordingly, microplastics made of PE and PP float in the water because of low densities when compared to the water (Cheng et al., 2020). The densities of PA, PS, PVC, and PET are higher than those of water. These plastic polymers (*i.e.*, PP, PE, PVC, PET, PS, and PA) have significant materials to develop various plastic-based products as well as wrapper materials, which actively eliminate microplastic polymers, particularly PA, PE, and PP into wastewater (Koelmans et al., 2019; Suaria et al., 2016). Further, microplastics containing wastewater migrate to water reservoirs. Currently, various techniques have been employed to detect microplastics in wastewater (Bilgin et al., 2020). Pharmaceuticals and personal care products (cosmetics, medicine, face scrubs, various toothpastes, etc.) can release a substantial amount of PE (Barboza and Gimenez, 2015; Madhumitha et al., 2022). On the other hand, the use of plastic products that include PP, PE, and PET can pollute aquatic environment (Chia et al., 2020; Wu et al., 2017). The various medical products, thermal garments, and building materials release a great amount of PP (Blair et al., 2019). Recently, the hazardous microplastic materials that are broadly noticed in oceans, estuaries, and lakes are well documented (Kanhai et al., 2018). The surface of the water contains a significant amount of microplastic particles (Xiong et al., 2018). The research carried out by (Eo et al., 2018) indicated that smaller (0.02–1 mm) microplastics are more widely dispersed than large (1–5 mm) microplastics. Further, the presence of microplastics in sediments and seawater demonstrates great variances.

14.6 TOXICOLOGICAL AND RISK ASSESSMENT OF MICROPLASTICS IN AQUATIC ENVIRONMENT

Microplastics have been investigated more for their impact on aquatic habitats than for their widespread prevalence in the environment (Chaukura et al., 2021). Recent researches indicate that plastic pollution is equally prevalent in the world's freshwater systems as it is in the oceans (Silva-Cavalcanti et al., 2023). The majority of plastic waste originates terrestrially but ultimately ends up in waterways, causing issues for aquatic creatures and the ecosystems on which they rely (Thushari and Senevirathna, 2020). Due to their propensity to become entangled in plastic rubbish or become unwell after consuming it, marine animals and birds have been the subject of much research (Smith et al., 2018; Tanaka et al., 2013). The consumption of microplastics by invertebrates has far-reaching biological repercussions, from the cellular to the population level (Prinz and Korez, 2020). Limited studies have been conducted on the effects of microplastics on aquatic organisms, as well as the mechanisms of absorption other than ingestion (de Sá et al., 2018).

It has been demonstrated that consuming microplastics impairs the body's ability to absorb vital nutrients (Guzzetti et al., 2018). Numerous animals incorrectly consume non-food items, such as bacteria and fungi, as food. There is evidence that both pelagic and benthic marine organisms ingest plastic (Corcoran, 2015; Ogunola et al., 2018). Marine invertebrates include amphipods, lugworms, barnacles, mussels, crabs, and sea cucumbers, among others (Lusher, 2015). The long-term consequences of microplastics have been studied and confirmed (Lusher et al., 2017). The effects of microplastics on newborn *Daphnia magna*, *Hyalella azteca*, and *Calanus helgolandicus* were negative. *D. magna* treated with PP nanoplastic particles had smaller, more seriously deformed offspring compared to controls (Aljaibachi and Callaghan, 2018; Au et al., 2015; Cole et al., 2015).

Two months of exposure to polystyrene microplastics had a detrimental effect on oyster fertility and embryo development. Reduced egg production (38%), oocyte size (5%), and sperm velocity (23%) all contributed to an 18% slowdown in oyster larval development (Sussarellu et al., 2016). The development of *Tigriopus japonicus* nauplii, copepodites, and adults was stunted by the presence of polystyrene microplastics (0.005–0.5 m) during both generations (Lee et al., 2013). It is likely that the number of offspring produced and the range of these species might expand if creatures began laying eggs on microplastics and other plastic debris (Goldstein et al., 2012). Plastic does not biodegrade like wood or kelp, allowing offspring to migrate to safer locations (Zhang et al., 2021). Paint chips include copious amounts of polymeric (PP, PE, PET, expanded PS, and polyurethane) components, and additives also affect reproduction in aquatic organisms (Engler, 2012). The exposure of *Assiminea grayana* snails to unspecified-size particles impaired the development of the snails' reproductive organs (i.e., necrotic oocytes and atypical spermatozoa) (Watermann et al., 2017). Other pollutants in the microplastics (diuron, tributyltin, and other metals) may have been more responsible for the abnormal development of reproductive organs (Besseling et al., 2014; Cole and Galloway, 2015; Falandysz et al., 2006).

As a result of the leaking of endocrine-disrupting chemicals, the developmental toxicity of old PE microplastics was shown to be greater in Japanese medaka (Amereh et al., 2020). The gene expression of oestrogen receptor (ER), vitellogenin (Vtg 1), and choriogenin H (Chg H) in Japanese medaka was inhibited by a two-month diet of aged PE microplastics (*Oryzias latipes*) (Kannan and Vimalkumar, 2021; Wang et al., 2022). Changes in fertility can encourage life cycle adaptations across populations because different species within a group rely on different food sources at various stages of development (Cocci et al., 2022; Park et al., 2020). In juvenile brown mussels (*Perna perna*) exposed to aged PE microplastics, developmental defects and mortality were more prevalent than in juveniles exposed to virgin microplastics (Hou et al., 2021; Ma et al., 2020).

14.7 NEED FOR A CHANGE

Recently, various technologies have been developed and suggested to mitigate microplastic pollution (Bilal and Iqbal, 2020). For example, photocatalysis is one of the suitable techniques for the remediation of microplastics, but there is the availability of minimum evidence regarding the elimination of byproducts (Ebrahimbabaie et al., 2022; Uheida et al., 2021). Electrooxidation is a suitable technique for the mitigation of various microplastic wastes, where microplastic effectively transforms into several gases (Kiendrebeogo et al., 2021). Nonetheless, eliminated gases can interact with other toxic gases. In addition, membrane-based technology has received great attention from environmental workers due to its effectiveness in removing microplastics (Du et al., 2021; Jin et al., 2021; Kundu et al., 2021). But, membrane fouling and blocking are one of the problematic issues. At present, no precise novel remediation method is associated with microplastics (Malankowska et al., 2021). Therefore, there is a requirement for more investigations on the remediation of hazardous microplastic wastes. It may be possible to carbonize the microplastics that have been collected from the various wastewaters in order to produce the high-cost carbon-related materials that may be used as a catalyst for the removal of various pollutants (Nguyen et al., 2022; Wang et al., 2022). This is a promising way to reduce the amount of pollution caused by microplastics. The development of innovative equipment that can identify microfibers during the washing process and other industrial processes could lead to a reduction in the amount of microfibers that end up in the sewage system.

14.8 BIOPLASTICS AS AN EMINENT ALTERNATE

At present, thermoplastics including PS, PP, and PE account for 60% of Europe's total demand for plastic (Kawecki et al., 2021). The majority of plastic materials are able to be produced using petrochemical chemicals. This category of plastics is responsible for a wide range of environmental problems, the most notable of which is the pollution of water and soil. Plastic compounds can also cause serious health problems in humans. As a result, alternative materials for plastics are required. Bioplastics are one material that has great potential for use in developing new goods (Figure 14.4).

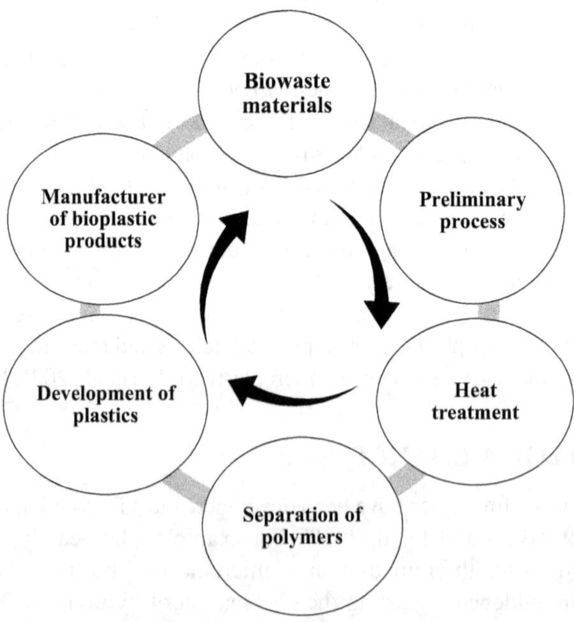

FIGURE 14.4 Process of bio-plastic manufacture.

On the other hand, not all of them are decomposable (Li et al., 2015; Narancic and O'Connor, 2017). Bioplastics are members of the plastic family, which can be separated into two distinct groups: (i) biodegradable and (ii) non-biodegradable bioplastics. The biodegradable related bioplastics such as polyhydroxyalkanoates (PHA), polylactic acid (PLA), starch, and cellulose. Likewise, both biobased and oil-based plastics are capable of being recycled (or) incinerated, but biobased and oil-based plastics are not currently recycled on a large scale due to the presence of contaminates. Certain types of plastics can be effectively broken down through processes such as microbial decomposition, anaerobic digestion, and home composting (Emadian et al., 2017; Narancic et al., 2018). Bioplastics that are not easily decomposed (non-biodegradable) in natural environments, include bio-polyethylene, polyol-polyurethane, and bio-polyethylene terephthalate.

14.9 CONCLUSIONS

The freshwater and marine environment plays a pivotal role in the food supply for many people, particularly in developing countries. In recent decades, a huge amount of plastic-based materials have been used globally, and unscientific disposal of plastic materials which releases microplastics, resulting in generate serious environmental pollution and affecting aquatic organisms. These microplastics are a great source of water pollution in rivers as well as oceans, and they are predicted to rise by 2025. According to recent reports, microplastics were noticed in various aquatic food materials and water. Food products polluted with microplastics can cause a variety of diseases and problems in humans, such as endocrine disruption, cancer, neurobehavioral variations, neurotoxicity, and reproductive toxicity. The authors strongly suggest that the development of bioplastic materials is a feasible way to reduce microplastic contamination. In particular, biomass-related polymers are appropriate substitutions for non-biodegradable plastics. The authors recommended that different waste materials including animal wastes, plant residues, and other biomass suitable for the development of environment-friendly bioplastics.

ACKNOWLEDGMENTS

Ponnuchamy Kumar acknowledges the Science Engineering and Research Board (SERB), New Delhi, India, for providing financial assistance from the major research project (EEQ/2017/000135; dated: March 23, 2018). The authors also thank the support of the RUSA – Phase 2.0 grant (F. 24-51/2014U).

REFERENCES

Ahmed, T., Shahid, M., Azeem, F., Rasul, I., Shah, A. A., Noman, M., Hameed, A., Manzoor, N., Manzoor, I., and Muhammad, S. (2018). Biodegradation of plastics: Current scenario and future prospects for environmental safety. *Environmental Science and Pollution Research*, 25(8), 7287–7298.

Akanyange, S. N., Zhang, Y., Zhao, X., Adom-Asamoah, G., Ature, A.-R. A., Anning, C., Tianpeng, C., Zhao, H., Lyu, X., and Crittenden, J. C. (2022). A holistic assessment of microplastic ubiquitousness: Pathway for source identification in the environment. *Sustainable Production and Consumption*, 33, 113–145.

Akarsu, C., Kumbur, H., Gökdağ, K., Kıdeyş, A. E., and Sanchez-Vidal, A. (2020). Microplastics composition and load from three wastewater treatment plants discharging into Mersin Bay, north eastern Mediterranean Sea. *Marine Pollution Bulletin*, 150, 110776.

Albignac, M., Ghiglione, J. F., Labrune, C., and ter Halle, A. (2022). Determination of the microplastic content in Mediterranean benthic macrofauna by pyrolysis-gas chromatography-tandem mass spectrometry. *Marine Pollution Bulletin*, 181, 113882.

Alfaro-Núñez, A., Astorga, D., Cáceres-Farías, L., Bastidas, L., Soto Villegas, C., Macay, K. C., and Christensen, J. H. (2021). Microplastic pollution in seawater and marine organisms across the tropical Eastern Pacific and Galápagos. *Scientific Reports*, 11(1), 6424. https://doi.org/10.1038/s41598-021-85939-3

Alimi, O. S., Farner Budarz, J., Hernandez, L. M., and Tufenkji, N. (2018). Microplastics and nanoplastics in aquatic environments: Aggregation, deposition, and enhanced contaminant transport. *Environmental Science & Technology*, 52(4), 1704–1724.

Aljaibachi, R., and Callaghan, A. (2018). Impact of polystyrene microplastics on Daphnia magna mortality and reproduction in relation to food availability. *PeerJ*, 6, e4601.

Amereh, F., Babaei, M., Eslami, A., Fazelipour, S., and Rafiee, M. (2020). The emerging risk of exposure to nano(micro)plastics on endocrine disturbance and reproductive toxicity: From a hypothetical scenario to a global public health challenge. *Environmental Pollution*, 261, 114158.

Ammala, A. (2013). Biodegradable polymers as encapsulation materials for cosmetics and personal care markets. *International Journal of Cosmetic Science*, 35(2), 113–124.

Amrutha, K., and Warrier, A. K. (2020). The first report on the source-to-sink characterization of microplastic pollution from a riverine environment in tropical India. *Science of The Total Environment*, 739, 140377. https://doi.org/10.1016/j.scitotenv.2020.140377

Andrady, A. L. (2017). The plastic in microplastics: A review. *Marine Pollution Bulletin*, 119(1), 12–22.

Andrady, A. L., and Neal, M. A. (2009). Applications and societal benefits of plastics. *Philosophical Transactions of the Royal Society B: Biological Sciences*, 364(1526), 1977–1984.

Anu Pavithran, V. (2021). Study on microplastic pollution in the coastal seawaters of selected regions along the northern coast of Kerala, southwest coast of India. *Journal of Sea Research*, 173, 102060. https://doi.org/10.1016/j.seares.2021.102060

Arun Kumar, A., Sivakumar, R., Sai Rutwik, Y., Nishanth, T., Revanth, V., and Kumar, S. (2019). Marine debris in India: Quantifying type and abundance of beach litter along Chennai, east coast of India. In M. L. Kolhe, P. K. Labhasetwar, and H. M. Suryawanshi (Eds.), *Smart Technologies for Energy, Environment and Sustainable Development* (pp. 217–230). Springer, Singapore. https://doi.org/10.1007/978-981-13-6148-7_23

Ashwini, S. K., and Varghese, G. K. (2020). Environmental forensic analysis of the microplastic pollution at "Nattika" Beach, Kerala Coast, India. *Environmental Forensics*, 21(1), 21–36. https://doi.org/10.1080/15275922.2019.1693442

Au, S. Y., Bruce, T. F., Bridges, W. C., and Klaine, S. J. (2015). Responses of Hyalella azteca to acute and chronic microplastic exposures: Effects of Microplastic Exposure on Hyalella azteca. *Environmental Toxicology and Chemistry*, 34(11), 2564–2572.

Avio, C. G., Gorbi, S., and Regoli, F. (2017). Plastics and microplastics in the oceans: From emerging pollutants to emerged threat. *Marine Environmental Research*, 128, 2–11.

Barboza, L. G. A., and Gimenez, B. C. G. (2015). Microplastics in the marine environment: Current trends and future perspectives. *Marine Pollution Bulletin*, 97(1–2), 5–12. 08

Barrett, J., Chase, Z., Zhang, J., Holl, M. M. B., Willis, K., Williams, A., Hardesty, B. D., and Wilcox, C. (2020). Microplastic pollution in deep-sea sediments from the great Australian bight. *Frontiers in Marine Science*, 7, 576170.

Bayo, J., Olmos, S., and López-Castellanos, J. (2020). Microplastics in an urban wastewater treatment plant: The influence of physicochemical parameters and environmental factors. *Chemosphere*, 238, 124593. https://doi.org/10.1016/j.chemosphere.2019.124593

Becucci, M., Mancini, M., Campo, R., and Paris, E. (2022). Microplastics in the Florence wastewater treatment plant studied by a continuous sampling method and Raman spectroscopy: A preliminary investigation. *Science of The Total Environment*, 808, 152025.

Besseling, E., Wang, B., Lürling, M., and Koelmans, A. A. (2014). Nanoplastic affects growth of S. obliquus and reproduction of D. magna. *Environmental Science & Technology*, 48(20), 12336–12343.

Bharath, K. M., Srinivasalu, S., Natesan, U., Ayyamperumal, R., Kalam, S. N., Anbalagan, S., Sujatha, K., and Alagarasan, C. (2021). Microplastics as an emerging threat to the freshwater ecosystems of Veeranam lake in south India: A multidimensional approach. *Chemosphere*, 264, 128502. https://doi.org/10.1016/j.chemosphere.2020.128502

Bilal, M., and Iqbal, H. M. N. (2020). Transportation fate and removal of microplastic pollution - A perspective on environmental pollution. *Case Studies in Chemical and Environmental Engineering*, 2, 100015.

Bilgin, M., Yurtsever, M., and Karadagli, F. (2020). Microplastic removal by aerated grit chambers versus settling tanks of a municipal wastewater treatment plant. *Journal of Water Process Engineering*, 38, 101604.

Blair, R. M., Waldron, S., and Gauchotte-Lindsay, C. (2019). Average daily flow of microplastics through a tertiary wastewater treatment plant over a ten-month period. *Water Research*, 163, 114909. https://doi.org/10.1016/j.watres.2019.114909

Blair, R. M., Waldron, S., Phoenix, V. R., and Gauchotte-Lindsay, C. (2019). Microscopy and elemental analysis characterisation of microplastics in sediment of a freshwater urban river in Scotland, UK. *Environmental Science and Pollution Research*, 26(12), 12491–12504.

Brandt, J., Mattsson, K., and Hassellöv, M. (2021). Deep learning for reconstructing low-quality FTIR and Raman Spectra—A case study in microplastic analyses. *Analytical Chemistry*, 93(49), 16360–16368.

Browne, M. A., Crump, P., Niven, S. J., Teuten, E., Tonkin, A., Galloway, T., and Thompson, R. (2011). Accumulation of microplastic on shorelines woldwide: Sources and sinks. *Environmental Science & Technology*, 45(21), 9175–9179.

Campanale, C., Massarelli, C., Bagnuolo, G., Savino, I., and Uricchio, V. F. (2019). The problem of microplastics and regulatory strategies in Italy. In F. Stock, G. Reifferscheid, N. Brennholt, and E. Kostianaia (Eds.), *Plastics in the Aquatic Environment-Part II* (Vol. 112, pp. 255–276). Springer International Publishing.

Carr, S. A., Liu, J., and Tesoro, A. G. (2016). Transport and fate of microplastic particles in wastewater treatment plants. *Water Research*, 91, 174–182. https://doi.org/10.1016/j.watres.2016.01.002

Chaisrikhwun, B., Ekgasit, S., and Pienpinijtham, P. (2023). Size-independent quantification of nanoplastics in various aqueous media using surfaced-enhanced Raman scattering. *Journal of Hazardous Materials*, 442, 130046.

Chamas, A., Moon, H., Zheng, J., Qiu, Y., Tabassum, T., Jang, J. H., Abu-Omar, M., Scott, S. L., and Suh, S. (2020). Degradation rates of plastics in the environment. *ACS Sustainable Chemistry & Engineering*, 8(9), 3494–3511.

Chaudhry, A. K., and Sachdeva, P. (2021). Microplastics' origin, distribution, and rising hazard to aquatic organisms and human health: Socio-economic insinuations and management solutions. *Regional Studies in Marine Science*, 48, 102018.

Chaukura, N., Kefeni, K. K., Chikurunhe, I., Nyambiya, I., Gwenzi, W., Moyo, W., Nkambule, T. T. I., Mamba, B. B., and Abulude, F. O. (2021). Microplastics in the aquatic environment-The occurrence, sources, ecological impacts, fate, and remediation challenges. *Pollutants*, 1(2), 95–118.

Cheng, H., Luo, H., Hu, Y., and Tao, S. (2020). Release kinetics as a key linkage between the occurrence of flame retardants in microplastics and their risk to the environment and ecosystem: A critical review. *Water Research*, 185, 116253.

Chia, W. Y., Ying Tang, D. Y., Khoo, K. S., Kay Lup, A. N., and Chew, K. W. (2020). Nature's fight against plastic pollution: Algae for plastic biodegradation and bioplastics production. *Environmental Science and Ecotechnology*, 4, 100065.

Choi, J. S., Hong, S. H., and Park, J.-W. (2020). Evaluation of microplastic toxicity in accordance with different sizes and exposure times in the marine copepod Tigriopus japonicus. *Marine Environmental Research*, 153, 104838.

Cocci, P., Gabrielli, S., Pastore, G., Minicucci, M., Mosconi, G., and Palermo, F. A. (2022). Microplastics accumulation in gastrointestinal tracts of Mullus barbatus and Merluccius merluccius is associated with increased cytokine production and signaling. *Chemosphere*, 307, 135813.

Cole, M., and Galloway, T. S. (2015). Ingestion of nanoplastics and microplastics by pacific oyster larvae. *Environmental Science & Technology*, 49(24), 14625–14632.

Cole, M., Lindeque, P., Fileman, E., Halsband, C., and Galloway, T. S. (2015). The impact of polystyrene microplastics on feeding, function and fecundity in the marine copepod Calanus helgolandicus. *Environmental Science & Technology*, 49(2), 1130–1137.

Cole, M., Webb, H., Lindeque, P. K., Fileman, E. S., Halsband, C., and Galloway, T. S. (2015). Isolation of microplastics in biota-rich seawater samples and marine organisms. *Scientific Reports*, 4(1), 4528. https://doi.org/10.1038/srep04528

Corcoran, P. L. (2015). Benthic plastic debris in marine and fresh water environments. *Environmental Science: Processes & Impacts*, 17(8), 1363–1369.

Corradini, F., Meza, P., Eguiluz, R., Casado, F., Huerta-Lwanga, E., and Geissen, V. (2019). Evidence of microplastic accumulation in agricultural soils from sewage sludge disposal. *Science of the Total Environment*, 671, 411–420.

Crawford, C. B., and Quinn, B. (2017). Plastic production, waste and legislation. In *Microplastic Pollutants* (pp. 39–56). Elsevier, https://doi.org/10.1016/B978-0-12-809406-8.00003-7

Cressey, D. (2016). Bottles, bags, ropes and toothbrushes: The struggle to track ocean plastics. *Nature*, 536(7616), 263–265.

Cutroneo, L., Reboa, A., Besio, G., Borgogno, F., Canesi, L., Canuto, S., Dara, M., Enrile, F., Forioso, I., Greco, G., Lenoble, V., Malatesta, A., Mounier, S., Petrillo, M., Rovetta, R., Stocchino, A., Tesan, J., Vagge, G., and Capello, M. (2020). Microplastics in seawater: Sampling strategies, laboratory methodologies, and identification techniques applied to port environment. *Environmental Science and Pollution Research*, 27(9), 8938–8952.

Daniel, D. B., Thomas, S. N., and Thomson, K. T. (2020). Assessment of fishing-related plastic debris along the beaches in Kerala Coast, India. *Marine Pollution Bulletin*, 150, 110696. https://doi.org/10.1016/j.marpolbul.2019.110696

De Frond, H., Rubinovitz, R., and Rochman, C. M. (2021). MATR-FTIR spectral libraries of plastic particles (FLOPP and FLOPP-e) for the analysis of microplastics. *Analytical Chemistry*, 93(48), 15878–15885.

De Matteis, A., Turkmen Ceylan, F. B., Daoud, M., and Kahuthu, A. (2022). A systemic approach to tackling ocean plastic debris. *Environment Systems and Decisions*, 42(1), 136–145.

de Sá, L. C., Oliveira, M., Ribeiro, F., Rocha, T. L., and Futter, M. N. (2018). Studies of the effects of microplastics on aquatic organisms: What do we know and where should we focus our efforts in the future? *Science of the Total Environment*, 645, 1029–1039.

Desidery, L., and Lanotte, M. (2022). Polymers and plastics: Types, properties, and manufacturing. In *Plastic Waste for Sustainable Asphalt Roads* (pp. 3–28). Woodhead Publishing, https://doi.org/10.1016/B978-0-323-85789-5.00001-0.

Dierkes, G., Lauschke, T., Becher, S., Schumacher, H., Földi, C., and Ternes, T. (2019). Quantification of microplastics in environmental samples via pressurized liquid extraction and pyrolysis-gas chromatography. *Analytical and Bioanalytical Chemistry*, 411(26), 6959–6968.

Dierkes, G., Lauschke, T., and Földi, C. (2021). Analytical methods for plastic (microplastic) determination in environmental samples. In F. Stock, G. Reifferscheid, N. Brennholt, and E. Kostianaia (Eds.), *Plastics in the Aquatic Environment-Part I* (Vol. 111, pp. 43–67). Springer International Publishing. http://dx.doi.org/10.1007/698_2021_744.

Dong, X., Liu, X., Hou, Q., and Wang, Z. (2023). From natural environment to animal tissues: A review of microplastics(nanoplastics) translocation and hazards studies. *Science of the Total Environment*, 855, 158686.

Dowarah, K., and Devipriya, S. P. (2019). Microplastic prevalence in the beaches of Puducherry, India and its correlation with fishing and tourism/recreational activities. *Marine Pollution Bulletin*, 148, 123–133. https://doi.org/10.1016/j.marpolbul.2019.07.066

Du, H., Xie, Y., and Wang, J. (2021). Microplastic degradation methods and corresponding degradation mechanism: Research status and future perspectives. *Journal of Hazardous Materials*, 418, 126377.

Duis, K., and Coors, A. (2016). Microplastics in the aquatic and terrestrial environment: Sources (with a specific focus on personal care products), fate and effects. *Environmental Sciences Europe*, 28(1), 2.

Dyachenko, A., Mitchell, J., and Arsem, N. (2017). Extraction and identification of microplastic particles from secondary wastewater treatment plant (WWTP) effluent. *Analytical Methods*, 9(9), 1412–1418.

Ebrahimbabaie, P., Yousefi, K., and Pichtel, J. (2022). Photocatalytic and biological technologies for elimination of microplastics in water: Current status. *Science of the Total Environment*, 806, 150603.

Ekvall, M. T., Lundqvist, M., Kelpsiene, E., Šileikis, E., Gunnarsson, S. B., and Cedervall, T. (2019). Nanoplastics formed during the mechanical breakdown of daily-use polystyrene products. *Nanoscale Advances*, 1(3), 1055–1061.

El Hadri, H., Gigault, J., Maxit, B., Grassl, B., and Reynaud, S. (2020). Nanoplastic from mechanically degraded primary and secondary microplastics for environmental assessments. *NanoImpact*, 17, 100206.

Emadian, S. M., Onay, T. T., and Demirel, B. (2017). Biodegradation of bioplastics in natural environments. *Waste Management*, 59, 526–536.

Engler, R. E. (2012). The complex interaction between marine debris and toxic chemicals in the ocean. *Environmental Science & Technology*, 46(22), 12302–12315.

Eo, S., Hong, S. H., Song, Y. K., Lee, J., Lee, J., and Shim, W. J. (2018). Abundance, composition, and distribution of microplastics larger than 20 μm in sand beaches of South Korea. *Environmental Pollution*, 238, 894–902.

Eriksen, M., Lebreton, L. C. M., Carson, H. S., Thiel, M., Moore, C. J., Borerro, J. C., Galgani, F., Ryan, P. G., and Reisser, J. (2014). Plastic pollution in the world's oceans: More than 5 trillion plastic pieces weighing over 250,000 tons afloat at sea. *PLoS ONE*, 9(12), e111913.

Evode, N., Qamar, S. A., Bilal, M., Barceló, D., and Iqbal, H. M. N. (2021). Plastic waste and its management strategies for environmental sustainability. *Case Studies in Chemical and Environmental Engineering*, 4, 100142.

Falandysz, J., Albanis, T., Bachmann, J., Bettinetti, R., Bochentin, I., Boti, V., Bristeau, S., Daehne, B., Dagnac, T., Galassi, S., Jeannot, R., Oehlmann, J., Orlikowska, A., Sakkas, V., Szczerski, R., Valsamaki, V., and Schulte-Oehlmann, U. (2006). Some chemical contaminant of surface sediments at the Baltic Sea Coastal Region with special emphasis on androgenic and anti-androgenic compounds. *Journal of Environmental Science and Health, Part A*, 41(10), 2127–2162.

Frehland, S., Kaegi, R., Hufenus, R., and Mitrano, D. M. (2020). Long-term assessment of nanoplastic particle and microplastic fiber flux through a pilot wastewater treatment plant using metal-doped plastics. *Water Research*, 182, 115860.

Geyer, R., Jambeck, J. R., and Law, K. L. (2017). Production, use, and fate of all plastics ever made. *Science Advances*, 3(7), e1700782.

Gigault, J., Pedrono, B., Maxit, B., and Ter Halle, A. (2016). Marine plastic litter: The unanalyzed nano-fraction. *Environmental Science: Nano*, 3(2), 346–350.

Goldstein, M. C., Rosenberg, M., and Cheng, L. (2012). Increased oceanic microplastic debris enhances oviposition in an endemic pelagic insect. *Biology Letters*, 8(5), 817–820.

Gomiero, A., Øysæd, K. B., Palmas, L., and Skogerbø, G. (2021). Application of GCMS-pyrolysis to estimate the levels of microplastics in a drinking water supply system. *Journal of Hazardous Materials*, 416, 125708.

Gopinath, K., Seshachalam, S., Neelavannan, K., Anburaj, V., Rachel, M., Ravi, S., Bharath, M., and Achyuthan, H. (2020). Quantification of microplastic in Red Hills Lake of Chennai city, Tamil Nadu, India. *Environmental Science and Pollution Research*, 27(26), 33297–33306. https://doi.org/10.1007/s11356-020-09622-2

Goswami, P., Vinithkumar, N. V., and Dharani, G. (2020). First evidence of microplastics bioaccumulation by marine organisms in the Port Blair Bay, Andaman Islands. *Marine Pollution Bulletin*, 155, 111163. https://doi.org/10.1016/j.marpolbul.2020.111163

Guo, X., Lin, H., Xu, S., and He, L. (2022). Recent advances in spectroscopic techniques for the analysis of microplastics in food. *Journal of Agricultural and Food Chemistry*, 70(5), 1410–1422.

Guo, X., and Wang, J. (2019). The chemical behaviors of microplastics in marine environment: A review. *Marine Pollution Bulletin*, 142, 1–14.

Guzzetti, E., Sureda, A., Tejada, S., and Faggio, C. (2018). Microplastic in marine organism: Environmental and toxicological effects. *Environmental Toxicology and Pharmacology*, 64, 164–171.

Hanvey, J. S., Lewis, P. J., Lavers, J. L., Crosbie, N. D., Pozo, K., and Clarke, B. O. (2017). A review of analytical techniques for quantifying microplastics in sediments. *Analytical Methods*, 9(9), 1369–1383.

Harley-Nyang, D., Memon, F. A., Jones, N., and Galloway, T. (2022). Investigation and analysis of microplastics in sewage sludge and biosolids: A case study from one wastewater treatment works in the UK. *Science of the Total Environment*, 823, 153735. https://doi.org/10.1016/j.scitotenv.2022.153735

Hongprasith, N., Kittimethawong, C., Lertluksanaporn, R., Eamchotchawalit, T., Kittipongvises, S., and Lohwacharin, J. (2020). IR microspectroscopic identification of microplastics in municipal wastewater treatment plants. *Environmental Science and Pollution Research*, 27(15), 18557–18564. https://doi.org/10.1007/s11356-020-08265-7

Hossain, S., Rahman, M. A., Ahmed Chowdhury, M., and Kumar Mohonta, S. (2020). Plastic pollution in Bangladesh: A review on current status emphasizing the impacts on environment and public health. *Environmental Engineering Research*, 26(6), 200535–0.

Hou, B., Wang, F., Liu, T., and Wang, Z. (2021). Reproductive toxicity of polystyrene microplastics: In vivo experimental study on testicular toxicity in mice. *Journal of Hazardous Materials*, 405, 124028.

Hou, P., Xu, Y., Taiebat, M., Lastoskie, C., Miller, S. A., and Xu, M. (2018). Life cycle assessment of end-of-life treatments for plastic film waste. *Journal of Cleaner Production*, 201, 1052–1060.

Huang, D., Tao, J., Cheng, M., Deng, R., Chen, S., Yin, L., and Li, R. (2021). Microplastics and nanoplastics in the environment: Macroscopic transport and effects on creatures. *Journal of Hazardous Materials*, 407, 124399.

Imhof, H. K., Sigl, R., Brauer, E., Feyl, S., Giesemann, P., Klink, S., Leupolz, K., Löder, M. G. J., Löschel, L. A., Missun, J., Muszynski, S., Ramsperger, A. F. R. M., Schrank, I., Speck, S., Steibl, S., Trotter, B., Winter, I., and Laforsch, C. (2017). Spatial and temporal variation of macro-, meso- and microplastic abundance on a remote coral island of the Maldives, Indian Ocean. *Marine Pollution Bulletin*, 116(1–2), 340–347. https://doi.org/10.1016/j.marpolbul.2017.01.010

Iñiguez, M. E., Conesa, J. A., and Fullana, A. (2017). Microplastics in Spanish table salt. *Scientific Reports*, 7(1), 8620.

Ivleva, N. P. (2021). Chemical analysis of microplastics and nanoplastics: Challenges, advanced methods, and perspectives. *Chemical Reviews*, 121(19), 11886–11936.

Jambeck, J. R., Geyer, R., Wilcox, C., Siegler, T. R., Perryman, M., Andrady, A., Narayan, R., and Law, K. L. (2015). Plastic waste inputs from land into the ocean. *Science*, 347(6223), 768–771.

Jayasiri, H. B., Purushothaman, C. S., and Vennila, A. (2013). Plastic litter accumulation on high-water strandline of urban beaches in Mumbai, India. *Environmental Monitoring and Assessment*, 185(9), 7709–7719. https://doi.org/10.1007/s10661-013-3129-z

Jehanno, C., and Sardon, H. (2019). Dynamic polymer network points the way to truly recyclable plastics. *Nature*, 568(7753), 467–468.

Jiang, Y., Yang, F., Zhao, Y., and Wang, J. (2020). Greenland Sea Gyre increases microplastic pollution in the surface waters of the Nordic Seas. *Science of the Total Environment*, 712, 136484.

Jin, T., Peydayesh, M., and Mezzenga, R. (2021). Membrane-based technologies for per- and poly-fluoroalkyl substances (PFASs) removal from water: Removal mechanisms, applications, challenges and perspectives. *Environment International*, 157, 106876.

Kane, I. A., and Clare, M. A. (2019). Dispersion, accumulation, and the ultimate fate of microplastics in deep-marine environments: A review and future directions. *Frontiers in Earth Science*, 7, 80.

Kanhai, L. D. K., Gårdfeldt, K., Lyashevska, O., Hassellöv, M., Thompson, R. C., and O'Connor, I. (2018). Microplastics in sub-surface waters of the Arctic Central Basin. *Marine Pollution Bulletin*, 130, 8–18.

Kannan, K., and Vimalkumar, K. (2021). A review of human exposure to microplastics and insights into microplastics as Obesogens. *Frontiers in Endocrinology*, 12, 724989.

Käppler, A., Fischer, M., Scholz-Böttcher, B. M., Oberbeckmann, S., Labrenz, M., Fischer, D., Eichhorn, K.-J., and Voit, B. (2018). Comparison of μ-ATR-FTIR spectroscopy and py-GCMS as identification tools for microplastic particles and fibers isolated from river sediments. *Analytical and Bioanalytical Chemistry*, 410(21), 5313–5327.

Karthik, R., Robin, R. S., Purvaja, R., Ganguly, D., Anandavelu, I., Raghuraman, R., Hariharan, G., Ramakrishna, A., and Ramesh, R. (2018). Microplastics along the beaches of southeast coast of India. *Science of the Total Environment*, 645, 1388–1399. https://doi.org/10.1016/j.scitotenv.2018.07.242

Kawecki, D., Wu, Q., Gonçalves, J. S. V., and Nowack, B. (2021). Polymer-specific dynamic probabilistic material flow analysis of seven polymers in Europe from 1950 to 2016. *Resources, Conservation and Recycling*, 173, 105733.

Kazemi, M., Faisal Kabir, S., and Fini, E. H. (2021). State of the art in recycling waste thermoplastics and thermosets and their applications in construction. *Resources, Conservation and Recycling*, 174, 105776.

Keerthika, K., Padmavathy, P., Rani, V., Jeyashakila, R., Aanand, S., and Kutty, R. (2022). Spatial, seasonal and ecological risk assessment of microplastics in sediment and surface water along the Thoothukudi, south Tamil Nadu, south east India. *Environmental Monitoring and Assessment*, 194(11), 820. https://doi.org/10.1007/s10661-022-10468-z

Khoironi, A., Hadiyanto, H., Anggoro, S., and Sudarno, S. (2020). Evaluation of polypropylene plastic degradation and microplastic identification in sediments at Tambak Lorok coastal area, Semarang, Indonesia. *Marine Pollution Bulletin*, 151, 110868.

Kiendrebeogo, M., Karimi Estahbanati, M. R., Khosravanipour Mostafazadeh, A., Drogui, P., and Tyagi, R. D. (2021). Treatment of microplastics in water by anodic oxidation: A case study for polystyrene. *Environmental Pollution*, 269, 116168.

King, C. A., Shamshina, J. L., Zavgorodnya, O., Cutfield, T., Block, L. E., and Rogers, R. D. (2017). Porous Chitin microbeads for more sustainable cosmetics. *ACS Sustainable Chemistry & Engineering*, 5(12), 11660–11667.

Koelmans, A. A., Mohamed Nor, N. H., Hermsen, E., Kooi, M., Mintenig, S. M., and De France, J. (2019). Microplastics in freshwaters and drinking water: Critical review and assessment of data quality. *Water Research*, 155, 410–422.

Kovač Viršek, M., Palatinus, A., Koren, Š., Peterlin, M., Horvat, P., and Kržan, A. (2016). Protocol for microplastics sampling on the sea surface and sample analysis. *Journal of Visualized Experiments*, 118, 55161.

Krishnakumar, S., Srinivasalu, S., Saravanan, P., Vidyasakar, A., and Magesh, N. S. (2018). A preliminary study on coastal debris in Nallathanni Island, Gulf of Mannar Biosphere Reserve, Southeast coast of India. *Marine Pollution Bulletin*, 131, 547–551. https://doi.org/10.1016/j.marpolbul.2018.04.026

Kristanti, R. A., Hadibarata, T., Wulandari, N. F., Sibero, M. T., Darmayati, Y., and Hatmanti, A. (2022). Overview of microplastics in the environment: Type, source, potential effects and removal strategies. *Bioprocess and Biosystems Engineering*, 46, 429–441, (2023). https://doi.org/10.1007/s00449-022-02784-y.

Kundu, A., Shetti, N. P., Basu, S., Raghava Reddy, K., Nadagouda, M. N., and Aminabhavi, T. M. (2021). Identification and removal of micro- and nano-plastics: Efficient and cost-effective methods. *Chemical Engineering Journal*, 421, 129816.

Lear, G., Kingsbury, J. M., Franchini, S., Gambarini, V., Maday, S. D. M., Wallbank, J. A., Weaver, L., and Pantos, O. (2021). Plastics and the microbiome: Impacts and solutions. *Environmental Microbiome*, 16(1), 2.

Lee, K.-W., Shim, W. J., Kwon, O. Y., and Kang, J.-H. (2013). Size-dependent effects of micro polystyrene particles in the Marine Copepod Tigriopus japonicus. *Environmental Science & Technology*, 47(19), 11278–11283.

Lenaker, P. L., Corsi, S. R., and Mason, S. A. (2021). Spatial distribution of microplastics in surficial Benthic Sediment of Lake Michigan and Lake Erie. *Environmental Science & Technology*, 55(1), 373–384.

Li, C., Gao, Y., He, S., Chi, H.-Y., Li, Z.-C., Zhou, X.-X., and Yan, B. (2021). Quantification of nanoplastic uptake in cucumber plants by pyrolysis gas chromatography/mass spectrometry. *Environmental Science & Technology Letters*, 8(8), 633–638.

Li, J., Yang, D., Li, L., Jabeen, K., and Shi, H. (2015). Microplastics in commercial bivalves from China. *Environmental Pollution*, 207, 190–195.

Li, X., Chen, L., Mei, Q., Dong, B., Dai, X., Ding, G., and Zeng, E. Y. (2018). Microplastics in sewage sludge from the wastewater treatment plants in China. *Water Research*, 142, 75–85.

Liu, X., Yuan, W., Di, M., Li, Z., and Wang, J. (2019). Transfer and fate of microplastics during the conventional activated sludge process in one wastewater treatment plant of China. *Chemical Engineering Journal*, 362, 176–182. https://doi.org/10.1016/j.cej.2019.01.033

Liu, Y., Li, R., Yu, J., Ni, F., Sheng, Y., Scircle, A., Cizdziel, J. V., and Zhou, Y. (2021). Separation and identification of microplastics in marine organisms by TGA-FTIR-GC/MS: A case study of mussels from coastal China. *Environmental Pollution*, 272, 115946.

Lu, H.-C., Ziajahromi, S., Neale, P. A., and Leusch, F. D. L. (2021). A systematic review of freshwater microplastics in water and sediments: Recommendations for harmonisation to enhance future study comparisons. *Science of the Total Environment*, 781, 146693.

Luo, H., Zeng, Y., Zhao, Y., Xiang, Y., Li, Y., and Pan, X. (2021). Effects of advanced oxidation processes on leachates and properties of microplastics. *Journal of Hazardous Materials*, 413, 125342.

Lusher, A. (2015). Microplastics in the marine environment: Distribution, interactions and effects. In M. Bergmann, L. Gutow, and M. Klages (Eds.), *Marine Anthropogenic Litter* (pp. 245–307). Springer International Publishing.

Lusher, A., Hollman, P., and Mendoza-Hill, J. (2017). Microplastics in fisheries and aquaculture: status of knowledge on their occurrence and implications for aquatic organisms and food safety. In *FAO Fisheries and Aquaculture Technical Paper, Rome, Italy* (Vol. 615, pp. 1–147).

Lusher, A. L., Welden, N. A., Sobral, P., and Cole, M. (2017). Sampling, isolating and identifying microplastics ingested by fish and invertebrates. *Analytical Methods*, 9(9), 1346–1360.

Ma, H., Pu, S., Liu, S., Bai, Y., Mandal, S., and Xing, B. (2020). Microplastics in aquatic environments: Toxicity to trigger ecological consequences. *Environmental Pollution*, 261, 114089.

Madhumitha, C. T., Karmegam, N., Biruntha, M., Arun, A., Al Kheraif, A. A., Kim, W., and Kumar, P. (2022). Extraction, identification, and environmental risk assessment of microplastics in commercial toothpaste. *Chemosphere*, 296, 133976.

Majewsky, M., Bitter, H., Eiche, E., and Horn, H. (2016). Determination of microplastic polyethylene (PE) and polypropylene (PP) in environmental samples using thermal analysis (TGA-DSC). *Science of the Total Environment*, 568, 507–511.

Malankowska, M., Echaide-Gorriz, C., and Coronas, J. (2021). Microplastics in marine environment: A review on sources, classification, and potential remediation by membrane technology. *Environmental Science: Water Research & Technology*, 7(2), 243–258.

Mintenig, S. M., Bäuerlein, P. S., Koelmans, A. A., Dekker, S. C., and van Wezel, A. P. (2018). Closing the gap between small and smaller: Towards a framework to analyse nano- and microplastics in aqueous environmental samples. *Environmental Science: Nano*, 5(7), 1640–1649.

Mintenig, S. M., Int-Veen, I., Löder, M. G. J., Primpke, S., and Gerdts, G. (2017). Identification of microplastic in effluents of waste water treatment plants using focal plane array-based micro-Fourier-transform infrared imaging. *Water Research*, 108, 365–372. https://doi.org/10.1016/j.watres.2016.11.015

Möller, J. N., Heisel, I., Satzger, A., Vizsolyi, E. C., Oster, S. D. J., Agarwal, S., Laforsch, C., and Löder, M. G. J. (2022). Tackling the challenge of extracting microplastics from soils: A protocol to purify soil samples for spectroscopic analysis. *Environmental Toxicology and Chemistry*, 41(4), 844–857.

Möller, J. N., Löder, M. G. J., and Laforsch, C. (2020). Finding microplastics in soils: A review of analytical methods. *Environmental Science & Technology*, 54(4), 2078–2090.

Munno, K., De Frond, H., O'Donnell, B., and Rochman, C. M. (2020). Increasing the accessibility for characterizing microplastics: Introducing new application-based and spectral libraries of plastic particles (SLoPP and SLoPP-E). *Analytical Chemistry*, 92(3), 2443–2451.

Murphy, F., Ewins, C., Carbonnier, F., and Quinn, B. (2016). Wastewater treatment works (WwTW) as a source of microplastics in the aquatic environment. *Environmental Science & Technology*, 50(11), 5800–5808. https://doi.org/10.1021/acs.est.5b05416

Nanda, S., and Berruti, F. (2021). Thermochemical conversion of plastic waste to fuels: A review. *Environmental Chemistry Letters*, 19(1), 123–148.

Napper, I. E., Baroth, A., Barrett, A. C., Bhola, S., Chowdhury, G. W., Davies, B. F. R., Duncan, E. M., Kumar, S., Nelms, S. E., Hasan Niloy, M. N., Nishat, B., Maddalene, T., Thompson, R. C., and Koldewey, H. (2021). The abundance and characteristics of microplastics in surface water in the transboundary Ganges River. *Environmental Pollution*, 274, 116348. https://doi.org/10.1016/j.envpol.2020.116348

Narancic, T., and O'Connor, K. E. (2017). Microbial biotechnology addressing the plastic waste disaster. *Microbial Biotechnology*, 10(5), 1232–1235.

Narancic, T., Verstichel, S., Reddy Chaganti, S., Morales-Gamez, L., Kenny, S. T., De Wilde, B., Babu Padamati, R., and O'Connor, K. E. (2018). Biodegradable plastic blends create new possibilities for end-of-life management of plastics but they are not a Panacea for plastic pollution. *Environmental Science & Technology*, 52(18), 10441–10452.

Narmatha Sathish, M., Immaculate Jeyasanta, K., and Patterson, J. (2020). Monitoring of microplastics in the clam Donax cuneatus and its habitat in Tuticorin coast of Gulf of Mannar (GoM), India. *Environmental Pollution*, 266, 115219. https://doi.org/10.1016/j.envpol.2020.115219

Nguyen, B., Claveau-Mallet, D., Hernandez, L. M., Xu, E. G., Farner, J. M., and Tufenkji, N. (2019). Separation and analysis of microplastics and nanoplastics in complex environmental samples. *Accounts of Chemical Research*, 52(4), 858–866.

Nguyen, M. K., Hadi, M., Lin, C., Nguyen, H.-L., Thai, V.-B., Hoang, H.-G., Vo, D.-V. N., and Tran, H.-T. (2022). Microplastics in sewage sludge: Distribution, toxicity, identification methods, and engineered technologies. *Chemosphere*, 308, 136455.

Ogunola, O. S., Onada, O. A., and Falaye, A. E. (2018). Mitigation measures to avert the impacts of plastics and microplastics in the marine environment (a review). *Environmental Science and Pollution Research*, 25(10), 9293–9310.

Padervand, M., Lichtfouse, E., Robert, D., and Wang, C. (2020). Removal of microplastics from the environment. A review. *Environmental Chemistry Letters*, 18(3), 807–828.

Park, E.-J., Han, J.-S., Park, E.-J., Seong, E., Lee, G.-H., Kim, D.-W., Son, H.-Y., Han, H.-Y., and Lee, B.-S. (2020). Repeated-oral dose toxicity of polyethylene microplastics and the possible implications on reproduction and development of the next generation. *Toxicology Letters*, 324, 75–85.

Parton, K. J., Godley, B. J., Santillo, D., Tausif, M., Omeyer, L. C. M., and Galloway, T. S. (2020). Investigating the presence of microplastics in demersal sharks of the North-East Atlantic. *Scientific Reports*, 10(1), 12204.

Patchaiyappan, A., Ahmed, S. Z., Dowarah, K., Jayakumar, S., and Devipriya, S. P. (2020). Occurrence, distribution and composition of microplastics in the sediments of South Andaman beaches. *Marine Pollution Bulletin*, 156, 111227. https://doi.org/10.1016/j.marpolbul.2020.111227

Patra, B. C., Shit, P. K., Bhunia, G. S., and Bhattacharya, M. (2022). *River Health and Ecology in South Asia: Pollution, Restoration, and Conservation*. Springer Nature, pp. 1–100.

Patterson, J., Jeyasanta, K. I., Sathish, N., Booth, A. M., and Edward, J. K. P. (2019). Profiling microplastics in the Indian edible oyster, Magallana bilineata collected from the Tuticorin coast, Gulf of Mannar, Southeastern India. *Science of the Total Environment*, 691, 727–735. https://doi.org/10.1016/j.scitotenv.2019.07.063

Perumal, K., and Muthuramalingam, S. (2021). Microplastics pollution studies in India: A recent review of sources, abundances and research perspectives [Preprint]. In Review. https://doi.org/10.21203/rs.3.rs-535083/v1

Prinz, N., and Korez, Š. (2020). Understanding how microplastics affect marine Biota on the cellular level is important for assessing ecosystem function: A review. In S. Jungblut, V. Liebich, and M. Bode-Dalby (Eds.), *YOUMARES 9-The Oceans: Our Research, Our Future* (pp. 101–120). Springer. https://doi.org/10.1007/978-3-030-20389-4_6.

Qiu, Q., Tan, Z., Wang, J., Peng, J., Li, M., and Zhan, Z. (2016). Extraction, enumeration and identification methods for monitoring microplastics in the environment. *Estuarine, Coastal and Shelf Science*, 176, 102–109.

Rahimi, A., and García, J. M. (2017). Chemical recycling of waste plastics for new materials production. *Nature Reviews Chemistry*, 1(6), 0046.

Renner, G., Schmidt, T. C., and Schram, J. (2018). Analytical methodologies for monitoring micro(nano)plastics: Which are fit for purpose? *Current Opinion in Environmental Science & Health*, 1, 55–61.

Robin, R. S., Karthik, R., Purvaja, R., Ganguly, D., Anandavelu, I., Mugilarasan, M., and Ramesh, R. (2020). Holistic assessment of microplastics in various coastal environmental matrices, southwest coast of India. *Science of The Total Environment*, 703, 134947. https://doi.org/10.1016/j.scitotenv.2019.134947

Rodríguez Chialanza, M., Sierra, I., Pérez Parada, A., and Fornaro, L. (2018). Identification and quantitation of semi-crystalline microplastics using image analysis and differential scanning calorimetry. *Environmental Science and Pollution Research*, 25(17), 16767–16775.

Sait, S. T. L., Sørensen, L., Kubowicz, S., Vike-Jonas, K., Gonzalez, S. V., Asimakopoulos, A. G., and Booth, A. M. (2021). Microplastic fibres from synthetic textiles: Environmental degradation and additive chemical content. *Environmental Pollution*, 268, 115745.

Samanta, P., Dey, S., Kundu, D., Dutta, D., Jambulkar, R., Mishra, R., Ghosh, A. R., and Kumar, S. (2022). An insight on sampling, identification, quantification and characteristics of microplastics in solid wastes. *Trends in Environmental Analytical Chemistry*, 36, December 2022, e00181. https://doi.org/10.1016/j.teac.2022.e00181.

Sarkar, D. J., Das Sarkar, S., Das, B. K., Manna, R. K., Behera, B. K., and Samanta, S. (2019). Spatial distribution of meso and microplastics in the sediments of river Ganga at eastern India. *Science of the Total Environment*, 694, 133712. https://doi.org/10.1016/j.scitotenv.2019.133712

Sathish, M. N., Jeyasanta, I., and Patterson, J. (2020). Occurrence of microplastics in epipelagic and mesopelagic fishes from Tuticorin, Southeast coast of India. *Science of the Total Environment*, 720, 137614. https://doi.org/10.1016/j.scitotenv.2020.137614

Sathish, N., Jeyasanta, K. I., and Patterson, J. (2019). Abundance, characteristics and surface degradation features of microplastics in beach sediments of five coastal areas in Tamil Nadu, India. *Marine Pollution Bulletin*, 142, 112–118. https://doi.org/10.1016/j.marpolbul.2019.03.037

Selvam, K., Xavier, K. A. M., Shivakrishna, A., Bhutia, T. P., Kamat, S., and Shenoy, L. (2021). Abundance, composition and sources of marine debris trawled-up in the fishing grounds along the north-east Arabian coast. *Science of the Total Environment*, 751, 141771. https://doi.org/10.1016/j.scitotenv.2020.141771

Shabaka, S. H., Marey, R. S., Ghobashy, M., Abushady, A. M., Ismail, G. A., and Khairy, H. M. (2020). Thermal analysis and enhanced visual technique for assessment of microplastics in fish from an Urban Harbor, Mediterranean Coast of Egypt. *Marine Pollution Bulletin*, 159, 111465.

Shi, B., Patel, M., Yu, D., Yan, J., Li, Z., Petriw, D., Pruyn, T., Smyth, K., Passeport, E., Miller, R. J. D., and Howe, J. Y. (2022). Automatic quantification and classification of microplastics in scanning electron micrographs via deep learning. *Science of the Total Environment*, 825, 153903.

Shim, W. J., Hong, S. H., and Eo, S. E. (2017). Identification methods in microplastic analysis: A review. *Analytical Methods*, 9(9), 1384–1391.

Silva-Cavalcanti, J. S., Silva, J. C. P., de Andrade, F. M., Brito, A. M. S. S., and Costa, M. F. da. (2023). Microplastic pollution in sediments of tropical shallow lakes. *Science of the Total Environment*, 855, 158671.

Smith, M., Love, D. C., Rochman, C. M., and Neff, R. A. (2018). Microplastics in seafood and the implications for human health. *Current Environmental Health Reports*, 5(3), 375–386.

Sobhani, Z., Lei, Y., Tang, Y., Wu, L., Zhang, X., Naidu, R., Megharaj, M., and Fang, C. (2020). Microplastics generated when opening plastic packaging. *Scientific Reports*, 10(1), 4841.

Song, Y. K., Hong, S. H., Jang, M., Han, G. M., Rani, M., Lee, J., and Shim, W. J. (2015). A comparison of microscopic and spectroscopic identification methods for analysis of microplastics in environmental samples. *Marine Pollution Bulletin*, 93(1–2), 202–209.

Sruthy, S., and Ramasamy, E. V. (2017). Microplastic pollution in Vembanad Lake, Kerala, India: The first report of microplastics in lake and estuarine sediments in India. *Environmental Pollution*, 222, 315–322. https://doi.org/10.1016/j.envpol.2016.12.038

Stanton, T., Johnson, M., Nathanail, P., MacNaughtan, W., and Gomes, R. L. (2020). Freshwater microplastic concentrations vary through both space and time. *Environmental Pollution*, 263, 114481.

Suaria, G., Avio, C. G., Mineo, A., Lattin, G. L., Magaldi, M. G., Belmonte, G., Moore, C. J., Regoli, F., and Aliani, S. (2016). The mediterranean plastic soup: Synthetic polymers in mediterranean surface waters. *Scientific Reports*, 6(1), 37551.

Sun, J., Dai, X., Wang, Q., van Loosdrecht, M. C. M., and Ni, B.-J. (2019). Microplastics in wastewater treatment plants: Detection, occurrence and removal. *Water Research*, 152, 21–37.

Sussarellu, R., Suquet, M., Thomas, Y., Lambert, C., Fabioux, C., Pernet, M. E. J., Le Goïc, N., Quillien, V., Mingant, C., Epelboin, Y., Corporeau, C., Guyomarch, J., Robbens, J., Paul-Pont, I., Soudant, P., and Huvet, A. (2016). Oyster reproduction is affected by exposure to polystyrene microplastics. *Proceedings of the National Academy of Sciences*, 113(9), 2430–2435.

Tagg, A. S., Sapp, M., Harrison, J. P., Sinclair, C. J., Bradley, E., Ju-Nam, Y., and Ojeda, J. J. (2020). Microplastic monitoring at different stages in a wastewater treatment plant using reflectance micro-FTIR imaging. *Frontiers in Environmental Science*, 8, 145. https://doi.org/10.3389/fenvs.2020.00145

Tanaka, K., Takada, H., Yamashita, R., Mizukawa, K., Fukuwaka, M., and Watanuki, Y. (2013). Accumulation of plastic-derived chemicals in tissues of seabirds ingesting marine plastics. *Marine Pollution Bulletin*, 69(1–2), 219–222.

Teichert, S., Löder, M. G. J., Pyko, I., Mordek, M., Schulbert, C., Wisshak, M., and Laforsch, C. (2021). Microplastic contamination of the drilling bivalve Hiatella arctica in Arctic rhodolith beds. *Scientific Reports*, 11(1), 14574.

Thushari, G. G. N., and Senevirathna, J. D. M. (2020). Plastic pollution in the marine environment. *Heliyon*, 6(8), e04709.

Tiwari, M., Rathod, T. D., Ajmal, P. Y., Bhangare, R. C., and Sahu, S. K. (2019). Distribution and characterization of microplastics in beach sand from three different Indian coastal environments. *Marine Pollution Bulletin*, 140, 262–273. https://doi.org/10.1016/j.marpolbul.2019.01.055

Toapanta, T., Okoffo, E. D., Ede, S., O'Brien, S., Burrows, S. D., Ribeiro, F., Gallen, M., Colwell, J., Whittaker, A. K., Kaserzon, S., and Thomas, K. V. (2021). Influence of surface oxidation on the quantification of polypropylene microplastics by pyrolysis gas chromatography mass spectrometry. *Science of The Total Environment*, 796, 148835.

Tong, H., Jiang, Q., Hu, X., and Zhong, X. (2020). Occurrence and identification of microplastics in tap water from China. *Chemosphere*, 252, 126493.

Trainic, M., Flores, J. M., Pinkas, I., Pedrotti, M. L., Lombard, F., Bourdin, G., Gorsky, G., Boss, E., Rudich, Y., Vardi, A., and Koren, I. (2020). Airborne microplastic particles detected in the remote marine atmosphere. *Communications Earth & Environment*, 1(1), 64.

Uheida, A., Mejía, H. G., Abdel-Rehim, M., Hamd, W., and Dutta, J. (2021). Visible light photocatalytic degradation of polypropylene microplastics in a continuous water flow system. *Journal of Hazardous Materials*, 406, 124299.

Valerio, O., Muthuraj, R., and Codou, A. (2020). Strategies for polymer to polymer recycling from waste: Current trends and opportunities for improving the circular economy of polymers in South America. *Current Opinion in Green and Sustainable Chemistry*, 25, 100381.

Varnava, C. K., and Patrickios, C. S. (2021). Polymer networks one hundred years after the macromolecular hypothesis: A tutorial review. *Polymer*, 215, 123322.

Veerasingam, S., Ranjani, M., Venkatachalapathy, R., Bagaev, A., Mukhanov, V., Litvinyuk, D., Mugilarasan, M., Gurumoorthi, K., Guganathan, L., Aboobacker, V. M., and Vethamony, P. (2021). Contributions of Fourier transform infrared spectroscopy in microplastic pollution research: A review. *Critical Reviews in Environmental Science and Technology*, 51(22), 2681–2743.

Veerasingam, S., Saha, M., Suneel, V., Vethamony, P., Rodrigues, A. C., Bhattacharyya, S., and Naik, B. G. (2016). Characteristics, seasonal distribution and surface degradation features of microplastic pellets along the Goa coast, India. *Chemosphere*, 159, 496–505. https://doi.org/10.1016/j.chemosphere.2016.06.056

Wang, C., Huang, R., Sun, R., Yang, J., and Dionysiou, D. D. (2022). Microplastics separation and subsequent carbonization: Synthesis, characterization, and catalytic performance of iron/carbon nanocomposite. *Journal of Cleaner Production*, 330, 129901.

Wang, J., Li, X., Gao, M., Li, X., Zhao, L., and Ru, S. (2022). Polystyrene microplastics increase estrogenic effects of 17α-ethynylestradiol on male marine medaka (Oryzias melastigma). *Chemosphere*, 287, 132312.

Wang, X.-S., Song, H., Liu, Y.-L., Pan, X.-R., Zhang, H.-C., Gao, Z., Kong, D.-Z., Wang, R., Wang, L., and Ma, J. (2021). Quantitively analyzing the variation of micrometer-sized microplastic during water treatment with the flow cytometry-fluorescent beads method. *ACS ES&T Engineering*, 1(12), 1668–1677.

Watermann, B. T., Löder, M., Herlyn, M., Daehne, B., Thomsen, A., and Gall, K. (2017). Long-term 2007-2013 monitoring of reproductive disturbance in the dun sentinel Assiminea grayana with regard to polymeric materials pollution at the coast of Lower Saxony, North Sea, Germany. *Environmental Science and Pollution Research*, 24(4), 3352–3362.

Wong, J. K. H., Lee, K. K., Tang, K. H. D., and Yap, P.-S. (2020). Microplastics in the freshwater and terrestrial environments: Prevalence, fates, impacts and sustainable solutions. *Science of the Total Environment*, 719, 137512.

Woo, H., Seo, K., Choi, Y., Kim, J., Tanaka, M., Lee, K., and Choi, J. (2021). Methods of analyzing microsized plastics in the environment. *Applied Sciences*, 11(22), 10640.

Woodall, L. C., Sanchez-Vidal, A., Canals, M., Paterson, G. L. J., Coppock, R., Sleight, V., Calafat, A., Rogers, A. D., Narayanaswamy, B. E., and Thompson, R. C. (2014). The deep sea is a major sink for microplastic debris. *Royal Society Open Science*, 1(4), 140317.

Wu, W.-M., Yang, J., and Criddle, C. S. (2017). Microplastics pollution and reduction strategies. *Frontiers of Environmental Science & Engineering*, 11(1), 6.

Xiong, X., Bond, T., Saboor Siddique, M., and Yu, W. (2021). The stimulation of microbial activity by microplastic contributes to membrane fouling in ultrafiltration. *Journal of Membrane Science*, 635, 119477.

Xiong, X., Chen, X., Zhang, K., Mei, Z., Hao, Y., Zheng, J., Wu, C., Wang, K., Ruan, Y., Lam, P. K. S., and Wang, D. (2018). Microplastics in the intestinal tracts of East Asian finless porpoises (Neophocaena asiaeorientalis sunameri) from Yellow Sea and Bohai Sea of China. *Marine Pollution Bulletin*, 136, 55–60.

Xu, E. G., Cheong, R. S., Liu, L., Hernandez, L. M., Azimzada, A., Bayen, S., and Tufenkji, N. (2020). Primary and secondary plastic particles exhibit limited acute toxicity but chronic effects on Daphnia magna. *Environmental Science & Technology*, 54(11), 6859–6868.

Xu, X., Hou, Q., Xue, Y., Jian, Y., and Wang, L. (2018). Pollution characteristics and fate of microfibers in the wastewater from textile dyeing wastewater treatment plant. *Water Science and Technology*, 78(10), 2046–2054.

Yang, T., Luo, J., and Nowack, B. (2021). Characterization of nanoplastics, fibrils, and microplastics released during washing and abrasion of polyester textiles. *Environmental Science & Technology*, 55(23), 15873–15881.

Yuan, Z., and Nag, R. E. R. E. (2022). Cummins, human health concerns regarding microplastics in the aquatic environment-from marine to food systems. *The Science of the Total Environment*, 7, 153730.

Yue, L., Bonab, V. S., Yuan, D., Patel, A., Karimkhani, V., and Manas-Zloczower, I. (2019). Vitrimerization: A novel concept to reprocess and recycle thermoset waste via dynamic chemistry. *Global Challenges*, 3(7), 1800076.

Zamboulis, A., Papadopoulos, L., Terzopoulou, Z., Bikiaris, D. N., Patsiaoura, D., Chrissafis, K., Gazzano, M., Lotti, N., and Papageorgiou, G. Z. (2019). Synthesis, thermal properties and decomposition mechanism of poly(Ethylene vanillate) polyester. *Polymers*, 11(10), 1672.

Zhang, D., Liu, X., Huang, W., Li, J., Wang, C., Zhang, D., and Zhang, C. (2020). Microplastic pollution in deep-sea sediments and organisms of the Western Pacific Ocean. *Environmental Pollution*, 259, 113948.

Zhang, J., Chevali, V. S., Wang, H., and Wang, C.-H. (2020). Current status of carbon fibre and carbon fibre composites recycling. *Composites Part B: Engineering*, 193, 108053.

Zhang, L., Xie, Y., Liu, J., Zhong, S., Qian, Y., and Gao, P. (2020). An overlooked entry pathway of microplastics into agricultural soils from application of sludge-based fertilizers. *Environmental Science & Technology*, 54(7), 4248–4255. https://doi.org/10.1021/acs.est.9b07905

Zhang, S., Wang, J., Yan, P., Hao, X., Xu, B., Wang, W., and Aurangzeib, M. (2021). Non-biodegradable microplastics in soils: A brief review and challenge. *Journal of Hazardous Materials*, 409, 124525.

Ziajahromi, S., Neale, P. A., Rintoul, L., and Leusch, F. D. L. (2017). Wastewater treatment plants as a pathway for microplastics: Development of a new approach to sample wastewater-based microplastics. *Water Research*, 112, 93–99. https://doi.org/10.1016/j.watres.2017.01.042

Zilinskaite, E., Futter, M., and Collentine, D. (2022). Stakeholders' perspectives on microplastics in sludge applied to agricultural land. *Frontiers in Sustainable Food Systems*, 6, 830637.

Index

Note: **Bold** page numbers refer to tables; *italic* page numbers refer to figures.

For Product Safety Concerns and Information please contact our EU representative GPSR@taylorandfrancis.com
Taylor & Francis Verlag GmbH, Kaufingerstraße 24, 80331 München, Germany

www.ingramcontent.com/pod-product-compliance
Lightning Source LLC
LaVergne TN
LVHW010831120826
845149LV00016B/465